FUNDAMENTALS OF DIGITAL SYSTEMS DESIGN

V. THOMAS RHYNE

Department of Electrical Engineering
Texas A&M University

PRENTICE-HALL, INC., *Englewood Cliffs, New Jersey*

Library of Congress Cataloging in Publication Data

RHYNE, V THOMAS,
   Fundamentals of digital systems design.
   (Prentice-Hall computer applications in electrical
engineering series)

   Includes bibliographical references.
   1. Switching theory. I. Title.
TK7868.S9R5   621.3815'37   72-6903
ISBN 0-13-336156-X

PRENTICE-HALL COMPUTER APPLICATIONS
IN ELECTRICAL ENGINEERING SERIES

Franklin F. Kuo, *Editor*

10 9 8 7 6 5 4 3 2 1

Printed in the United States of America

PRENTICE-HALL INTERNATIONAL, INC., *London*
PRENTICE-HALL OF AUSTRALIA, PTY. LTD., *Sydney*
PRENTICE-HALL OF JAPAN, INC., *Tokyo*
PRENTICE-HALL OF CANADA, LTD., *Toronto*
PRENTICE-HALL OF INDIA PRIVATE LTD., *New Delhi*

This book is dedicated to my father,
VERNON T. RHYNE, JR.

Contents

Preface

The subject matter that pertains to most topics can be separated into two general areas: theoretical aspects and practical aspects. This distinction may be illustrated by contrasting a discussion of the theory of finite-state machines to a course dealing with soldering tips, wiring suggestions, etc. My personal conviction is that a good engineer knows both theory *and* application; he can bridge the gap between a concept and its realization in some working system.

Engineering teachers often find that the available texts on many subjects deal primarily with theoretical aspects. In recent years, however, there has been a discernible shift toward practicality in engineering teaching. The COSINE Committee of the Commission on Education, in a March 1971 report* stated that:

> One of the major problems facing an electrical engineering faculty is that of making a student aware of how theoretical material presented in lecture courses can be applied to solve realistic engineering problems. In particular, it is extremely important for a student to be introduced to the practice of engineering as an integral part of his academic program.

In keeping with this emphasis, this text is an attempt to bridge the gap between the blackboard and the breadboard in the area of digital systems design. The intent is to teach a consistent approach to designing digital

---

Digital Systems Laboratory Courses and Laboratory Developments, Cosine Committee Report, Commission on Education of the National Academy of Engineering, Washington, D.C., March, 1971, p. 1.

devices, i.e., a design sequence which the student can follow, leading him from the description of what a digital device must do all the way to the interconnection of the components that will actually perform the desired functions. Design techniques have been selected for usefulness and universality. None of the techniques given are dated; they all should be applicable to new technology as it emerges.

This text is not intended to be a departure from the current texts in this field. Rather, it is meant to be a companion, teaching application techniques to accompany the theoretical techniques that are presented elsewhere. Basic theoretical concepts are included, however, so that a single course taught from this text can serve to introduce a student to digital systems design. He then can expand his theoretical background by further study.

The general emphasis herein is on presenting at least one good method for solving each type of design problem, rather than giving many alternative approaches. Consistency, organization, and reasonableness of design effort are the aims of this presentation. The development proceeds from introductory concepts to techniques for complete systems design. Practical examples are used wherever possible. Reference is made to many other texts and articles that are more theoretical in nature, however, so that interested students can augment the basic concepts that are given in this text.

The first seven chapters are organized for use in a one-semester course. These chapters form a logical sequence, with each new section building upon the concepts that are introduced in the previous sections. Throughout, detailed electronic considerations are minimized, thereby allowing this text to be used quite early in the Electrical Engineering curriculum and with non-EE students. Here at Texas A&M, where this material has been taught to several hundred students, it has found wide acceptance by students from the sophomore level through the graduate level, students in engineering, mathematics, physics, computer science, and several other academic departments, as well as by practicing engineers.

Chapter 1 provides introductory information; it may be omitted for students who are already familiar with binary numbering. Chapters 2–4 present a coordinated approach to combinational design. Chapter 2 introduces the Boolean algebra. Chapter 3 introduces the voltage symbolism technique for relating Boolean expressions to the gating circuits that implement them. The common gating devices are first described as idealized models, with subsequent discussion covering deviations from the ideal, e.g., threshold effects, delays, and loading. The material in Chapter 3 is presented early in the text so that coordinated laboratory work with actual gating devices can be started as soon as possible. Chapter 4 discusses the minimization of combinational expressions. Graphical, tabular, and computer-aided techniques for minimizing both single- and multiple-output functions are presented. Many practical combinational design examples, including the use of medium-scale integrated-circuit devices in combinational networks, are presented.

Chapters 5 and 6 cover sequential design. Chapter 5 introduces the common syntax that is used to describe sequential systems, the problems that can arise in sequential operation, and various techniques for solving these problems. The common sequential storage devices are also described in Chapter 5. Chapter 6 describes and illustrates design methods for use with sequential devices. The transition map technique for designing counters is presented first, with subsequent sections covering generalized sequential design and related sequential devices (shift registers, ring counters, etc.).

Chapter 7 completes the presentation of digital system design by introducing control techniques and devices. Several alternative approaches to designing controllers are presented. Examples of the complete system design process are included in Chapter 7.

The remaining chapters cover related concepts that are of general interest to digital systems designers. These last chapters also contain many combinational, sequential, and systems design examples.

Laboratory practice is essential for a digital systems course. Many of the design problems that are given at the ends of the chapters are suitable as laboratory exercises. Especially suitable problems of this type are marked with an asterisk (*). Laboratory experiments can be performed with a minimum of hardware, as is described in the COSINE report quoted earlier. Experiments using modern gating circuits, sequential devices, and MSI devices (full-adders, counters, etc.) are easy to develop and serve to reinforce the design techniques that are presented in the text.

Where time and curriculum constraints allow, it is expected that *two* courses in digital systems design should be taught: one from this text introducing basic switching theory and design techniques, and one covering advanced switching theory. The key to this text is that it teaches the student how to put theoretical concepts into action.

It is impossible to acknowledge or reference adequately all of the ideas, techniques, and examples that are presented in this text. Many of these have come from the fine engineers and technicians who have been my co-workers. I am also indebted to the many good teachers who have helped me in the past, notably Mr. Robert Guyton at Mississippi State University and Dr. John Peatman at Georgia Institute of Technology. Also, I want to acknowledge the support of Texas A&M University, in particular that of Dean Fred Benson and Dr. W. B. Jones, Jr.

I truly believe that this text should be a help to digital designers both new and old. I hope that I am correct.

V. T. RHYNE

College Station, Texas

1 Basic Concepts

1.0 What Is a Digital System?

Since this textbook is intended to introduce its readers to modern techniques for designing digital systems, the above question is a valid beginning. To answer it, the meaning of the term *digital* in this application must be established. The world we live in is, for the most part, a continuum of physical variables such as time, temperature, distance, weight, and so on. In conversation, in data processing, in record keeping, and in most other activities, the exact values of these kinds of variables cannot be used. Rather, one must measure the variable and make use of some numerical value that is representative of the actual value. Room temperature, for example, may be measured and recorded as 72°F, while in reality it is somewhere between 72°F and 73°F. A woman may give her age as thirty-nine, while in reality it is something quite different, and she is growing older even as she states it.

The measurement process, which combines the concepts of accuracy, resolution, and the time at which the numerical value is determined, is commonly referred to as *digitizing* the variable. This points to the fact that the original variable is replaced by a numerical value whose digits represent the size of the variable at a specific point in time. Once it has been converted into a digital form, a value can be processed or stored with no further loss of accuracy or resolution.

Thus, to answer the original question, a digital system is one that processes information that is in a digital (or numerical) form, rather than processing the continuous variables themselves. Systems that perform the conversion between the continuous variables and their digitized representations (and the reverse process) are also classified as digital systems.

1

This continuous-to-digital conversion process is familiar to us all. A driver is asked, "How fast are you going?" He looks at his speedometer (a continuous device) and says, "Sixty miles per hour" (a digital response). He is probably not going at that exact rate, but he must make the conversion to numerical form in order to state his speed.

1.1 Accuracy and Resolution

In the above example the correctness of the driver's stated speed is related to two factors: the basic accuracy with which the speedometer indicates the speed of the vehicle and the resolution with which the observer reads the indicated value. Accuracy is related to the quality of the measurement process. Increased accuracy usually requires improved measurement techniques or devices. Measuring the diameter of a ball bearing with a tape measure will certainly not give the same degree of accuracy as that obtained by using a micrometer. The accuracy of a given measurement is of great concern in many scientific and engineering applications.

The resolution of the continuous/digital conversion process is related to the minimum separation between successive numerical values that can result from the measurement process. This minimum separation, commonly called the *unit of resolution*, limits the accuracy of the conversion. Continuous values falling between successive numerical values will have to be given an approximate digital value that is either greater than or less than their true value. The unit of resolution in the U.S. currency is the penny. Thus, for example, if two children find a nickel and seek to divide their prize, one will have to receive only two cents while the other will get three. An exact division is not possible. The error in both amounts will be one-half cent, or one-half of the unit of resolution. Generally, the accuracy with which a digital value represents a continuous variable can be no better than plus or minus one-half of the unit of resolution, for an error range of one full unit.

The accuracy of a measurement can be increased by narrowing the interval between successive numerical values. This increase in resolution necessitates that more numerical information be obtained during the conversion process. For example, if a metal bar is measured to be 12 in. long, the measurement implies that the bar is between 11.5 and 12.5 inches in length. A more precise measurement of 12.00 in. places the length between 11.995 and 12.005 in.

Just as accuracy cannot be improved beyond the limit imposed by the unit of resolution, it is useless to increase resolution beyond the limit of the basic accuracy of the conversion process. Some people can perform a multiplication on a slide rule (a continuous device) and announce the answer to five or six decimal places or more. This excessive resolution is not truly useful, since the accuracy of the slide rule is generally limited to about three digits.

The accuracy and the resolution of the continuous/digital conversion process go hand in hand. An increase in either will generally increase either (or both) the cost and complexity of the conversion system. Increasing the resolution, thereby increasing the numerical information that makes up the digital values resulting from the conversion process, will increase the data storage and information processing tasks that the digital system will have to perform. It is desirable to use only the resolution and accuracy that is actually required by the application.

1.2 Continuous Systems Versus Digital Systems

Not all systems are digital in nature. Many computational and control systems process information which is in a continuous form. Mercury thermometers, most bathroom scales, slide rules, and electronic analog computers are devices that operate in a continuous mode. The contrast between a continuous system and a digital system can be illustrated by comparing a mercury thermometer and a digital temperature measuring system such as those incorporated in many signs advertising banks or savings associations. The thermometer is a continuous device. The mercury in it will rise to a height that corresponds to the temperature of the bulb of the thermometer. Any change in temperature will bring about a corresponding change in the height of the mercury column.

The digital system, on the other hand, will periodically convert the temperature to a digital value and display that value on its sign, usually to the nearest degree. A change in temperature will not be indicated until the change is large enough to move the digital value to the next higher or lower degree. Otherwise, the indicated value will remain the same. Like a stopped clock that is only correct twice a day, the digital system will rarely indicate the exact temperature, but it should always be correct within a resolution of one degree.

Digital systems are finding wider application for a variety of reasons. First, by narrowing the unit of resolution, digital systems can have more accuracy than continuous systems. This leads to repeatability in system operation, an important consideration in such areas as computation and numerically controlled machine tooling. Also, digital data can be stored and/or transmitted without deterioration. Consider the difference between measuring a voltage and recording the measured value and trying to store the voltage level on a capacitance. Surely changes in environment and time

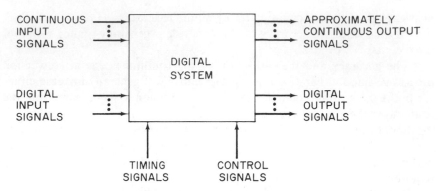

Figure 1-0. A Generalized Digital System

will have less effect on the digitally recorded voltage than on the voltage level that is stored on the capacitance. In addition, digital data can be encoded in new ways, leading to efficient, error-protected data handling. Also, operating electronic devices in a digital mode leads to reliable, trouble-free operation, even when low-cost components with loose tolerances are used. These and other reasons have led to widespread application of digital systems in new areas and in areas where continuous systems were previously used.

A generalized digital system is shown in Fig. 1-0. The inputs to the system may be in both continuous and digital form. In an electronic digital system, the continuous inputs must be converted into a suitable form, commonly voltage or current, by appropriate signal transducers. Thermocouples, strain gauges, and photocells are typical devices used for this purpose. The continuous output signals from such transducers are the electrical *analogs* of the original variables. These electrical signals are converted into digital form by digital subsystems called *analog-to-digital converters* (ADCs).

Direct digital input signals need no such conversion. Entries from a cash register or keyboard and data read from a punched card are examples of this. Also, some transducers convert their input signals directly to a digital form. A turbine flowmeter which generates an electrical pulse for each of its revolutions is an example. The pulses can be accumulated in a counter to obtain information about total flow or flow rates. Internal to the digital system, however, all signals are digital in form.

Direct digital input signals and the digital signals that are generated by analog-to-digital conversion are processed by the digital system to obtain the desired output information. Where necessary, digital output information can be converted back to an approximately continuous form by a *digital-to-analog converter* (DAC). The design of analog-to-digital and digital-to-analog converters is covered in Chapter 11. Outputs may be presented in a digital form as well, with scoreboards and line printers being examples.

1.3 The Design of Digital Systems

Digital design may be separated into three general aspects: component design, functional design, and systems design. *Component design* centers around the electronic devices that perform the simplified logical operations within a digital system. These devices have unusual names (gates, flip-flops, etc.), and their design and use is somewhat different from other aspects of electronics or circuit analysis. The goal of digital component design is to make cheap, reliable digital devices. A wide variety of digital components is available from various electronic manufacturers, and this facet of digital design is covered only briefly in this text. Readers specifically interested in digital electronics should see Refs. [3], [4], and [6].

Functional design is the assembly of digital components into working subsystems such as counters or adders. These functional subsystems are then combined, along with the appropriate timing, control, and interface systems, into the overall digital system that performs the desired operations. The last task is, of course, the *systems design* phase. Typical digital systems include traffic light controllers, numerically controlled milling machines, and digital computers. This textbook is mainly concerned with the functional and systems aspects of digital design. Functional design techniques are covered in Chapters 2 through 6. Chapter 7 covers system design, and the remainder of the text deals with selected applications of digital systems technology.

EXAMPLE 1-0: A Digital Rotational Speed Measurement System

It is desired to measure the speed of a rotating shaft to the nearest revolution per minute. The digital system shown in Fig. E1-0 is developed for this purpose. A black band is painted on the shaft, leaving one white timing

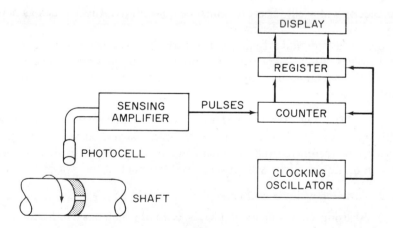

Figure E1-0

mark. A photocell senses the presence of the timing mark, generating an electrical pulse for each revolution of the shaft. The pulses are totalized in a counter that is reset once per minute. The counter thus records the number of revolutions per minute. Just prior to each periodic reset, the value in the counter is transferred to a storage register, where it remains, unchanging, for the next full minute. The value in the storage register is used to drive a numerical display, indicating the shaft speed in revolutions per minute.

This system may be broken down into its various design aspects:

Component Design: design of the photocell and its electronics, the oscillator that provides periodic resetting, the storage and counting devices, and the numerical display devices and their associated electronics.

Functional Design: design of the counter, the storage register, and the numerical display unit.

Systems Design: assembly of the system and provision of the necessary electronic interfaces.

1.4 Information and Quantity

Once converted into digital form, the items that are processed or stored by a digital system may be separated into two general classes: items of information and items of quantity. Information items give facts and may be thought of in a YES/NO or TRUE/FALSE sense, although many information items have more than two values. Quantitative items are numerical in nature. In a computer-based personnel file, for example, there will be informative entries such as sex (male or female) and marital status (single, married, divorced, or widowed). There will also be quantitative entries such as height, weight, and number of dependents.

Insofar as a digital system is concerned, there is much similarity between the two types of items. The next two subsections discuss both types, showing how all the operands used by a digital system can be reclassified into very simple forms.

1.4.1. The Representation of Information.

Information is in general multivalued. To repeat the previous example, marital status can be single, married, divorced, widowed, or other (bigamist). Such multivalued information can be stored, processed, or displayed in either of two ways:

1. By using a single multistate device.

2. By using a multiplicity of devices, each of which has a reduced number of states.

The total number of states, or conditions into which the device (or devices) can be set, must equal or exceed the number of possible values that the item of information can take on.

To illustrate the above choices, consider a scoreboard that must indicate the final outcome of a sporting event. Four illuminated displays are to be provided, labeled *WIN, LOSE, TIE,* and *FORFEIT*. The selection switching for these displays may be provided by a single four-position rotary switch, four single-pole, single-throw (SPST) toggle switches, or two single-pole, double throw (SPDT) toggle switches, as shown in Fig. 1-1. The last case in

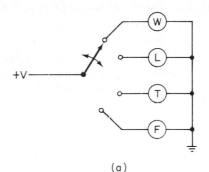

(a)

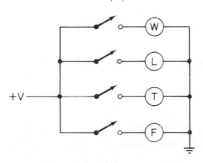

(b)

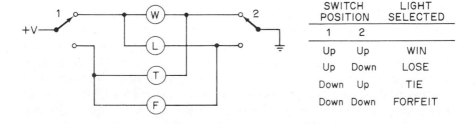

(c)

SWITCH POSITION		LIGHT SELECTED
1	2	
Up	Up	WIN
Up	Down	LOSE
Down	Up	TIE
Down	Down	FORFEIT

Figure 1-1. Alternative Switching Arrangements for Simple Scoreboard

Fig. 1-1 illustrates a vital point. The particular indicator light that is selected depends on the setting of *both* the control switches, each of which has two positions. Thus, there are $2^2 = 4$ possible combinations that can be selected with a pair of two-valued devices.

The selection of an optimum information representation system has been considered by Richards [5], showing that devices which can take on e (approximately 2.7) states each are optimum, based upon minimizing the number of devices and the number of different states per device. This leads to three-state devices as the nearest integral choice, with two-state devices as a next-to-optimum selection. The availability of two-valued electrical and electronic devices, coupled with the ease and reliability of operating electronic circuits in an ON/OFF mode, has led to the use of the two-valued information representation system in almost all electronic digital systems.

Thus, given an informative item that can take on N states, M two-valued variables may be used to represent the item, where M is the smallest integer such that

$$2^M \geqq N \qquad\qquad (1\text{-}0)$$

EXAMPLE 1-1:   Given a variable that can take on ten different values, how many two-valued variables must be used to represent it?

Answer: Four, since 2^4 is 16, while 2^3 is only 8.
Is there any redundancy?

Answer: Yes, 6 of the 16 possible combinations of the 4 variables are not used.

The replacement of multivalued items of information by several two-valued items simplifies the operation of a digital system. The system need consider only one of two choices at its inputs and outputs that are informative in nature. What of quantitative items, however?

1.4.2. The Representation of Quantity.   As has been stated, quantitative information in a digital system is bounded on the lower end by a resolution limit that is fixed by the numerical conversion process. This limit establishes the smallest quantitative value that can be carried by the system. Also, owing to the finite size of all realizable digital systems, an upper limit must be established on the size that a numerical variable can have. This, of course, may necessitate the use of numerical scaling, should quantities that exceed that limit arise. The odometer of a car illustrates this point, since it has a resolution of 0.1 miles and a maximum capacity of six digits (99999.9 miles).

The upper and lower limits combine to establish a finite number of possible quantitative values for any quantitative item, e.g., 1 million different mileage readings for the odometer. Thus, a quantitative item in a digital system can be thought of as a multivalued item of information telling which of a finite number of quantitative values most closely describes the desired quantity.

EXAMPLE 1-2:   A digital system has a resolution of 0.001 units and a maximum capacity of four digits. How many possible quantitative values can a variable take on?

   Answer: 10,000

What is their range?

   Answer: 0.000–9.999

Just as a multivalued item of information can be reexpressed in other ways, quantitative information can be represented in, various ways within a digital system. For example, consider the representation of an integral quantity that falls in the range from 0 to 999. At one end of the representation spectrum is the use of only *one* item of information to describe the quantity, with this one item, or symbol, having 1000 different possible values. Clearly, the specification of 1000 different symbols that are easy to write, print, interpret, etc., is a difficult task. Imagine the mechanical complexity of a single-pole, 1000-position rotary switch. This extreme is obviously a poor choice for quantitative description.

The decimal number system is one approach to the problem of representing numerical values. It permits a numerical quantity to be reexpressed in terms of multivalued coefficients, a_i, as

$$N = \sum_{i=0}^{M-1} a_i \times 10^i \qquad \text{(integers only, } 0 \le N < 10^M) \qquad (1\text{-}1)$$

where each coefficient can take on only *ten* values. Thus, more items of information are used, but each has a more limited range of possible values. With M symbols, each taking on ten different values, 10^M different combinations are possible. In the decimal system these combinations represent the numerical values from 0 to $10^M - 1$, unless scaling, in the form of a displaced decimal point, is used.

Numerical representations in the form of Eq. (1-1) are referred to as *positional number systems*. The summation shown in Eq. (1-1) is usually written in shorthand form as

$$N = a_{M-1} \cdots a_3 a_2 a_1 a_0 \qquad (1\text{-}2)$$

where the *position* of each coefficient tells which of the powers of 10 acts as its multiplier.

EXAMPLE 1-3:   Express the decimal number 102693 in the summation form of Eq. (1-1).

Answer:

$$102693 = 1 \times 10^5 + 0 \times 10^4 + 2 \times 10^3 + 6 \times 10^2$$
$$+ 9 \times 10^1 + 3 \times 10^0$$

The decimal system is a single case of a positional numbering system. In general, given a numerical quantity, N, and a *radix* or *base*, R, the quantity can be expressed in terms of M coefficients, where

$$M = [\log_R (N)] \tag{1-3}$$

The brackets imply that M is the smallest integer equal to or greater than the quantity within the brackets. The general form for expressing the quantity is

$$N = \sum_{i=0}^{M-1} a_i R^i \tag{1-4}$$

where each coefficient can take on R values, ranging from 0 to $R - 1$. The possible values for N range from 0 to $R^M - 1$, a total of R^M possibilities.

In Sec. 1.4.1 it was shown that reducing informative items to their simplest form, the two-valued or YES/NO case, was the most convenient way for a digital system to operate. Similarly, the representation of quantitative information in a base-2, or *binary*, numbering system, has advantages in digital applications involving the processing of numerical data. In the binary numbering system the radix is 2, restricting the coefficients in Eq. (1-4) to two values, zero and 1. Thus, each coefficient may be considered as a two-valued item of information, e.g., FALSE when the quantitative equivalent is zero and TRUE when it is 1. The number of symbols required to represent a quantity in the binary numbering system will exceed that required by the decimal system (or any other numbering system having a radix greater than 2, for that matter), but the simplicity of having two-valued coefficients generally offsets the increase in their number.

Certainly not all digital systems process numerical information, but the frequency with which the binary number system and other number systems closely related to it appear in digital systems practice makes a familiarity with binary notation desirable. The next section briefly introduces this system for representing quantitative information. Arithmetic operations are covered in more detail in Chapters 8 and 9.

1.5   The Binary Number System

The binary number system uses the quantity 2 as its base. Integral quantities are expressed as

$$N = a_0 2^0 + a_1 2^1 + a_2 2^2 + a_3 2^3 + \cdots$$
$$= a_0 + 2a_1 + 4a_2 + 8a_3 + \cdots \qquad (1\text{-}5)$$

Each coefficient can take on only the values zero or 1, making binary numbers look like decimal numbers with the symbols for 2–9 missing. Basic to understanding the binary number system is a familiarity with the powers of 2. Table 1-0 gives this information for the first ten powers. These powers should become as familiar to digital systems designers as the powers of 10 are to school children.

Table 1-0:   The Powers of 2

i	2^i
0	1
1	2
2	4
3	8
4	16
5	32
6	64
7	128
8	256
9	512
10	1024

Of note is the last entry in Table 1-0, showing that 2^{10} is approximately 1000, an approximation which is useful in estimating the number of binary coefficients required to represent large decimal numbers.

EXAMPLE 1-4:   Approximately how many coefficients will the binary equivalent of 100,000 (base 10) have?

Answer:

$$100,000 = 100 \times 10^3$$
$$\simeq 128 \times 10^3$$
$$\simeq 2^7 \times 2^{10}$$
$$\simeq 2^{17}$$

Thus, approximately 17 binary coefficients will be required.

The coefficients of a binary number correspond to the digits of a decimal number and are commonly called *bits*, a contraction of *binary digits.**

*The extension of this contracted form to other numbering systems leads to problems. Consider the selection of a shortened name for the coefficients of the base-3, or *ternary*, system.

The number of bits present in a binary number establishes the maximum value that the number can take on. For N bits, the number can take on 2^N values ranging from 0 to $2^N - 1$. The binary equivalents for the decimal numbers from 0 to 15 are given in Table 1-1. These, too, should be learned.

Table 1-1: Binary/Decimal Equivalence

Decimal Number	Binary Number
0	0000
1	0001
2	0010
3	0011
4	0100
5	0101
6	0110
7	0111
8	1000
9	1001
10	1010
11	1011
12	1100
13	1101
14	1110
15	1111

The representation of quantities greater than 15 requires at least a five-bit binary number.

1.5.1. Binary/Decimal Conversion.   Human beings are conditioned to work with the decimal number system rather than the binary system. Who wants to get a paycheck made out for $1000 only to find out that it is worth only $8? Digital systems operate best with two-valued variables, however, making the binary number system most useful for digital processing. Thus, to obtain numerical data from or return data to the "real world," digital systems must often provide conversion between decimal and binary numbers. This is strictly a digital process, and no loss in accuracy or resolution is implied. Actual systems for making these conversions are designed in later chapters; at this point only manual conversion processes are introduced.

Decimal-to-Binary Conversion

Decimal-to-binary conversion may be made in several ways. For small decimal values (15 or less), familiarity with Table 1-1 is suggested. For larger decimal values, two methods are useful:

1. Extraction of powers of 2.
2. The *dibble-dabble* method [3].

The first method involves a decomposition of the decimal number into a summation of powers of 2 from which the binary equivalent can be found.

EXAMPLE 1-5: Convert 273 to binary by using the extraction of powers of 2.

    Procedure:

$$273 = 256 + 16 + 1$$
$$= 2^8 + 2^4 + 2^0$$
$$= 1 \times 2^8 + 0 \times 2^7 + 0 \times 2^6 + 0 \times 2^5 + 1 \times 2^4$$
$$+ 0 \times 2^3 + 0 \times 2^2 + 0 \times 2^1 + 1 \times 2^0$$

Thus $273_{10} = 100010001_2$. (The subscript indicates the value of the radix.)

EXAMPLE 1-6: Convert 496_{10} into binary in the same way.

    Procedure (left to right):

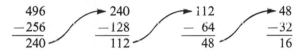

Thus

$$496_{10} = 256 + 128 + 64 + 32 + 16$$
$$= 2^8 + 2^7 + 2^6 + 2^5 + 2^4$$

Hence $496_{10} = 111110000_2$.

———

The second method for decimal-to-binary conversion, the dibble-dabble technique, is an algorithmic approach that is useful with larger decimal numbers. The decimal number is repeatedly divided by 2, yielding a succession of remainders of zero or 1. These remainders, when read in reverse order, yield the binary equivalent to the binary number.

EXAMPLE 1-7: Convert 273_{10} to binary by using the dibble-dabble method.

    Procedure:

$$273 \div 2 = 136 \text{ and a remainder of } 1$$
$$136 \div 2 = 68 \text{ and a remainder of } 0$$
$$68 \div 2 = 34 \text{ and a remainder of } 0$$
$$34 \div 2 = 17 \text{ and a remainder of } 0$$
$$17 \div 2 = 8 \text{ and a remainder of } 1$$
$$8 \div 2 = 4 \text{ and a remainder of } 0$$
$$4 \div 2 = 2 \text{ and a remainder of } 0$$
$$2 \div 2 = 1 \text{ and a remainder of } 0$$
$$1 \div 2 = 0 \text{ and a remainder of } 1$$

Reading the remainders from bottom to top gives

$$273_{10} = 100010001_2$$

EXAMPLE 1-8: Convert 496 to binary.

Procedure:

$$496 \div 2 = 248 \text{ and } 0$$
$$248 \div 2 = 124 \text{ and } 0$$
$$124 \div 2 = 62 \text{ and } 0$$
$$62 \div 2 = 31 \text{ and } 0$$
$$31 \div 2 = 15 \text{ and } 1$$
$$15 \div 2 = 7 \text{ and } 1$$
$$7 \div 2 = 3 \text{ and } 1$$
$$3 \div 2 = 1 \text{ and } 1$$
$$1 \div 2 = 0 \text{ and } 1$$

Thus $496_{10} = 111110000_2$.

EXAMPLE 1-9: Convert 1000 to binary.

Procedure:

$$1000 \div 2 = 500 \text{ and } 0$$
$$500 \div 2 = 250 \text{ and } 0$$
$$250 \div 2 = 125 \text{ and } 0$$
$$125 \div 2 = 62 \text{ and } 1$$
$$62 \div 2 = 31 \text{ and } 0$$
$$31 \div 2 = 15 \text{ and } 1$$
$$15 \div 2 = 7 \text{ and } 1$$
$$7 \div 2 = 3 \text{ and } 1$$
$$3 \div 2 = 1 \text{ and } 1$$
$$1 \div 2 = 0 \text{ and } 1$$

Thus $1000_{10} = 1111101000_2$.

It should be remembered that a binary number consisting of N consecutive 1s is equal to $2^N - 1$. Thus, for example,

$$1111111_2 = 2^7 - 1 = 127_{10} \tag{1-6}$$

Binary-to-Decimal Conversion

Binary-to-decimal conversion can be accomplished by reversing the two techniques given above, yielding

1. The summation of powers of 2.

2. The *dabble-dibble* algorithm (what else?).

The first technique is best illustrated by example.

EXAMPLE 1-10:　Convert 101011_2 to decimal.

Procedure:

$$101011_2 = 1 \times 2^5 + 1 \times 2^3 + 1 \times 2^1 + 1 \times 2^0$$
$$= 32 + 8 + 2 + 1$$
$$= 43_{10}$$

EXAMPLE 1-11:　Convert 1111101_2 to decimal.

Procedure:

$$1111101_2 = 2^6 + 2^5 + 2^4 + 2^3 + 2^2 + 2^0$$
$$= 64 + 32 + 16 + 8 + 4 + 1$$
$$= 125_{10}$$

Note that since $1111111_2 = 127_{10}$, 1111101_2 *must be* 125_{10}, since the 2^1 factor is the only one missing.

For longer numbers the summation method becomes tedious. Then the reverse of the dibble-dabble method, the dabble-dibble* method, is useful. This is implemented by multiplying the most-significant bit of the binary number by 2, adding the next bit to the product, multiplying by 2 again, and repeating the process. No multiplication by 2 takes place after the last bit (the one in the 2^0 position) is added to the accumulated product.

EXAMPLE 1-12:　Convert 1111111_2 to decimal by using the dabble-dibble algorithm.

Procedure:　Begin with leftmost bit, as

$$1 \times 2 = 2$$

Add the next bit:

$$2 + 1 = 3$$

*The term is an alliterative perversion of the *double-add* sequence used by the algorithm.

Multiply and repeat:

$$3 \times 2 = 6, \qquad 6 + 1 = 7$$
$$7 \times 2 = 14, \qquad 14 + 1 = 15$$
$$15 \times 2 = 30, \qquad 30 + 1 = 31$$
$$31 \times 2 = 62, \qquad 62 + 1 = 63$$
$$63 \times 2 = 126, \qquad 126 + 1 = 127$$

The last bit has been reached; hence

$$1111111_2 = 127_{10}$$

EXAMPLE 1-13:   Convert 101101_2 to decimal.

Procedure:

$$1 \times 2 = 2, \qquad 2 + 0 = 2$$
$$2 \times 2 = 4, \qquad 4 + 1 = 5$$
$$5 \times 2 = 10, \qquad 10 + 1 = 11$$
$$11 \times 2 = 22, \qquad 22 + 0 = 22$$
$$22 \times 2 = 44, \qquad 44 + 1 = 45$$

Thus

$$101101_2 = 45_{10}$$

As a check,

$$45_{10} = 32 + 8 + 4 + 1$$
$$= 2^5 + 2^3 + 2^2 + 2^0$$
$$= 101101_2$$

EXAMPLE 1-14:   Convert 1111101000_2 to decimal.

Procedure:

$$1 \times 2 = 2, \qquad 2 + 1 = 3$$
$$3 \times 2 = 6, \qquad 6 + 1 = 7$$
$$7 \times 2 = 14, \qquad 14 + 1 = 15$$
$$15 \times 2 = 30, \qquad 30 + 1 = 31$$
$$31 \times 2 = 62, \qquad 62 + 0 = 62$$
$$62 \times 2 = 124, \qquad 124 + 1 = 125$$
$$125 \times 2 = 250, \qquad 250 + 0 = 250$$
$$250 \times 2 = 500, \qquad 500 + 0 = 500$$
$$500 \times 2 = 1000, \qquad 1000 + 0 = 1000$$

Thus $1111101000_2 = 1000_{10}$.

Octal/Binary Relationships

As a convenient, shorthand method for expressing a lengthy binary number, the base-8 or *octal* number system can be used. Since 8 is 2^3, each octal digit can be determined by examining a group of three consecutive bits in the binary number. The three bits will give a value from 0 to 7, and that value will correspond to the octal coefficient. Conversion to and from octal can be used as an intermediate step in binary/decimal conversions. It is also useful in displaying and printing binary information, such as when listing the contents of successive memory locations within the core memory of a digital computer.

EXAMPLE 1-15:   Convert 1011101_2 to octal and then to decimal.

Procedure: Divide the binary number into groups of three bits:

$$1011101 = \underset{1}{001} \quad \underset{3}{011} \quad \underset{5}{101} \quad \text{(Note that high-order zeros are added if needed.)}$$

Thus $1011101_2 = 135_8$.

Conversion to decimal can then be accomplished by using the summation of powers method:

$$135_8 = 1 \times 8^2 + 3 \times 8^1 + 5 \times 8^0$$
$$= 1 \times 64 + 3 \times 8 + 5 \times 1$$
$$= 64 + 24 + 5$$
$$= 93_{10}$$

Note that octal numbers will always appear to be *larger* than their decimal equivalents.

EXAMPLE 1-16:   Convert 110101001001_2 to octal.

Answer: $110101001001_2 = 6511_8$

EXAMPLE 1-17:   Convert 1762 from octal to binary.

Answer: $1762_8 = 1111110010_2$

Hexadecimal/Binary Relationships

An alternative to the octal shorthand used to represent a binary number is the *hexadecimal* or "hex" representation. This method uses groups of four binary bits to describe the coefficients of a base-16 (2^4) number. Since the coefficients must range from 0 to 15 in value, the normal decimal digits must be augmented in order to express the values above 9. The six letters, *A–F*, are commonly used for this purpose. The 16 hex symbols and their binary, decimal, and octal equivalents are shown in Table 1-2. Octal representations use one-third as many characters as the equivalent binary number; hex uses one-fourth as many.

Table 1-2: Hexadecimal, Binary, Octal, and Decimal Equivalents

Hex	Binary	Octal	Decimal
0	0000	0	0
1	0001	1	1
2	0010	2	2
3	0011	3	3
4	0100	4	4
5	0101	5	5
6	0110	6	6
7	0111	7	7
8	1000	10	8
9	1001	11	9
A	1010	12	10
B	1011	13	11
C	1100	14	12
D	1101	15	13
E	1110	16	14
F	1111	17	15

EXAMPLE 1-18:   Convert 10110111_2 to hex and then to decimal.

Procedure: Separate the binary number into four-bit groups to get the hex coefficients.

$$10110111_2 = \underset{\underset{B}{\uparrow}}{1011} \quad \underset{\underset{7}{\uparrow}}{0111}$$

$$= B7_{16}$$

The decimal equivalent can be found by using the summation of powers once again:

$$B7_{16} = B \times 16^1 + 7 \times 16^0$$
$$= 11 \times 16 + 7 \times 1$$
$$= 176 + 7$$
$$= 183_{10}$$

EXAMPLE 1-19:   Convert 100101101_2 to hex.

Answer: 12D

EXAMPLE 1-20:   Convert BAD to binary.

Answer: 101110101101_2

1.5.2. Basic Binary Arithmetic.   The usual arithmetic operations of addition, subtraction, multiplication, and division can be performed in the binary numbering system and in octal and hexadecimal as well. Binary addition and subtraction are introduced in this subsection; Chapter 8 covers this subject in more detail. Readers interested in more complete discussion of arithmetic operations in digital systems are referred to Refs. [1], [2], and [5].

Binary Addition

In all positional numbering systems, the addition of two numbers may be performed column by column (referred to as decade by decade in the decimal system) rather than by operating on the numbers in their entirety. Thus, only two coefficients need be considered at a time. School children learn the addition table for the decimal system and by applying the table to each decade of the numbers learn to add decimal numbers of any length.

Whenever the sum of two coefficients equals or exceeds the radix, R, the *carry* operation is used to propagate the excess into the next higher column, since that column has a weight which is R times greater than the column being added. The carry adds one unit to the next higher column. Thus, for all columns except the least significant, the addition process involves two coefficients and a carry from the next lower column. The carry can only have the values of zero or 1, no matter what radix is used by the numbering system. In the binary system the coefficients can only have values of zero or 1, as well, making the binary addition table much simpler than that used for decimal addition.

In the least-significant position of a binary number only *two* operands are added. This process, commonly called *half-addition*, is described in Table 1-3, where all possible combinations for the two coefficients are considered.

Table 1-3:   **Tabulation of the Half-addition Operation**

Addend Coefficient	Augend Coefficient	Carry to Next Column	Sum in this Column
0	0	0	0
0	1	0	1
1	0	0	1
1	1	1	0

The values of the resultant sum and carry for each combination are shown. The last row shows that $1 + 1$ yields a sum of zero and a carry of 1, or 10_2, which is the way the quantity 2 must be represented in binary.

For higher-ordered columns, the binary addition process involves three operands. This operation is called *full addition*. There are eight possible combinations of the three two-valued inputs, as shown in Table 1-4. Table 1-4

Table 1-4:   Tabulation of the Full-addition Operation

Addend Coefficient	Augend Coefficient	Carry from Lower Column	Carry to Next Column	Sum in this Column
0	0	0	0	0
0	0	1	0	1
0	1	0	0	1
0	1	1	1	0
1	0	0	0	1
1	0	1	1	0
1	1	0	1	0
1	1	1	1	1

shows that whenever the sum of the three operands is 2 or more, a carry is propagated to the next higher column. The sum then becomes two less than the total of the operands, reducing to zero if the precarry sum was 2 and becoming 1 if this sum was 3.

The complete binary addition operation is illustrated by the following examples.

EXAMPLE 1-21:   Add 5_{10} and 9_{10}, using binary notation.

Procedure:

$$5_{10} = 0101_2$$
$$9_{10} = 1001_2$$

Thus

```
0 0 1 ← carries
  0101
 +1001
 ─────
  1110 = 14₁₀
```
$$\quad 1110 = 14_{10}$$

EXAMPLE 1-22:   Add 1011_2 and 0111_2.

Answer:

```
1 1 1 1 ← carries
   1011
  +0111
  ─────
  10010
```

(This is $11 + 7 = 18$ in decimal.)

EXAMPLE 1-23: $11111 + 1 = ?$

Answer:

$$
\begin{array}{r}
1\,1\,1\,1\,1 \leftarrow \text{carries} \\
11111 \\
+ \qquad 1 \\
\hline
100000
\end{array}
$$

Familiarity with binary addition is expected of a digital systems designer, for the binary addition operation is at the heart of almost every processing step in computational applications of digital systems. Further examples of the binary addition operation are presented in the problems given in Sec. 1-9.

Binary Subtraction

The performance of binary subtraction is similar to that of binary addition, except that *borrows*, rather than carries, are propagated to higher-ordered columns. The borrow operation occurs in a column whenever the coefficient of the subtrahend exceeds that of the minuend. The borrow, if unaccounted for in the next higher column, must then be propagated further. The two-operand subtraction operation, called *half-subtraction*, is defined by Table 1-5. Note that the borrow is propagated only when the coefficient of

Table 1-5: Tabulation of the Half-subtraction Operation

Minuend Coefficient	Subtrahend Coefficient	Borrow to Next Column	Difference in this Column
0	0	0	0
0	1	1	1
1	0	0	1
1	1	0	0

the minuend is zero and the coefficient of the subtrahend is 1. The borrow reduces the next higher column by 1, bringing a value of 2 into the column that originated the borrow. Thus, since $2 - 1 = 1$, the difference in that column becomes 1, as shown in Table 1-5.

EXAMPLE 1-24:   Subtract 1_2 from 10_2.

Procedure: Begin with the rightmost column:

$$
\begin{array}{r}
0 \\
-1 \\
\hline
\end{array}
$$

This combination requires a borrow from the next higher column:

$$1 \leftarrow \text{borrow}$$
$$10$$
$$\underline{- \ 1}$$

The borrow reduces the second column by 1, making it have a net value of zero. The borrow also adds 2 to the first column:

$$\begin{array}{c} 10 \\ \underline{- \ 1} \end{array} \longrightarrow \begin{array}{c} 02 \\ \underline{- \ 1} \end{array}$$

which clearly gives a difference of 1 as the answer; i.e.,

$$10_2 - 1_2 = 1_2$$

For all columns except the lowest ordered, the subtraction operation involves three operands: the coefficients of the minuend and the subtrahend and the borrow from the next lower column. The borrow acts as a -1, just as does the coefficient of the subtrahend. The *full-subtraction* operation is described by Table 1-6. The fourth row of Table 1-6 shows that whenever the coefficient of the minuend is zero and both the coefficient of the subtrahend and the incoming borrow are 1, a net value of -2 results in the column. A borrow is then made, bringing down a value of 2 to cancel the -2

Table 1-6.   Tabulation of the Full-subtraction Operation

Minuend Coefficient	*Subtrahend Coefficient*	*Borrow from Lower Column*	*Borrow to Next Column*	*Difference in this Column*
0	0	0	0	0
0	0	1	1	1
0	1	0	1	1
0	1	1	1	0
1	0	0	0	1
1	0	1	0	0
1	1	0	0	0
1	1	1	1	1

value and making the difference in the column have a net value of zero. Of interest is the similarity between the Sum column of Table 1-4 and the Difference column of Table 1-6. Binary addition and subtraction differ only in the logic behind the propagation of carries or borrows.

Several examples of binary subtraction follow; other examples appear in the problems given in Sec. 1-9 and in Chapter 8.

EXAMPLE 1-25: Subtract 3_{10} from 7_{10}, using binary notation.

Procedure:

$$3_{10} = 11_2$$
$$7_{10} = 111_2$$

Thus

0 0 ← borrows
111
$$-011$$
$$100 = 4_{10}$$

EXAMPLE 1-26: Subtract 1 from 100 (binary).

Answer:

1 1 ← borrows
100
$$-001$$
$$11$$

EXAMPLE 1-27: What is $110_2 - 11_2$?

Answer:

1 1 ← borrows
110
$$-011$$
$$011$$

Binary subtraction is somewhat more difficult to perform manually than binary addition because of the borrowing process. For this reason, checking the results of the subtraction by adding the difference and the subtrahend to see that the value of the original minuend is obtained is strongly recommended.

To illustrate a second difficulty that may arise during subtractive operations, consider the following example.

EXAMPLE 1-28: Subtract 111_2 from 100_2.

Procedure:

1 1 1 0 ← borrows
100
$$-111$$
$$101$$

Note that the net difference appears as 101, with a borrow into the fourth column unresolved. This high-order borrow occurs whenever the subtrahend is larger than the minuend. It has a value, in this example, of -2^3 or -8. The difference appears in a *complemented* form, with the unresolved borrow pointing to the fact that the complementary form has arisen. The negative answer is really expressed as $-8 + 5 = -3$. The use of complementary forms for expressing negative numbers and for implementing binary subtraction through the addition of complements is discussed in detail in Chapter 8.

As an alternative, it may be recognized that the subtrahend is larger than the minuend, in which case the subtraction can be reversed, yielding

$$\begin{array}{r} {\scriptstyle 0\ 0\ 0\leftarrow\text{borrows}} \\ 111 \\ -100 \\ \hline 011 = 3_{10} \end{array}$$

The difference is now 3, and a minus sign must be assigned:

$$100_2 - 111_2 = -11_2$$

The binary number system is to quantitative information what the reduction to TRUE/FALSE variables is to multivalued items of information. The coefficients of a binary number are, in fact, two-valued items of information themselves, since they can describe only a value of 1 or zero. By using binary notation to represent quantity and by reducing informative data to TRUE/FALSE form, digital systems need process only two-valued variables. For this reason, digital systems are often called *logical* systems, since they operate in a TRUE/FALSE mode. This simplification is the key to the usefulness of digital systems.

1.6 Summary

Digital systems process, store, and display information that is originally in a discrete, digitized form or that is a digitized approximation of a continuous variable. Both informative and quantitative items can be processed, but both are converted into items of information that can take on only two possible values. Quantitative data are expressed in the binary numbering system. Simple binary arithmetic operations using this system are introduced. By using binary numbers and TRUE/FALSE expressions for items of information, digital systems design can be reduced to the manipulation of two-valued variables.

1.7 Did you Learn?

1. The contrast between *continuous* and *digital* systems?

2. What analog-to-digital conversion is and why it is needed?

3. How to represent multivalued information by using two-valued variables?

4. What a positional numbering system is?

5. How to convert numerical information from binary to octal, decimal, and hexadecimal? The reverse processes?

6. Binary addition and subtraction?

1.8 References

[1] Chu, Yoahan, *Digital Computer Design Fundamentals.* New York: McGraw-Hill Book Company, 1962.

[2] Flores, Ivan, *The Logic of Computer Arithmetic.* Englewood Cliffs, N.J.: Prentice-Hall, Inc., 1963.

[3] Hurley, R. B., *Transistor Logic Circuits.* New York: John Wiley & Sons, Inc., 1961.

[4] Millman, Jacob, and Herbert Taub, *Pulse, Digital, and Switching Waveforms.* New York: McGraw-Hill Book Company, 1965.

[5] Richards, R. K., *Arithmetic Operations in Digital Computers.* New York: Van Nostrand Reinhold Company, 1956.

[6] Sifferlen, Thomas, and Vartan Vartanian, *Digital Electronics with Engineering Applications.* Englewood Cliffs, N.J.: Prentice-Hall, Inc., 1970.

1.9 Problems

1-1. Example 1-0 described a digital rpm measuring system. How could information as to direction of rotation be included in this system? Discuss the component, functional, and systems design aspects of the system proposed. *Suggestion:* Consider painting two bands on the shaft.

1-2. A digital voltmeter has a range of 100 V and a resolution of 0.1 V. How many different readings are possible? How many two-valued variables are necessary to represent all possible readings? Is there any redundancy? How many three-valued variables would be needed?

1-3. A decimal number is to be indicated on a scoreboard by using a 4×6 grid of lamps as shown in Fig. P1-3. List the selection of bulbs that must be turned on in order to form each digit. Refer to the bulbs by letter and number. This could be done by a 10-position, 24-pole rotary switch. What other choices are possible?

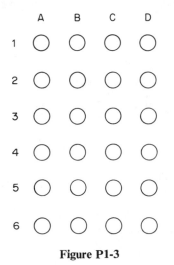

Figure P1-3

1-4. Discuss the component, functional, and systems design aspects of a traffic-light control system at a complex intersection.

1-5. A class has 32 students. Each is to be given a plastic card with a personal code combination (holes or no-holes) punched into it. How many hole locations must be specified? List all the possible combinations that can be generated. Is there any redundancy? Suppose that there were 33 students.

1-6. Convert the following binary numbers to octal, hexadecimal, and decimal. Use both the summation-of-powers and the dibble-dabble algorithm to make the binary-to-decimal conversion.

a. 101010
b. 111101101
c. 11110000

d. 10010111
e. 1001001001
f. 1111111110

1-7. Convert the following octal and hexadecimal numbers to binary and to decimal.

a. 777_8
b. 777_{16}
c. $B7A_{16}$

d. 123_8
e. 123_{16}
f. $FC1_{16}$

1-8. Convert the following decimal numbers to binary, octal, and hexadecimal. Use both the extraction-of-powers and the dabble-dibble algorithm to make the decimal-to-binary conversion.

a. 1000	d. 123	g. 27
b. 2000	e. 65	h. 270
c. 777	f. 63	i. 17

1-9. Estimate (by using the technique given in Ex. 1-4) the number of bits required to represent each of the following decimal numbers. Reexpress the answers in terms of octal and hexadecimal coefficients.

a. 10^4	d. 10^7
b. 10^5	e. 10^{15}
c. 10^6	f. 10^{20}

1-10. Estimate the decimal magnitude of each of the binary numbers which follow.

a. 10000000000	d. 2^{30}
b. 1000000000000000	e. 2^{40}
c. 2^{20}	f. $11111111111111111111 = 2^{18} - 1$

1-11. Develop an octal equivalent to the dibble-dabble and dabble-dibble algorithms. Use it to perform the octal/decimal conversions of Probs. 1-7 and 1-8.

1-12. Repeat Prob. 1-11 for the hexadecimal case.

1-13. The base-3 (ternary) system has three-valued coefficients. The values are 0, 1, and 2.

    a. List a decimal-to-ternary conversion table for the decimal values from 0 to 80.

    b. Convert the decimal numbers given in Prob. 1-8 into ternary form by using the extraction-of-powers method.

    c. Develop a ternary equivalent to the dabble-dibble algorithm and use it to convert the decimal numbers given in Prob. 1-8 to ternary.

    d. Develop a ternary version of the dibble-dabble algorithm. Using both the algorithm and the summation-of-powers method, convert the following ternary numbers into decimal form.

(1) 12021	(4) 20
(2) 112211	(5) 1010
(3) 210	(6) 10000

1-14. Perform the following binary arithmetic operations. Convert the operands into decimal and verify the answers.

a. $110101 + 110101$	d. $110101 - 100111$
b. $111111 + 100101$	e. $1110000 - 100011$
c. $110101101 + 10111111$	f. $10101111 - 1001001$

1-15. The addition table for octal numbers can be expressed as in Table P1-15. The sum that results when the numbers of a row and column are added is entered at their intersection. Boldface indicates that a carry is generated, with the carry going into the next higher octal column. The table on the left corresponds to octal addition with no carry from the lower column. The table on the right applies when an incoming carry is present. The tables are only partially completed.

Table P1-15

Without Incoming Carry

	0	1	2	3	4	5	6	7
0	0	1	2	3	4	5	6	7
1	1	2	3					
2	2	3	4					
3	3	4	5					
4	4	5	6					
5	5	6	7					
6	6	7	**0**					
7	7	**0**	**1**					

With Incoming Carry

	0	1	2	3	4	5	6	7
0	1	2	3	4	5	6	7	**0**
1	2							
2	3							
3	4							
4	5							
5	6							
6	7							
7	**0**							

   a. Complete the tables and, by using them, solve the following octal
      addition problems. Verify the results by using binary equivalents.
      (1) 723 + 156.
      (2) 165 + 165
      (3) 301 + 376
   b. Develop the tables that apply to octal subtraction with and without
      an incoming borrow. By using these tables, solve the octal subtraction
      problems given below. Check the results by using binary equivalents.
      (1) 723 − 156
      (2) 162 − 77
      (3) 300 − 65

1-16. Develop addition and subtraction tables for hexadecimal numbers. By using them, solve the following hexadecimal arithmetic problems. Check the results by using binary equivalents.

    a. A1 + B2   d. E00 − FF
    b. AA + BB   e. 95 − C
    c. 7A + 191   f. 100 − 10

1-17. Repeat Prob. 1-16 for the ternary numbering system. Perform the following ternary arithmetic operations. Verify the results by using decimal equivalents.

    a. 122 + 211   d. 200 − 122
    b. 200 + 200   e. 100 − 22
    c. 20120 + 11222   f. 1122 − 200

1-18. Assume that an M-valued item of information is to be represented in a digital system. Prove that if the cost of representing the item is defined as the product of the number of variables that are required and the number of values that each variable can take on, then the cheapest number of variables is

$$N = \ln (M)$$

where each variable can take on e values. (See Richards [5], pp. 7–11.)

2 Introduction to Boolean Algebra

2.0 The Nature of Two-Valued Variables

The manipulation of variables that have only two possible values has been systematized by Shannon [6], using ideas that were first expressed by an English mathematician, George Boole [1]. This branch of mathematical theory is given the name Boolean (Boo′·lē·ŭn) algebra. Unlike the usual algebraic variables, a Boolean variable, say A, can take on only two values, which are commonly referred to as TRUE and FALSE. Other values may be assigned, however, such as hot/cold, male/female, or tall/short. The symbols 0 and 1 are used to represent the two possible values of Boolean variables. $A = 1$ usually means that A is TRUE in a Boolean sense, making $A = 0$ indicate that A is FALSE. A Boolean variable may then be related to some piece of information, e.g., $A = 1$ meaning that the switch associated with A is open and $A = 0$ signifying that the same switch is closed. Another variable, B, may relate to room temperature, being TRUE when the temperature exceeds 70°F and FALSE otherwise.

Boolean variables do not take on quantitative values, but they may be used to represent quantitative information. For example, a four-bit binary number can be represented by four Boolean variables. Each variable can be related to one of the coefficients of the binary number, indicating that the coefficient has a value of 1 when the variable is TRUE and a value of 0 when the variable is FALSE (or the reverse of this). In this way the 16 possible TRUE/FALSE combinations of the variables can be related to the quantities $0-15_{10}$ that the binary number can take on. Knowing the TRUE/FALSE values of each of the variables will enable the quantity that they represent to be calculated.

Boolean variables can be manipulated by using operators that are similar to the usual algebraic operators. These Boolean operators are commonly called *logical connectives*. The major connectives are discussed in Secs. 2.1–2.7.

2.1 The AND Connective

The AND connective is defined for two or more Boolean arguments and can be related, somewhat incorrectly, to multiplication. The common symbols for the AND operation are

$$f(A,B) = A \cdot B = A.B = AB = A \cap B = A \wedge B \qquad (2\text{-}0)$$

where A and B are Boolean variables. The first and third symbols (the center-ed dot and the *implied* AND) are used in this text.

In a word statement, the AND of several Boolean variables is TRUE if and only if all the variables are TRUE. An alternative way of expressing the behavior of a logical connective (or any Boolean expression, for that matter) is the *truth table*. Since each of the arguments of the connective can take on only two values, only a finite number of different input combinations is possible. For an m-argument connective this number is 2^m. A truth table lists all these possible input combinations and gives the TRUE/FALSE value of the connective for each of them. Thus, the table completely specifies the operation of the logical connective. In truth table form, the two- and three-argument AND operations are

A	B	A·B
0	0	0
0	1	0
1	0	0
1	1	1

A	B	C	ABC
0	0	0	0
0	0	1	0
0	1	0	0
0	1	1	0
1	0	0	0
1	0	1	0
1	1	0	0
1	1	1	1

Note that $A \cdot B$ is TRUE only when both A *and* B are TRUE.

The AND operation can be related to the electrical operation of two or more toggle switches connected in series. The switch equivalent to the two-argument AND operation is illustrated in Fig. 2-0. The light indicates when-ever the output signal is TRUE. Note that the light can be illuminated only when switch A *and* switch B are both closed.

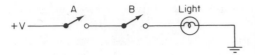

Figure 2-0. The Switch-Equivalent to the Two-Argument AND Connective

There are many different symbolic representations for the AND operation. The Institute of Electrical and Electronics Engineers and the American Standards Association have proposed the "shape distinctive" IEEE-ASA symbol set. These symbols are used throughout this text. The IEEE-ASA symbol for the AND operation is shown in Fig. 2-1. This symbol is commonly called an *AND gate*. The term *gate* is a carry-over from early switching systems that used relays to perform logical functions. The arm of the relay was opened or closed to "gate" the electrical signals.

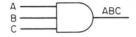

Figure 2-1. The AND Gate Symbol

2.2 The OR Connective

Like the AND connective, the OR connective is defined for two or more Boolean arguments. The OR connective can be loosely related to normal algebraic addition. The common symbols for the OR operation are

$$f(A,B) = A + B = A \cup B = A \vee B \tag{2-1}$$

The first notation is used herein. The use of the word *plus* should be avoided when working with the OR function.

In a word statement, the OR of several Boolean variables is TRUE if *any* of the variables is TRUE. In truth table form, the two-variable and three-variable OR operations are

A	B	A + B
0	0	0
0	1	1
1	0	1
1	1	1

A	B	C	A + B + C
0	0	0	0
0	0	1	1
0	1	0	1
0	1	1	1
1	0	0	1
1	0	1	1
1	1	0	1
1	1	1	1

showing that $A + B$ is TRUE if either *A or B* is TRUE.

The normal OR function is TRUE even when both *A* and *B* are TRUE, and is sometimes called the INCLUSIVE OR. A related function, the EXCLUSIVE OR, is FALSE in this case. Its truth table (for two variables) is

A	B	A EX-OR B
0	0	0
0	1	1
1	0	1
1	1	0

The EXCLUSIVE-OR function arises frequently in Boolean algebra and has been given its own symbol, the *ring sum*:

$$A \text{ EX-OR } B = A \oplus B \qquad\qquad (2\text{-}2)$$

While the INCLUSIVE OR is TRUE if any one or more of its variables is TRUE, the EXCLUSIVE OR of two or more variables is TRUE only if an *odd number* of the variables are true. The three-argument EXCLUSIVE-OR function, expressed in truth table form, is

A	B	C	$A \oplus B \oplus C$
0	0	0	0
0	0	1	1
0	1	0	1
0	1	1	0
1	0	0	1
1	0	1	0
1	1	0	0
1	1	1	1

The EXCLUSIVE-OR function is often related to the *odd-parity* property of a group of Boolean variables, since it indicates when an odd number of the variables are TRUE. This property is useful in error-detection encoding and decoding. Some other interesting properties of the EXCLUSIVE-OR function are shown in Prob. 2-2.

In common usage the term OR always refers to the INCLUSIVE-OR operation. The switch equivalent to the normal OR function is a parallel connection of toggle switches, as shown in Fig. 2-2a. The light is illuminated if one or more of the toggle switches is closed. The switch equivalent to the two-argument EXCLUSIVE-OR function requires *double-throw* switches, as shown in Fig. 2-2b. This type of connection is common to the two-way

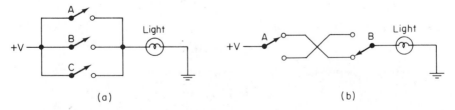

(a)   (b)

Figure 2-2. The Switch-Equivalents to the OR and the EXCLU-SIVE-OR Connectives

Figure 2-3. The OR Gate Symbol

switching arrangement found in many hallways. The light is illuminated if one switch is up and one is down. It is off if both are up or both are down.

The IEEE-ASA symbol for the OR operation is given in Fig. 2-3. The symbol for the EXCLUSIVE-OR function is given in Fig. 2-4.

Figure 2-4. The EXCLUSIVE-OR Gate Symbol

2.3 The NOT Operative

The NOT operative is defined for only one argument and hence is not truly a logical connective. The NOT function changes the logical value of its argument; it is also called *complementation* and *inversion*. Common symbols for the NOT operation include

$$f(A) = \bar{A} = A' = A^* \tag{2-3}$$

In truth table form, the NOT operation is

A	$\bar{A}$
0	1
1	0

showing the change in logical value.

Figure 2-5 shows the IEEE-ASA symbolic representation for the NOT function. This symbol is called a NOT gate or an *inverter*. There is no switch equivalent to the NOT function. Implementation of the NOT operation requires an active element such as a relay or a transistor.

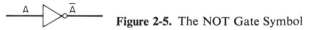

Figure 2-5. The NOT Gate Symbol

EXAMPLE 2-0:   How does the following expression relate to Shakespeare?

$$00101011 + (\overline{00101011})$$

Answer: Expressed in hex, the expression reads

$$\text{2B OR NOT (2B)}$$

2.4 The AND-OR-NOT Complete Set

The AND, OR, and NOT operators form a *complete set*, meaning that all other Boolean operations can be expressed in an equivalent form that involves only A-O-N connectives. It follows that if devices that can perform the AND, OR, and NOT functions are available, then any given logical operation can be implemented. General techniques for developing A-O-N forms for logical expressions are introduced in Secs. 2.8 and 2.9.

EXAMPLE 2-1:   Give an A-O-N form for the EXCLUSIVE-OR connective.

   Answer: $A \oplus B = (A \cdot \bar{B}) + (\bar{A} \cdot B)$

Show this symbolically. (See Fig. E2-1.)

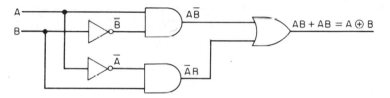

Figure E2-1

As proof of this expression, consider the following truth tabulation

A B	$(A \cdot \bar{B})$	$(\bar{A} \cdot B)$	$(A \cdot \bar{B}) + (\bar{A} \cdot B)$
0 0	0	0	0
0 1	0	1	1
1 0	1	0	1
1 1	0	0	0

The last column is $A \oplus B$.

2.5 The Fundamental Properties of Boolean Algebra

The A-O-N set of logical connectives (or any other complete set of connectives) may be used to manipulate Boolean variables in much the same way that normal algebraic operations are performed. There are several identities, laws, and theorems which are basic to Boolean algebra. These relationships are presented below in A-O-N form.

2.5.1. Elemental Propositions.   The following propositions, although incorrect (in part) in normal algebra, are correct in and basic to Boolean algebra:

$$a + \bar{a} = 1 \tag{2-4}$$

$$a \cdot \bar{a} = a\bar{a} = 0 \tag{2-5}$$

$$a + 1 = 1 \tag{2-6}$$

$$a \cdot 1 = a \tag{2-7}$$

$$a + a = a \tag{2-8}$$

$$a \cdot a = a \tag{2-9}$$

$$\overline{(\bar{a})} = a \tag{2-10}$$

$$a + 0 = a \tag{2-11}$$

$$a \cdot 0 = 0 \tag{2-12}$$

For the most part, the above propositions are intuitively obvious, but they all may be proved by using a truth table. These propositions can also be used in lieu of truth tabulation to prove other algebraic relationships.

EXAMPLE 2-2:   Prove Eq. (2-5) by truth tabulation.

   Proof:

a	$\bar{a}$	$(a \cdot \bar{a})$
0	1	0
1	0	0

The final column contains nothing but zeros (logically FALSE), proving that $a \cdot \bar{a} = 0$.

2.5.2. The Fundamental Laws.
The Associative law:

$$(a + b) + c = a + (b + c) \tag{2-13}$$

$$(ab)c = a(bc) \tag{2-14}$$

The Commutative law:

$$a + b = b + a \tag{2-15}$$

$$ab = ba \tag{2-16}$$

The Distributive law:

$$a(b + c) = ab + ac \tag{2-17}$$

$$a + bc = (a + b)(a + c) \tag{2-18}$$

Equation (2-18) appears incorrect if interpreted in normal algebraic terms. It can be proved, however, by truth tabulation or by using the previously shown relationships.

Proof of Eq. (2-18):

$$a + bc = (a + b)(a + c)$$
$$= aa + ac + ba + bc$$
$$= a + ab + ac + bc \qquad \text{from Eqs. (2-9) and (2-16)}$$
$$= a(1 + b + c) + bc \qquad \text{from Eqs. (2-7) and (2-17)}$$
$$= a(1) + bc \qquad \text{from Eq. (2-6)}$$
$$= a + bc \qquad \text{from Eq. (2-7)}$$

2.5.3 De Morgan's Theorem (Duality.)

De Morgan's theorem, in two-argument form, is

$$\overline{(a + b)} = \bar{a} \cdot \bar{b} \qquad \qquad (2\text{-}19)$$

$$\overline{(ab)} = \bar{a} + \bar{b} \qquad \qquad (2\text{-}20)$$

In a more general form, De Morgan's theorem states that whenever a Boolean expression in the A-O-N form is transformed by

1. Replacing all AND connectives by OR connectives,

2. Replacing all OR connectives by AND connectives,

3. Replacing all variables by their complements, noting that variables that originally appeared in complemented form thus become uncomplemented as per Eq. (2-10),

then the complement of the original expression is obtained.

De Morgan's theorem offers a simple process whereby Boolean expressions in the A-O-N form can be reexpressed in an equivalent form that, while still requiring only A-O-N connectives, is different from the original expression.

EXAMPLE 2-3:   Apply De Morgan's theorem to

$$F = A\bar{B} + C\bar{D}$$

Answer: $\bar{F} = (\bar{A} + B)(\bar{C} + D)$

Hence $F = \overline{(\bar{A} + B)(\bar{C} + D)}$.

EXAMPLE 2-4:   Obtain an alternative A-O-N expression for

$$F = A\bar{B} + \bar{A}B = A \oplus B$$

Answer: By De Morgan's theorem,
$$\bar{F} = (\bar{A} + B)(A + \bar{B})$$
$$= \bar{A}A + \bar{A}\bar{B} + AB + B\bar{B}$$
$$= \bar{A}\bar{B} + AB$$

Complementing both sides of the expression for $\bar{F}$ gives
$$F = \overline{\bar{A}\bar{B} + AB}$$

Applying De Morgan's theorem again gives
$$F = (A + B)(\bar{A} + \bar{B})$$

EXAMPLE 2-5:   Find the dual form for
$$F = \overline{a\bar{b}(c + \bar{d}(e + f))}.$$

Answer: $F = \bar{a} + b + \bar{c}(d + \bar{e} \cdot \bar{f})$

2.6 Reduction of Boolean Expressions by Algebraic Manipulation

Many of the Boolean expressions that arise in digital systems design are not in their simplest form. Some internal redundancy is present in such expressions, and, with cleverness and luck, the redundancy can be removed by using the previously shown propositions and laws. This reduction eliminates unnecessary gates, thereby saving cost, space, and weight. Such reduction is one of the major considerations of combinational design.

EXAMPLE 2-6:   Given $F = A + \bar{A}B$, reduce the expression by algebraic methods.

Solution:

$$F = A + \bar{A}B$$
$$= A \cdot 1 + \bar{A}B \qquad \text{from Eq. (2-7)}$$
$$= A(B + \bar{B}) + \bar{A}B \qquad \text{from Eq. (2-4)}$$
$$= AB + A\bar{B} + \bar{A}B$$
$$= AB + AB + A\bar{B} + \bar{A}B \qquad \text{from Eq. (2-8)}$$
$$= A(B + \bar{B}) + B(A + \bar{A})$$
$$= A + B \qquad \text{from Eqs. (2-4) and (2-7)}$$

The reduced expression can be implemented with only a two-input OR gate, while the original expression would require the same OR gate plus an inverter and a two-input AND gate.

The simplification of Boolean expressions via algebraic reduction is not always a straightforward process, especially at points where a factor must be duplicated in order to gain further reduction, as in the fifth step of Ex. 2-6. Even so, a familiarity with such reduction techniques, and with Boolean algebra in general, is desirable for digital designers. Practice with Boolean manipulation is essential; many of the problems given in Sec. 2-14 are devoted to this.

Fortunately, there are tools that the digital systems designer can use to help him in reducing Boolean expressions. These tools, involving both manual and computer-aided methods, are described in Chapter 4. An understanding of the Boolean algebraic principles that underlie such design techniques is necessary, however, and can best be gained by experience.

2.7 Other Complete Sets of Connectives

The A-O-N connectives are not the only complete set of connectives. At least two other complete sets are commonly used in applications of Boolean algebra. These sets are unusual in that each involves only a single logical connective. These two connectives form complete sets all by themselves. This property is of great usefulness in engineering practice since only one type of logical device needs to be purchased and stockpiled, rather than the separate AND, OR, and NOT gates required for A-O-N implementation. These two functionally complete connectives are described in the next two subsections.

2.7.1. The NOR Connective. The NOR connective is a combination of the NOT and the OR operations, and hence the name NOR. It can operate on one or more Boolean arguments. In A-O-N form, the two-variable NOR operation is

$$f(A,B) = (A \text{ NOR } B) = \overline{(A + B)} = A \downarrow B \qquad (2\text{-}21)$$

showing that the NOR is the NOT of the OR of its Boolean arguments. The last symbol is used in this text. In truth table form the two-argument NOR function is

A B	$A + B$	$\overline{(A + B)} = A \downarrow B$
0 0	0	1
0 1	1	0
1 0	1	0
1 1	1	0

Note that the NOR function is TRUE if and only if *all* its arguments are FALSE.

The IEEE-ASA symbol for the NOR gate is shown in Fig. 2-6. The dot on the output of the symbol indicates logical inversion.

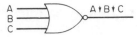

Figure 2-6. The NOR Gate Symbol

When operating on a single Boolean variable, the NOR function is equivalent to the NOT operation. The NOR function can be used to obtain the NOT operation in two ways. First, if a Boolean variable is combined with itself via the NOR operation, the result is

$$A \downarrow A = \overline{A + A} \qquad (2\text{-}22)$$

which is equivalent to $\overline{A}$. Second, if all but one of the arguments of a NOR operation are forced to be logically FALSE, the output of the connective will be the complement of the remaining argument. In other words,

$$(A \downarrow 0 \downarrow 0 \cdots) = \overline{A} \qquad (2\text{-}23)$$

The NOR gate will act as an inverter in either of the connections shown in Fig. 2-7.

Figure 2-7. The NOR Gate used as a NOT Gate

EXAMPLE 2-7:   Express the two-argument OR function in NOR terms.

Solution:

$$A + B = \overline{(\overline{A + B})} \qquad \text{(since two successive inversions cancel each other)}$$

$$= \overline{\overline{A \downarrow B}}$$

$$= (A \downarrow B) \downarrow 0$$

In symbolic form this is as shown in Fig. E2-7.

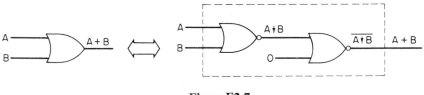

Figure E2-7

EXAMPLE 2-8:   Express the EXCLUSIVE-OR function in NOR form.

Answer: $A \oplus B = (A \downarrow B) \downarrow (\overline{A} \downarrow \overline{B})$

Here, $\overline{A} = A \downarrow A$ or $A \downarrow 0$, and $\overline{B}$ is obtained in the same fashion.

The answer that is given above can be proved by truth tabulation, remembering that the NOR function is TRUE only when all its arguments are false:

A B	$\bar{A}$ $\bar{B}$	$(\bar{A} \downarrow \bar{B})$	$(A \downarrow B)$	$[(\bar{A} \downarrow \bar{B}) \downarrow (A \downarrow B)]$
0 0	1 1	0	1	0
0 1	1 0	0	0	1
1 0	0 1	0	0	1
1 1	0 0	1	0	0

Examination of the last column shows that the function is equivalent to the EXCLUSIVE-OR connective. The NOR gating for the EXCLUSIVE-OR function is shown in Fig. E2-8. Generalized techniques for expressing Boolean operations in NOR form are given in Sec. 2-10.

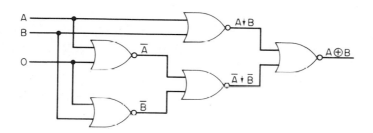

Figure E2-8

EXAMPLE 2-9:   Express the two-argument AND operation by using only the NOR operator.

Solution:

$$A \cdot B = \overline{(\bar{A})} \cdot \overline{(\bar{B})}$$
$$= \overline{\bar{A} + \bar{B}} \qquad \text{from Eq. (2-19)}$$
$$= \bar{A} \downarrow \bar{B} \qquad \text{by definition}$$
$$= (A \downarrow 0) \downarrow (B \downarrow 0) \qquad \text{from Eq. (2-23)}$$

In symbolic form this is as shown in Fig. E2-9.

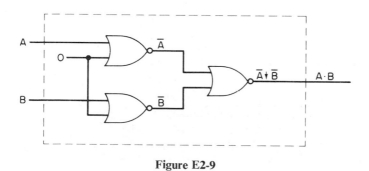

Figure E2-9

2.7.2. The NAND Connective. The NAND connective is a combination of the NOT and the AND operations and is defined for one or more Boolean arguments. The two-variable NAND operation in A-O-N form is

$$f(A,B) = (A \text{ NAND } B) = \overline{A \cdot B} = A \uparrow B \qquad (2\text{-}24)$$

indicating that the NAND is the NOT of the AND of its Boolean arguments. Hereafter, the upward arrow is used to represent the NAND operation. In truth table form,

A B	AB	$\overline{AB} = A \uparrow B$
0 0	0	1
0 1	0	1
1 0	0	1
1 1	1	0

showing that the NAND function is FALSE if and only if all its arguments are TRUE.

Like the NOR, the NAND operation can act as a NOT when given only one argument. To obtain the NOT operation from the NAND, the variable to be complemented can be NANDed with itself:

$$\bar{A} = A \uparrow A \uparrow A \uparrow \cdots \qquad (2\text{-}25)$$

or the variable can be NANDed with all TRUE arguments:

$$\bar{A} = A \uparrow 1 \uparrow 1 \uparrow \cdots \qquad (2\text{-}26)$$

The NAND, including its ability to act as a NOT, forms a complete set by itself. The IEEE-ASA NAND gate symbol is shown in Fig. 2-8.

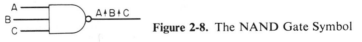

Figure 2-8. The NAND Gate Symbol

EXAMPLE 2-10: Using only the NAND connective, express the three-argument AND operation.

Solution:

$$A \cdot B \cdot C = \overline{(\overline{ABC})}$$
$$= \overline{A \uparrow B \uparrow C}$$
$$= (A \uparrow B \uparrow C) \uparrow 1$$

In symbolic form this is as shown in Fig. E2-10.

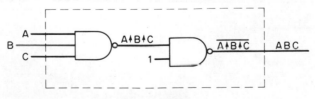

Figure E2-10

The NAND and NOR functions can be used to reexpress all the identities, laws, and theorems that were given in A-O-N form in Sec. 2.5. For example, Eq. (2-9), when reexpressed in NAND form, is

$$a \uparrow a = \overline{a \cdot a} = \bar{a} \tag{2-27}$$

Similarly, Eq. (2-15) becomes

$$a \downarrow b = b \downarrow a \tag{2-28}$$

when reexpressed in NOR form. Even De Morgan's theorem, Eqs. (2-19) and (2-20), can be expressed in NAND/NOR terms. This reexpression is the subject of Probs. 2-16 and 2-17.

2.8 Combinational Design: Minterm Decomposition

A general digital system was shown in Fig. 1-0. All such digital systems can be classified into two general categories: *combinational* systems and *sequential* systems. If the output signals from a system depend only on the current input signals, the system is combinational. The outputs from the system at a given instant can be produced by logical operation (combination) upon the input signals at that time. If, however, information concerning previous inputs or outputs is required in order to determine the present output signals, the system is sequential in nature. The design of sequential systems is covered in Chapters 5, 6, and 7.

In designing combinational digital systems, all input and output signals to and from the system are reexpressed in TRUE/FALSE form and represented by Boolean variables. Algebraic expressions defining the output variables in terms of the input variables are then defined according to the specifications of the digital system. These expressions are manipulated in order to simplify the amount of logical processing that is required to implement the system. The system is then constructed with suitable logical devices, such as electronic gating circuits, in accordance with the simplified logical equations that describe the system.

A standardized approach to this design process has been developed. This approach is useful in converting a combinational design problem into a set of logical expressions that "solve" the problem. These expressions are usually in one of two A-O-N forms. To illustrate this design procedure and to introduce these two common A-O-N forms for expressing a Boolean function, a combinational design example follows.

EXAMPLE 2-11. Code-to-Code Conversion: A smoke/fire alarm is installed in a remote pumping station. The alarm is sensitive to eight different environmental conditions and indicates these conditions by making different combinations of Boolean variables appear at its three output terminals, *A*, *B*, and *C*. The conditions and the codes that represent them are shown below:

Conditions	A B C
a. No smoke, no fire	0 1 0
b. Light smoke, no fire	1 1 1
c. Medium smoke, no fire	0 0 1
d. Heavy smoke, no fire	1 1 0
e. Light smoke, flame	0 0 0
f. Medium smoke, flame	1 0 0
g. Heavy smoke, flame	0 1 1
h. Disaster	1 0 1

It is desired to convert this series of output codes into normal ascending binary numbers. The device that will do this is a code-to-code converter that will accept A, B, and C and generate three new Boolean variables, X, Y, and Z, according to the following table:

Condition	Input A B C	Outputs X Y Z
a	010	000
b	111	001
c	001	010
d	110	011
e	000	100
f	100	101
g	011	110
h	101	111

Example 2-11 actually consists of three separate design problems, each involving the development of the Boolean equations for generating either X, Y, or Z from the given values of A, B, and C.

Consider the generation of X first. An examination of the X column shows that the Boolean variable X is TRUE for the last four combinations of A, B, and C. X can therefore be decomposed into four separate functions:

$$X = X_1 + X_2 + X_3 + X_4 \qquad (2\text{-}29)$$

where X_1, X_2, X_3, and X_4 are defined as shown below:

A B C	X_1	X_2	X_3	X_4
0 1 0	0	0	0	0
1 1 1	0	0	0	0
0 0 1	0	0	0	0
1 1 0	0	0	0	0
0 0 0	1	0	0	0
1 0 0	0	1	0	0
0 1 1	0	0	1	0
1 0 1	0	0	0	1

Examination of the above table shows that X_1 is TRUE *only* when A is FALSE, B is FALSE, and C is FALSE. Thus, X_1 may be written as

$$X_1 = \bar{A}\bar{B}\bar{C} \tag{2-30}$$

Similarly,

$$X_2 = A\bar{B}\bar{C}, \qquad X_3 = \bar{A}BC, \qquad X_4 = A\bar{B}C \tag{2-31}$$

The function X may thus be expressed in A-O-N form as

$$X = \bar{A}\bar{B}\bar{C} + A\bar{B}\bar{C} + \bar{A}BC + A\bar{B}C \tag{2-32}$$

Similarly, the expressions for Y and Z may be found to be

$$Y = \bar{A}\bar{B}C + A\bar{B}C + \bar{A}BC + A\bar{B}\bar{C} \tag{2-33}$$
$$Z = ABC + AB\bar{C} + A\bar{B}\bar{C} + A\bar{B}C$$

If the input variables in the Boolean expression for X are written as 1 when TRUE and 0 when FALSE, Eq. (2-32) becomes

$$X = 000 + 100 + 011 + 101 \tag{2-34}$$

If the right-hand side of Eq. (2-34) is interpreted as a group of binary numbers, X may be written as

$$X = \sum 0, 4, 3, 5 \tag{2-35}$$

Equation (2-35) is a compact form for writing the expression for X. The fact that X is formed from only *three* input variables must also be known, along with the order in which these variables appeared in the original expression. This leads to the following forms for describing X, Y, and Z:

$$X(A,B,C) = \sum 0, 3, 4, 5$$
$$Y(A,B,C) = \sum 1, 3, 5, 6 \tag{2-36}$$
$$Z(A,B,C) = \sum 4, 5, 6, 7$$

Terms such as those used to express X_1, X_2, X_3, and X_4 in Eq. (2-32) are called *minterms*. Such terms must contain *all* the input variables, with each variable in either its TRUE or complemented form. For a system with n input variables there will be 2^n different minterms. The title *min*term arises from the fact that each minterm has a *min*imum number of 1s in its output column, excluding, of course, the possibility of having *no* 1s (the exclusive-never function).

Minterms can be expressed in a shorthand notation by using a lowercase m along with the decimal number that relates to the TRUE and FALSE combination of its component Boolean variables. For a three-variable system the eight minterms are

$$m_0 = \bar{A}\bar{B}\bar{C}, \qquad m_1 = \bar{A}\bar{B}C, \qquad m_2 = \bar{A}B\bar{C}, \qquad m_3 = \bar{A}BC,$$
$$m_4 = A\bar{B}\bar{C}, \qquad m_5 = A\bar{B}C, \qquad m_6 = AB\bar{C}, \qquad m_7 = ABC \tag{2-37}$$

EXAMPLE 2-12:   How many minterms will a six-variable system have?

Answer: $2^6 = 64$

EXAMPLE 2-13: Given a digital system with four input variables, A, B, C, and D, what are the expressions for m_2, m_8, m_{11}, and m_{15}?

Answer:

$$m_2 = \bar{A}\bar{B}C\bar{D}$$
$$m_8 = A\bar{B}\bar{C}\bar{D}$$
$$m_{11} = A\bar{B}CD$$
$$m_{15} = ABCD$$

Can there be a minterm numbered m_{16}?

Answer: No: m_{16} would be defined only for a system having five or more input variables.

EXAMPLE 2-14: Give the minterm notation for the following:

$$AB\bar{C}DE, \qquad \bar{A}\bar{B}CDE, \qquad ABCDEF, \qquad \text{and} \qquad A\bar{B}C\bar{D}E\bar{F}G$$

Answers: m_{27}, m_7, m_{63}, and m_{85}

In minterm form, the functions for X, Y, and Z that were given in Eq. (2-36) can be rewritten as

$$X = m_0 + m_3 + m_4 + m_5$$
$$Y = m_1 + m_3 + m_5 + m_6 \qquad (2\text{-}38)$$
$$Z = m_4 + m_5 + m_6 + m_7$$

Such a form is called the *expanded sum-of-products* (SOP) form for a Boolean expression, the *sum-of-minterms* form, the *minterm decomposition*, or the *disjunctive canonical form* [2].

Often there are simplifications that can be made in expressions given in this form. Consider the expression for Z that is given in Eq. (2-33):

$$Z = A\bar{B}\bar{C} + A\bar{B}C + AB\bar{C} + ABC \qquad (2\text{-}33)$$
$$= A\bar{B}(\bar{C} + C) + AB(\bar{C} + C)$$
$$= A\bar{B} + AB$$
$$= A(\bar{B} + B)$$
$$= A$$

showing that the four minterms that make up Z can be combined into a very simple expression that requires no gating at all for its implementation. The identification and elimination of such redundancies from Boolean expressions that are in the expanded sum-of-products form is the subject of Chapter 4.

2.9 Maxterm Decomposition

Referring back to the table that initially defined X in terms of A, B, and C, the FALSE entries in the X column can be combined to obtain the minterm decomposition for the complement of X:

$$\bar{X} = \bar{A}B\bar{C} + ABC + \bar{A}\bar{B}C + AB\bar{C} \qquad (2\text{-}39)$$

or

$$\bar{X} = m_2 + m_7 + m_1 + m_6 = \sum 1, 2, 6, 7 \qquad (2\text{-}40)$$

Obviously, those minterms that did not appear in the minterm decomposition of X will appear in the minterm decomposition of $\bar{X}$. De Morgan's theorem may now be applied to the minterm decomposition of $\bar{X}$, yielding

$$X = \overline{\bar{A}B\bar{C} + ABC + \bar{A}\bar{B}C + AB\bar{C}}$$
$$= (A + \bar{B} + C)(\bar{A} + \bar{B} + \bar{C})(A + B + \bar{C})(\bar{A} + \bar{B} + C) \quad (2\text{-}41)$$

This form for expressing X is called the *expanded product of sums* and is an alternative to the sum-of-minterms expression. Terms of the form $(A + \bar{B} + C)$ are called *maxterms* and, like minterms, must contain all the variables, each in either the normal or complemented form. The title *max*term results from the fact that when an individual maxterm is truth tabulated, its output column contains only one zero. This is the *max*imum number of 1s, except for the all-1s column.

The shorthand notation for a maxterm uses a capital M and a subscript determined by the TRUE/FALSE states of the variables that make up the maxterm. For example, the term $(A + \bar{B} + C)$ is M_5, since its TRUE/FALSE variables form 101, the binary equivalent of five.*

In maxterm form, Eq. (2-41) may be written as

$$X = M_5 \cdot M_0 \cdot M_6 \cdot M_1 \qquad (2\text{-}42)$$

or, in a more compact notation,

$$X(A,B,C) = \prod 0, 1, 5, 6 \qquad (2\text{-}43)$$

The expanded product of sums (POS) is also referred to as the *product-of-maxterms* form, the *maxterm decomposition*, and the *conjunctive canonical form* [2]. In some cases a reduction based upon a maxterm decomposition will lead to a simpler form for expressing a function than will a reduction based upon the minterm decomposition. Both forms should be considered in designing combinational systems.

---

*There is some disagreement over the choice of the maxterm subscript. The notation shown above is found in refs. [2], [4], and [7] and other texts. An alternative determination for the subscript uses the subscript of the minterm that when complemented via De Morgan's theorem yields the particular maxterm as the maxterm subscript. This necessitates reading a maxterm such as $(A + \bar{B} + C)$ as 010, or M_2, rather than M_5. References [3], [5], and [8] use this notation. Both forms are found in the literature.

A short-cut method for converting from the sum-of-minterms form for $\bar{X}$, as in Eq. (2-40), to the product-of-maxterms form for X, as in Eq. (2-43), is described in Probs. 2-7 and 2-8 in Sec. 2-14.

2.10 The Eight Standard Forms

Sections 2.8 and 2.9 have shown a process whereby a combinational design problem can be converted into a set of Boolean expressions in either the sum-of-minterms form (commonly called the AND/OR form since the input variables are first ANDed to form the minterms and then ORed to form the output variables) or the product-of-maxterms form (OR/AND). Both forms require the NOT function, as well as the AND and OR. Algebraic reductions starting with either of these forms can be made, but the basic AND/OR or OR/AND structure of the combinational expressions will generally be preserved.

These two forms are not the only ways in which a Boolean expression can be represented. There are six other standard forms into which a given Boolean expression can be cast. All these forms are *two-level* gating forms, since they involve two levels of logical operations between their input and output variables (not counting any NOT operations that may be required). The eight standard forms are

1. AND/OR.
2. NAND/NAND.
3. OR/NAND.
4. NOR/OR.
5. AND/NOR.
6. NAND/AND.
7. OR/AND.
8. NOR/NOR.

Forms 1 and 7 result from the minterm and maxterm decompositions. Forms 2 and 8 point to the logical completeness of the NAND and NOR functions.

Figure 2-9 illustrates the initial phases of the combinational design process and shows the relationships that exist between the eight standard forms. As shown in Fig. 2-9, a combinational design problem normally begins as a set of descriptive statements that specify what the output Boolean variables are to be for each combination of input conditions. These statements are used to fill out a truth table. The truth table is used to express each output

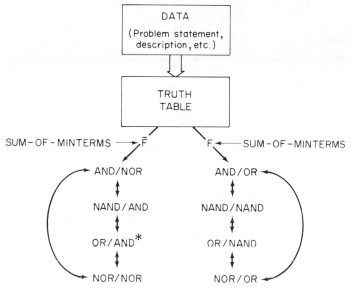

* PRODUCT - OF - MAXTERMS

Figure 2-9. The Eight Standard Forms for Boolean Expression

function and its complement in the sum-of-minterms form. From these two expressions, the sequence of standard forms shown in Fig. 2-9 can be obtained by successive applications of De Morgan's theorem. To illustrate this process, an example involving the two-argument EXCLUSIVE-OR function follows.

EXAMPLE 2-15: Develop the eight standard forms for expressing the EXCLUSIVE-OR function defined by the following truth table:

Inputs		Output
A	B	F
0	0	0
0	1	1
1	0	1
1	1	0

Procedure: From the truth table, express both F and $\bar{F}$ in sum-of-minterms form.

$$F = A\bar{B} + \bar{A}B = m_1 + m_2$$
$$\bar{F} = \bar{A}\bar{B} + AB = m_0 + m_3$$

Starting with the expression for F, proceed as follows:

$$F = A\bar{B} + \bar{A}B \qquad \text{(AND/OR)}$$
$$= \overline{(A\bar{B} + \bar{A}B)}$$
$$= \overline{(A\bar{B})} \cdot \overline{(\bar{A}B)} = (A \uparrow \bar{B}) \uparrow (\bar{A} \uparrow B) \qquad \text{(NAND/NAND)}$$
$$= (\bar{A} + B) \cdot (A + \bar{B}) = (\bar{A} + B) \uparrow (A + \bar{B}) \qquad \text{(OR/NAND)}$$
$$= \overline{(\bar{A} + B)} + \overline{(A + \bar{B})} = (\bar{A} \downarrow B) + (A \downarrow \bar{B}) \qquad \text{(NOR/OR)}$$

Beginning with $\bar{F}$ in sum-of-minterms form, the transformation process yields the other four forms:

$$\bar{F} = \bar{A}\bar{B} + AB$$
$$F = \overline{\bar{A}\bar{B} + AB} = (\bar{A}\bar{B}) \downarrow (AB) \qquad \text{(AND/NOR)}$$
$$= \overline{(\bar{A}\bar{B})} \cdot \overline{(AB)} = (\bar{A} \uparrow \bar{B}) \cdot (A \uparrow B) \qquad \text{(NAND/AND)}$$
$$= (A + B) \cdot (\bar{A} + \bar{B}) \qquad \text{(OR/AND)}$$
$$= \overline{\overline{(A + B) \cdot (\bar{A} + \bar{B})}}$$
$$= \overline{\overline{(A + B)} + \overline{(\bar{A} + \bar{B})}} = (A \downarrow B) \downarrow (\bar{A} \downarrow \bar{B}) \qquad \text{(NOR/NOR)}$$

The eight standard forms are separated into two groups of four related forms, as shown in Fig. 2-9. One group can be called the *AND/OR group*, since it contains that form for expressing a function. The other members of the AND/OR group are the NAND/NAND, OR/NAND, and NOR/OR forms. The second group is the OR/AND family. It contains the AND/NOR, NAND/AND, and NOR/NOR forms as well.

When generating the eight standard forms for a given Boolean expression, conversion between the Boolean forms that are within the same group is relatively easy. Intergroup conversion is not. One way that the group-to-group conversion can be accomplished is by truth tabulation of the original expression. Then, entry into the AND/OR group or the OR/AND group can be accomplished by using the sum-of-minterms expressions for the function and its complement, as illustrated in Fig. 2-9 and Ex. 2-15. A second method for making the jump from one group to the other involves algebraic expansion of the AND/OR or OR/AND expressions. This expansion process uses Eqs. (2-17) and (2-18) in reverse and converts an AND/OR form to an OR/AND form and vice versa.

EXAMPLE 2-16:   Convert $F = (A + \bar{B})(C + \bar{D})$ to AND/OR form.

Solution:

$$F = (A + \bar{B})C + (A + \bar{B})\bar{D} \qquad \text{from Eq. (2-17)}$$
$$= AC + \bar{B}C + A\bar{D} + \bar{B}\bar{D} \qquad \text{from Eq. (2-17) again}$$

The last expression is in AND/OR form.

EXAMPLE 2-17:   Convert $F = AB + CD$ into OR/AND form.

Solution:

$$F = (AB) + CD$$
$$= (AB + C)(AB + D) \qquad \text{from Eq. (2-18)}$$
$$= (C + A)(C + B)(D + A)(D + B) \qquad \text{from Eq. (2-18) again}$$

The last expression is in OR/AND form.

EXAMPLE 2-18:   Convert $F = (A \downarrow \bar{B}) \downarrow C$ into its eight standard forms.

Solution: The original expression is in the NOR/NOR form. This places it in the OR/AND group. Its related forms are obtained via De Morgan's theorem:

$$F = \overline{(A \downarrow \bar{B}) \downarrow C} \qquad \text{(NOR/NOR)}$$
$$= \overline{(A + \bar{B}) + C} \qquad \text{by definition}$$
$$= \bar{A}B + C = \bar{A}B \uparrow C \qquad \text{(AND/NOR)}$$
$$= \overline{(\bar{A}B)} \cdot \bar{C} = (\bar{A} \uparrow B) \cdot \bar{C} \qquad \text{(NAND/AND)}$$
$$= (A + \bar{B}) \cdot \bar{C} \qquad \text{(OR/AND)}$$

Expanding the OR/AND expression gives

$$F = A\bar{C} + \bar{B}\bar{C} \qquad \text{(AND/OR)}$$
$$= \overline{\overline{A\bar{C} + \bar{B}\bar{C}}}$$
$$= \overline{(A\bar{C}) \cdot (\bar{B}\bar{C})} = (A \uparrow \bar{C}) \uparrow (\bar{B} \uparrow \bar{C}) \qquad \text{(NAND/NAND)}$$
$$= \overline{(\bar{A} + C) \cdot (B + C)} = (\bar{A} + C) \uparrow (B + C) \qquad \text{(OR/NAND)}$$
$$= \overline{(\bar{A} + C)} + \overline{(B + C)} = (\bar{A} \downarrow C) + (B \downarrow C) \qquad \text{(NOR/OR)}$$

The Boolean expressions that result from the design process that is illustrated in Fig. 2-9 are also referred to as *switching functions* or *combinational equations*. The term switching is a carry-over from the relays that were used to perform the logical operations in early digital systems.

The AND/OR and OR/AND forms for expressing Boolean functions have found the most widespread use for several reasons. These forms arise most directly from the combinational design process. Also, the organization of human thought processes most closely parallels A-O-N forms. One may say, "I will be there if it doesn't rain and my car will start or if you will get me a date who is pretty." In AND/OR Boolean form, this statement is

$$A = \bar{R}S + DP \tag{2-44}$$

where A is TRUE if I arrive, R is TRUE if it rains, S is TRUE if the car starts, D is TRUE if you get me a date, and P is TRUE if she is pretty.

However, most modern electronic gating circuits do not operate in AND/OR fashion. Rather, for a variety of reasons, they act as NAND or NOR gates. The reasons for this are discussed briefly in Chapter 3 and in more detail in Refs. [4] and [9]. This logical mismatch can cause problems. Chapter 3 is addressed, in part, to the explanation and illustration of techniques for bridging the gap between Boolean equations that are preferably in A-O-N form and gating devices that are NAND or NOR in nature.

2.11 Summary

Chapter 1 has shown that informative and quantitative items can be re-expressed in two-valued form. Chapter 2 introduces the algebraic operations that may be used with such two-valued variables. The AND-OR-NOT complete set is described, as are the NAND and NOR functions, which form complete sets by themselves. The symbols for these various logical operations are shown, along with basic techniques for the algebraic simplification of Boolean expressions. The use of Boolean algebra in combinational design is introduced, as are the eight standard forms for Boolean expressions. The most important of these are the minterm and maxterm decompositions. These two forms for expressing a Boolean function require only AND, OR, and NOT connectives and are the starting point for designing most combinational digital systems.

2.12 Did You Learn?

1. What a Boolean variable is? How Boolean variables can be used to represent both information and quantity?

2. The definitions of the AND, OR, and NOT operators? The meaning of *complete set*? The AND-OR-NOT complete set?

3. The EXCLUSIVE OR and its differences from the normal (INCLUSIVE-) OR connective?

4. The NAND and NOR connectives? How to use them to perform the NOT operation? How they each form a complete set?

5. The symbols for the AND, OR, NOT, EXCLUSIVE-OR, NAND, and NOR gates?

6. The fundamental identities and laws that apply to Boolean algebra?

7. De Morgan's theorem?

8. How to simplify (reduce) a Boolean expression by using algebraic methods?

9. How to prove a Boolean equality by truth tabulation and by algebraic manipulation?

10. The definitions of the minterm and the maxterm?

11. How to express a Boolean function as a sum of minterms and a product of maxterms?

12. How to convert a Boolean expression into its eight standard forms?

13. Why AND-OR-NOT Boolean expressions are preferred to other forms?

2.13 References

[1] Boole, G., *An Investigation of the Laws of Thought, on which Are Founded the Mathematical Theories of Logic and Probability* (1849). Reprinted: New York: Dover Publications, Inc., 1954.

[2] Givone, Donald D., *Introduction to Switching Circuit Theory*. New York: McGraw-Hill Book Company, 1970.

[3] Hill, Fredrick J., and Gerald R. Peterson, *Introduction to Switching Theory and Logical Design*. New York: John Wiley & Sons, Inc., 1968.

[4] Hurley, R. B., *Transistor Logic Circuits*. New York: John Wiley & Sons, Inc., 1961.

[5] Kohavi, Zvi, *Switching and Finite Automata Theory*. New York: McGraw-Hill Book Company, 1970.

[6] Shannon, C. E., "Symbolic Analysis of Relay and Switching Circuits," *Trans. of AIEE*, 57 (1938), 713–723.

[7] Sifferlen, Thomas, and Vartan Vartanian, *Digital Electronics with Engineering Applications*. Englewood Cliffs, N. J.: Prentice-Hall, Inc., 1970.

[8] Sobel, Herbert S., *Introduction to Digital Computer Design*. Reading, Mass.: Addison-Wesley Publishing Company, Inc., 1970.

[9] Wicks, William E., *Logic Design with Integrated Circuits*. New York: John Wiley & Sons, Inc., 1968.

2.14 Problems

2-1. Prove the following Boolean equalities by using both truth tabulation and algebraic manipulation.

a. $ABC + AB\bar{C} + A\bar{B}C + A\bar{B}\bar{C} = A$

b. $(\bar{A} + \bar{B})(A + B) = A \oplus B$

c. $\bar{A} \uparrow \bar{B} = A + B$

d. $ABC + A\bar{B}C + AB\bar{C} = A(B + C)$

e. $\overline{\bar{A} \uparrow \bar{B}} = A \downarrow B$

f. $XY(X + Y) = XY$

g. $\bar{A} \downarrow \bar{B} = AB$

h. $XY(X + \bar{Y}) = XY$

i. $\overline{A \downarrow B} = A + B$

j. $(C \oplus D) \downarrow C = \bar{C}\bar{D}$

    k. $(a + b)(a + c)(a + d) = a + bcd$

    l. $x(x + y) = x$

2-2. Prove the following relations involving the EXCLUSIVE-OR operator.

    a. $\overline{A \oplus B} = AB + \bar{B}\bar{A}$

    b. $A \oplus \bar{B} = \bar{A} \oplus B = \overline{A \oplus B}$

    c. $\bar{A} \oplus \bar{B} = A \oplus B$

    d. $A \oplus (A \oplus B) = B$

    e. $(A \oplus B) \oplus (\bar{A} \oplus \bar{B}) = 0$

    f. Prove that if $D = A \oplus B \oplus C$ then, for all combinations of values of A, B, and C, the number of TRUE values found among A, B, C, and D will always be even.

2-3. Give the logic diagrams for the following Boolean expressions.

    a. $F = A\bar{B} + C\bar{D}$ d. $G = \overline{WX} \downarrow (Y \oplus Z)$

    b. $F = A \oplus (B \downarrow \bar{C})$ e. $G = W\bar{X} + WZ + (X \oplus Y)$

    c. $F = (\bar{A} + B) \uparrow C$ f. $F = [\bar{w}\bar{x}(y + \bar{z})] \oplus w$

2-4. Determine the Boolean expression that is implemented by each of the logic diagrams in Fig. P2-4.

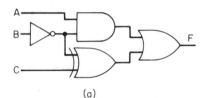

(a)

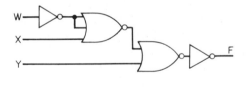

(b)

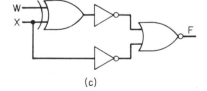

(c)

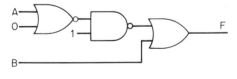

(d)

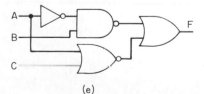

(e)

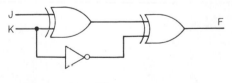

(f)

Figure P2-4

2-5. Express each of the following functions (given in truth table form) as a sum of minterms and a product of maxterms. Show a logic diagram for implementing each expression.

A B	F_1		C D E	F_2		W X Y Z	F_3
0 0	1		0 0 0	0		0 0 0 0	0
0 1	0		0 0 1	1		0 0 0 1	1
1 0	1		0 1 0	0		0 0 1 0	0
1 1	0		0 1 1	1		0 0 1 1	0
			1 0 0	0		0 1 0 0	0
			1 0 1	0		0 1 0 1	1
			1 1 0	0		0 1 1 0	1
			1 1 1	1		0 1 1 1	1
						1 0 0 0	0
						1 0 0 1	0
						1 0 1 0	0
						1 0 1 1	1
						1 1 0 0	1
						1 1 0 1	0
						1 1 1 0	0
						1 1 1 1	0

2-6. Give a truth table representation for each of the following functions.
   a. $F(A,B) = \sum 0, 1, 3$
   b. $F(A,B,C) = \sum 2, 5, 6, 7$
   c. $F(A,B,C,D) = \sum 0, 1, 4, 6, 9, 11, 13, 14$
   d. $F(P,Q,R,S,T) = \sum 0, 2, 9, 14, 15, 17, 23, 26, 28, 30, 31$
   e. $G(A,B,C,D,E,F) = \sum 1, 9, 15, 20, 27, 30, 38, 42, 49, 55, 60, 62, 63$

2-7. Section 2-9 showed that the procedure for expressing a Boolean function as a product of maxterms is
   a. Express the complement of the function as a sum of minterms.
   b. Convert to a product of maxterms by complementing both sides of that expression. De Morgan's theorem converts the sum of minterms to a product of maxterms.
Prove that if m_j is a minterm in the sum-of-minterms expression for $\bar{F}$, then the maxterm that is produced when m_j is complemented to form F as a product of maxterms is M_J, where

$$J = 2^n - 1 - j$$

and n is the number of Boolean variables. Illustrate by reworking the product-of-maxterms results of Prob. 2-5.

2-8. The results of Prob. 2-7 allow the procedure for finding the product-of-maxterms form for a function of n variables to be reduced to
   a. Express $\bar{F}$ as a sum of minterms.
   b. Subtract each of these minterm subscripts from $2^n - 1$.
This gives the maxterm coefficients for F in product-of-maxterms form. By

using this simplified procedure, reexpress each of the following functions in product-of-maxterms form (remember that minterms not present in the sum-of-minterms expression for F will appear in the sum-of-minterms expression for $\bar{F}$).

   a. $F(A,B,C) = \sum 0, 2, 5, 7$
   b. $F(A,B,C,D) = \sum 1, 3, 6, 7, 9, 10, 15$
   c. $F(A,B,C,D) = \sum 0, 2, 5, 9, 11, 14, 15$
   d. $F(W,X,Y,Z) = \sum 1, 3, 5, 7, 9, 11, 13, 15$
   e. $F(W,X,Y,Z) = \sum 0, 1, 2, 4, 6, 8, 9, 12, 14, 15$
   f. $G(A,B,C,D,E) = \sum 0, 2, 9, 10, 12, 14, 15, 17, 20, 23, 25, 30, 31$
   g. $F(v,w,x,y,z) = \sum 1, 3, 6, 10, 11, 13, 16, 18, 19, 22, 24, 26, 30, 31$

2-9.   Using the reverse of the procedure given in Prob. 2-8, reexpress the following functions in sum-of-minterms form.

   a. $F(A,B,C) = \prod 0, 2, 5, 7$
   b. $F(A,B,C,D) = \prod 0, 6, 7, 8, 10, 12, 14, 15$
   c. $F(A,B,C,D) = \prod 1, 5, 6, 9, 11, 13, 14$
   d. $F(W,X,Y,Z) = \prod 0, 1, 2, 3, 4, 5, 6, 7$
   e. $F(a,b,c,d,e) = \prod 2, 3, 5, 9, 11, 15, 16, 19, 21, 25, 29, 30$
   f. $G(v,w,x,y,z) = \prod 1, 3, 6, 10, 11, 13, 16, 18, 19, 22, 24, 26, 30, 31$

2-10.   Under what conditions will the sum-of-minterms and the product-of-maxterms expressions for a function have identical subscripts?

2-11.   A Boolean function M is to follow the logical condition of the majority of its three inputs, A, B, and C. Tabulate the behavior of M and express $M(A,B,C)$ as a sum of minterms and as a product of maxterms. Give a logic diagram for each expression.

2-12.   A function X is to represent the EXCLUSIVE OR of four input variables, A, B, C, and D. Express X as a sum of minterms and as a product of maxterms. Prove that both expressions are equivalent to

$$X(A,B,C,D) = A \oplus [B \oplus (C \oplus D)]$$

Give a logic diagram for the sum-of-minterms expression, the product-of-maxterms expression, and the equivalent EXCLUSIVE-OR expression.

2-13.   The block diagram given in Fig. P2-13 represents the *half-adder* (see

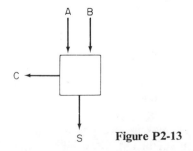

S   **Figure P2-13**

Table 1-3). Express the sum, S, and carry, C, in both sum-of-minterms and product-of-maxterms form. Give A-O-N logic diagrams for each. Can the EXCLUSIVE-OR function be used to simplify the implementation of either output? If so, give both the Boolean expression and a logic diagram.

2-14. Repeat the analysis procedure given in Prob. 2-13 for the *full adder* (Table 1-4), the *half-subtractor* (Table 1-5), and the *full subtractor* (Table 1-6).

2-15. Use De Morgan's theorem to reexpress each of the following expressions in uncomplemented form.

a. $\bar{F} = A + \bar{B}(CD + E)$
b. $\bar{G} = \bar{A}B + C\bar{D}(E + \bar{F})$
c. $\bar{G} = W\bar{X}(\bar{Y} + Z) + W\bar{X}$
d. $\bar{G} = wxyz + \bar{w}\bar{x}\bar{y}\bar{z}$
e. $\bar{G} = A + BC + \bar{A}(B + \bar{C})$
f. $\bar{F} = (w + x\bar{y})(z + \bar{x}y)$

2-16. Develop an analogy to De Morgan's theorem that involves conversions between NAND and NOR operations. Using the modified theorem, reexpress each of the following expressions in uncomplemented form.

a. $\bar{F} = A \uparrow B$
b. $\bar{F} = A \downarrow B$
c. $\bar{F} = A \downarrow (B \uparrow \bar{C})$
d. $\bar{G} = (W \downarrow \bar{X}) \uparrow (\bar{Y} \downarrow Z)$
e. $\bar{G} = (\bar{w} \downarrow y \downarrow \bar{z}) \uparrow \bar{x}$
f. $\bar{G} = (A \uparrow \bar{B}) \downarrow (C \downarrow \bar{D})$

2-17. Extend De Morgan's theorem and its modified form (Prob. 2-16) to include expressions that combine AND, OR, NAND, and NOR operators. Then reexpress each of the following expressions in uncomplemented form. Verify the validity of the new forms by using truth tabulation or algebraic manipulation.

a. $\bar{F} = (A + B) \uparrow (C + D)$
b. $\bar{F} = (A \downarrow B)(\bar{C} \uparrow \bar{D})$
c. $\bar{F} = AB \downarrow \overline{CD}$
d. $\bar{G} = W\bar{X} \uparrow (Y + \bar{Z})$
e. $\bar{G} = (w \downarrow xy) + \bar{y}z)$
f. $\bar{G} = (A\bar{B} + C) \uparrow (\bar{B} \downarrow C\bar{D})$

2-18. Convert each of the following expressions into OR/AND form.

a. $F = A\bar{B} + BC$
b. $F = \bar{W}\bar{X}Y + W\bar{X}\bar{Y} + \bar{W}X\bar{Y}$
c. $G = MNP + \bar{M}\bar{N}\bar{P}$
d. $G = (A \uparrow B \uparrow C) \uparrow (\bar{A} \uparrow D)$
e. $F = (W + \bar{X}) \uparrow (\bar{W} + Y + Z)$
f. $F = (\bar{w} \downarrow \bar{y}) + (x + y)$

2-19. Convert each of the following expressions into AND/OR form.

a. $F = A(\bar{B} + C + D)(B + \bar{D})$
b. $F = (A + \bar{B})(C + \bar{D} + E)$
c. $G = (W + Y)(W + Z)(W + \bar{Z})$
d. $G = (A\bar{B}) \downarrow (\bar{A}C)$
e. $F = (W \uparrow \bar{X})(Y \uparrow X)(\bar{W} \uparrow Y)$
f. $G = (w \downarrow y \downarrow z) \downarrow (\bar{x} \downarrow y)$

2-20.   Express each of the following functions in all eight standard forms.

a. $F = \bar{A}B\bar{C} + \bar{B}C$

b. $F = (X + Y)(\bar{Y} + Z)$

c. $G = \bar{X}(Y + Z)$

d. $G = (X \downarrow \bar{Y}) \downarrow (\bar{X} \downarrow Y)$

e. $F(A,B,C,D) = \sum 1, 2, 6, 7, 9, 14$

f. $G(W,X,Y,Z) = \sum 0, 2, 9, 10, 12, 15$

2-21.   Assume that a function is given in AND/OR or OR/AND form. Develop an algebraic procedure for converting it to NAND/NAND or NOR/NOR form. Use the procedure to reexpress each of the following expressions in NAND/NAND and NOR/NOR form. Show a logic diagram for each expression. Use NAND or NOR gates to perform any NOT operations that are required.

a. $F = \bar{A}B\bar{C} + \bar{B}C$

b. $F = (A + \bar{B})(B + \bar{C})$

c. $F = AB + \bar{C}\bar{D}$

d. $F = W(X + \bar{Y})$

e. $G = (\bar{w} + x)(\bar{w} + \bar{x} + y)$

f. $G = (\bar{w} + \bar{y} + z)(w + x)(\bar{x} + y + z)$

3 Combinational Logic Circuits and Voltage Symbolism

3.0 Introduction

The preceding chapter introduces a general technique for reducing a Boolean design problem to a set of combinational equations that require only AND, OR, and NOT connectives. In a working logical system some electrical or mechanical device must perform the logical operations specified by these Boolean equations. This text discusses the use of two-valued *electronic* gating circuits, although other gating devices (fluidic logic, for example) can be used in their place. Only the external behavior of the circuits is considered, with techniques for logical circuit design omitted. For information on that topic see Refs. [8], [11], [14], and [17].

3.1 Common Gating Circuits

All two-valued electronic gating circuits operate by using two different electrical conditions (voltage or current) to represent the logical states of their inputs and outputs. Typical values for the two conditions are 0 and 3.6 V, $+10$ and -10 V, or 5 mA into the circuit and 5 mA out of the circuit. The use of voltage-sensitive gating circuits is discussed herein, although the remarks can be generalized to current-sensitive gates or fluidic devices.

Logical gating circuits are commonly classified according to the types of electrical components that they contain. The early electronic gating circuits used only diodes and resistors and hence were called *diode-resistor logic* (DRL) *gates*. Different DRL circuits were required to perform the AND and the OR functions [8], with active-element inversion circuits acting as NOT gates. In practical applications DRL circuits proved to have serious limitations. Each successive stage of gating introduced a voltage drop associated with the forward-biased diodes within it, so that, after passing through several logical stages, a voltage level could become so degraded as to be logically meaningless. Also, since there was no power gain within a DRL gate, the logical input signals had to supply the power needs of all successive stages *and* the output loads. To circumvent these problems, other gating circuits were developed.

There are several types of modern, voltage-sensitive gating circuits; *diode-transistor logic* (DTL), *resistor-transistor logic* (RTL), *transistor-transistor logic* (TTL or T²L), and *emitter-coupled logic* (ECL) are examples. In large-scale integrated circuit logical networks, metal-oxide-semiconductor field-effect transistor (MOSFET) gating networks are also used. A complete discussion of these gating circuits is given in Refs. [6], [11], and [17]. In particular, Garrett [6, Part III] gives a very complete tabulation of comparative characteristics.

The DTL gate is the most popular form for gates constructed from discrete electronic components. However, the development of integrated circuit (IC) technology has made the use of other gate configurations economically feasible. Older ideas about saving cost by minimizing the number of active components in a gating system are not so applicable to devices that are manufactured by integrated-circuit techniques. As a result, the use of RTL, TTL, ECL, and other gating circuits, all of which contain more active elements than their discrete-component counterparts, has become prominent.

These various gate types have differing characteristics. From a purely logical standpoint, however, the gates fall into three general categories: those that implement the NAND function, those that implement the NOR function, and those that implement other functions (AND-OR-INVERT, EXCLUSIVE OR, etc.). There are crossovers between these types, as shown later in this chapter, but the basic differences remain.

The typical RTL gate acts as a NOR, while the basic TTL gate acts as a NAND. For example purposes, therefore, these two gate types are used throughout the remainder of this text. Practical implementation techniques are illustrated by using both RTL and TTL gates. A few examples of the use of other gate types are given as well. The techniques that are taught are not limited to RTL and TTL gates, however. Extension to any other gate type is easily accomplished, as is shown for the EXCLUSIVE-OR and AND-OR-INVERT gates in Sec. 3.9, and for other medium-scale-integration (MSI) combinational devices in Sec. 3.11.

3.2   The Voltage Truth Table

Just as a Boolean function can be specified by a logical truth table, the electrical operation of a gating circuit can be specified by a *voltage truth table* that gives the output voltage that the circuit will generate (ideally) for all combinations of input voltages. The idealized voltage truth table for a two-input RTL gate is shown in Fig. 3-0. The two voltage levels 0 V (ground)

A			A (V)	B (V)	C (V)
	RTL GATE	C	0	0	3.6
B			0	3.6	0
			3.6	0	0
			3.6	3.6	0

Figure 3-0. The RTL Gate and its Voltage Truth Table

and 3.6 V are the accepted industry standards for RTL devices. Note that only *ideal* conditions at inputs and outputs are considered in forming the voltage truth table. The voltage truth table can be obtained through analysis of the electronic circuitry of the gate, through experimental observation, or from the manufacturer of the circuit.

If the more positive of the two voltage conditions is chosen to represent the logically TRUE condition at both the inputs and the output of the gate shown in Fig. 3-0, the voltage truth table can be transformed into a logical truth table:

A	B	C
0	0	1
0	1	0
1	0	0
1	1	0

Comparison with the truth table of Sec. 2.7.1 shows that this is the NOR function. The RTL gate may thus be referred to as a *positive-TRUE* NOR gate, with the term positive-TRUE specifying that the more positive of the two voltage choices represents the TRUE logical condition at *all* the inputs and outputs of the circuit.*

*It should be noted that whenever the two voltage levels that are used with a particular gate type include 0 V the manufacturer of the gate often uses the 0-V level to represent logical 0 (FALSE) in referring to the gate and the logical functions it performs. As a result, the nonzero voltage is used to represent the TRUE condition. Thus, if a TTL manufacturer describes a gate having voltage levels of 0 and 5 V as being a NAND gate, he probably means that the gate performs the NAND function if 0 V represents FALSE and 5 V represents TRUE at *all* the inputs and outputs to the gate. There are exceptions to this generality, but this definition has appeared quite frequently in advertisement and descriptive literature.

It is significant to note that if the more negative of the two voltage choices (0 V) is used to represent the TRUE condition, the logical truth table becomes

A	B	C
1	1	0
1	0	1
0	1	1
0	0	1

which is the NAND function. Thus, the RTL gate is both a positive-TRUE NOR gate *and* a negative-TRUE NAND gate. This duality is true in general, in that any gate that performs the NOR function will always perform the NAND function if the logical meaning of all its inputs and outputs is inverted.

EXAMPLE 3-0:   Prove that $(\overline{\bar{A} \downarrow \bar{B}}) = A \uparrow B$.

Proof:

$$\begin{aligned}
(\overline{\bar{A} \downarrow \bar{B}}) &= \overline{(\bar{A} + \bar{B})} \\
&= \bar{\bar{A}} + \bar{\bar{B}} \\
&= \overline{AB} \qquad \text{from De Morgan's theorem, Eq. (2-20)} \\
&= A \uparrow B
\end{aligned}$$

EXAMPLE 3-1:   Prove that $(\overline{\bar{A} \uparrow \bar{B}}) = A \downarrow B$.

Proof:

$$\begin{aligned}
(\overline{\bar{A} \uparrow \bar{B}}) &= \overline{(\bar{\bar{A} \cdot \bar{B}})} \\
&= \bar{\bar{A}} \cdot \bar{\bar{B}} \\
&= \overline{A + B} \\
&= A \downarrow B
\end{aligned}$$

In actual use, the inputs and outputs to a gate circuit rarely equal the idealized values that are listed in the voltage truth table. The input voltages are generally *nearer* to either the high or the low voltage, however, and the gate circuit responds accordingly. Of great concern to the engineering user are the margins that are allowed about the high- and low-voltage values. "How low can an input voltage go and still appear to the gate as if it were at the high value?" is the type of question that must be answered.

Figure 3-1 shows a typical input characteristic for an inverter. Ideally the circuit acts as a logical NOT gate, changing a high-voltage input to low voltage and vice versa. Voltage levels of 0 and 5 V are assumed in the illustra-

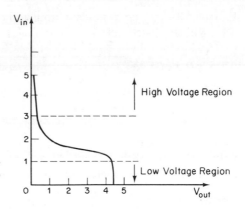

Figure 3-1. The Voltage-transfer Characteristic of an Example NOT Gate

tion. The output of the circuit remains close to 0 V as long as the input voltage is above 3 V. The output switches to nearly 5 V when the input goes below 1 V. In between, the output from the gate circuit is in a transition region, and a logical uncertainty arises.

The thresholds that apply to a given gate circuit can be determined by worst-case circuit analysis or by extensive experimental evaluation. The parameters that are usually determined include

1. The lowest input voltage that will still act as an ideal high value.
2. The highest input voltage that will still act as an ideal low voltage.
3. The highest output voltage that will be produced by a gate whose output is supposed to be low.
4. The lowest output voltage that will be produced by a gate whose output is supposed to be high.

Since gates are generally connected in sequence, it is necessary that the thresholds defined in 1 and 4 and in 2 and 3 overlap sufficiently so that the worst-case gate output voltage will act as the correct value at the inputs to all subsequent gates. These thresholds also specify the electrical noise immunity of the gate circuit, since the input voltage can vary within the allowable regions defined by the thresholds without causing any change in the output of the gate.

The output voltage from a gate may decrease under load, making it necessary to limit the load current that is supplied by the gate. This is easily accomplished by using normalized·*loading units*, as described in Sec. 3.10.4. The thresholds that apply to all common gating types are sufficiently wide to tolerate all normal operating conditions. As a result, the logical operation of the gates can usually be considered in an idealized fashion.

EXAMPLE 3-2:   The worst-case input and output voltage margins for a typical TTL gate are listed below:

	Input (V)	Output (V)
Highest-low	0.8	0.4
Lowest-high	2.0	2.4

There is a margin of 0.4 V in both cases; e.g., a gate input will consider all voltages below 0.8 V as being "low," and a "low" output from a gate will never exceed 0.4 V under allowable operating conditions.

The idealized voltage truth table for a two-input TTL gate is shown in Fig. 3-2. The standard voltage levels for TTL logic are 0 and 5 V. This type of gate acts as either a positive-TRUE NAND gate or a negative-TRUE NOR gate.

A (V)	B (V)	C (V)
0	0	+5
0	+5	+5
+5	0	+5
+5	+5	0

Figure 3-2. The TTL Gate and its Voltage Truth Table

The NAND/NOR gate circuit (whether DTL, RTL, TTL, ECL, or any other form) offers the electrical advantages of power gain, good noise immunity, and elimination of the voltage loss inherent in DRL. Also, since the NAND and NOR functions form complete sets by themselves, all combinational needs can be met with only one type of gate circuit. This simplifies inventory and purchasing problems. These economic and electrical considerations have made the NAND/NOR types of gating circuits the most useful form of construction for both discrete-component and integrated-circuit logical gates.

3.3 Voltage Symbolism Introduced

The use of the NAND/NOR types of logical gates does create a major difficulty for the digital systems designer: the implementation of AND-OR-NOT Boolean expressions with circuits that operate in a NAND or NOR logical form. Chapter 2 introduces the basic combinational design procedures (minterm or maxterm decomposition) which lead to A-O-N expressions. These expressions cannot be implemented directly with NAND/NOR gates, however. One approach to this problem is to logically manipulate these expressions into NAND/NAND or NOR/NOR form, using the techniques shown in Sec. 2-10. This process is tedious and offers ample opportunity for error. It also breaks the conceptual linkage between a Boolean design problem and its A-O-N expression.

EXAMPLE 3-3: Consider the NAND/NAND expression for "I will be there if it doesn't rain and my car will start or if you get me a date who is pretty" (repeated from Sec. 2-10). The A-O-N expression for the statement is given in Eq. (2-44):

$$A = \bar{R}S + DP \qquad \text{(2-44) repeated}$$

where A is TRUE if I arrive, R is TRUE if it rains, S is TRUE if the car starts, D is TRUE if you get me a date, and P is TRUE if she is pretty.

Converting Eq. (2-44) to NAND/NAND form gives

$$A = (\bar{R} \uparrow S) \uparrow (D \uparrow P)$$

This expression indicates that "I will arrive if it is not the case that both it did not happen that my car started and it did not rain and it did not happen that you got me a pretty date." The latter statement, although it is unfortunately very similar to many statements heard in classroom lectures, is not a very natural way to express the conditions under which arrival is expected.

Voltage symbolism (Refs. [1], [4], [9], [12], and [16]), the technique used for combinational implementation throughout the remainder of this text, is an easy method for bridging the gap between A-O-N expressions and NAND/NOR gates. There are other approaches to the resolution of this logical problem; for information on these see Refs. [7], [10], or [11]. Voltage symbolism is used herein because it offers direct conversion from A-O-N equations to NAND/NOR gating without losing the AND/OR or OR/AND structure of the equations. This technique works well in practice, is easy to learn and use, and can be extended to any type of gating circuit. In addition, voltage symbolism is useful in designing gating circuitry for logical systems that have electrical constraints upon their input and output signals, a significant utility that is not available with any other method. Voltage symbolism is also the only implementation technique whereby different types of gating circuits (NAND and EXCLUSIVE OR, for example) can be mixed with ease.

The voltage symbolism technique was first published by Washburn [16], who, in 1953, predicted that

. . . the synthesis of electronic switching circuits can be carried out in a routine manner using the Boolean algebra and other formal techniques. . . . However, the end result of this process in the present case is a circuit composed of more or less idealized elements. Considerable ingenuity may be necessary to replace these elements with their physical equivalents. Nevertheless, it may be anticipated that as new gates and memory devices appear, this final step in the design of a particular circuit may be accomplished in an almost routine fashion.

Washburn, in 1953, was concerned with relay logic, DRL, and vacuum-tube electronic logic circuits. With the widespread availability of prepackaged logical circuitry, the last statement of Washburn's prediction has become true. In particular, the use of voltage symbolism helps to reduce the implementation of a given combinational expression to a "routine" operation.

Voltage symbolism and several other closely related techniques have found great acceptance among practicing engineers. These techniques have found little acceptance in academic circles, however. The reasons for this omission are not clear. The technique is certainly useful, both to the designer and the technician. It works well in practice and is of great benefit during a checkout and/or troubleshooting procedure, since the use of voltage symbolism permits the A-O-N structure of the logical arrangement to be mainlined. In fact, one of the major assets of the voltage symbolism technique is that it changes the way in which the gating circuits are interpreted so as to make the gates correspond to the form of the A-O-N expressions that are to be implemented, rather than changing the equations to make them fit to the gates. Thus, voltage symbolism re-defines the behavior of the gating circuits to make them correspond to the way in which the user most naturally thinks, rather than making the user think in accordance with the behavior of the gating circuit.

The use of voltage symbolism involves the placing of a plus or minus sign at each input and output point in the logical circuit. This symbol indicates which of the two possible voltage levels represents the TRUE state of the Boolean variable at that point in the circuit. It specifies the relationship between the electronic characteristics of the gating circuit and the logical function that it is to perform. The use of the symbol is shown in Fig. 3-3.

The symbol is usually placed below the line. Note that neither of the voltage levels needs to be either positive or negative. In a circuit that operates with $+10$ and $+5$ V, $+5$ is the *more* negative and would represent TRUE with negative symbolism. Similarly, if the levels were -100 and -25 V, -25 would be the *more* positive and would represent TRUE with positive symbolism.

As an example, the RTL circuit (a negative-TRUE NAND gate) may be represented by the symbol shown in Fig. 3-4, where the minus signs drawn below the input and output lines indicate that the more negative of the voltage choices is used to represent the TRUE condition at these points. Simi-

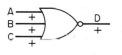

Figure 3-3. The Voltage Symbol and its Meaning

larly, the same RTL circuit can be used as a positive-TRUE NOR gate. The symbol for the gate when used in this manner is shown in Fig. 3-5.

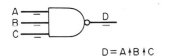

$$D = A \dagger B \dagger C$$

Figure 3-4. The RTL Gate When Used as a Negative-TRUE NAND Gate

$$D = A \dagger B \dagger C$$

Figure 3-5. The RTL Gate When Used as a Positive-TRUE NOR Gate

Figures 3-4 and 3-5 illustrate that the same circuit can be used to perform two different logical functions simply by changing the choice of which of the two voltage possibilities is used to represent the TRUE condition at its inputs and output. If a given gating circuit will perform the NAND function with a specified voltage symbolism, the same circuit will *always* perform the NOR function if the voltage symbolism at all its inputs and its output is reversed. This applies even in the case of a circuit with mixed symbolism, as shown in Fig. 3-6. The NAND and NOR operations form a *reversed-symbolism* pair.

Figure 3-6. The NAND/NOR Reversed-symbolism Pair

3.4 Voltage Symbolism and Inversion

Figure 3-7 shows eight fundamental relationships involving Boolean variables and voltage symbolism. The four cases on the left of Fig. 3-7 show the *logical* capability of a piece of wire. Note that an equivalent signal is obtained if the voltage symbolism is reversed *and* the Boolean function represented by the voltage on the wire is complemented. Thus, if a Boolean variable A is needed in negative-TRUE form, the signal $\bar{A}$ with positive-TRUE sense may be used in its place; the two are equivalent.

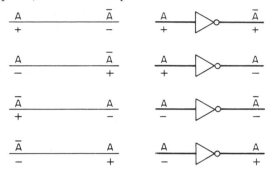

Figure 3-7. Voltage Symbolism and Logical Inversion

The four identities on the right of Fig. 3-7 show that an inverter (NOT circuit) can be used in either of two ways:

1. To *change the voltage symbolism* without changing the Boolean function.

2. To *complement the Boolean function* without changing the voltage symbolism.

EXAMPLE 3-4:   Given a Boolean variable X in positive-TRUE form, generate both X and $\bar{X}$ in positive-TRUE and negative-TRUE form.

Solution: Only one inverter is required, as shown in Fig. E3-4. When both X and $\bar{X}$ are available with the same voltage symbolism, the variable X is said to be in *double-rail* form.

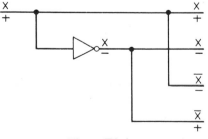

Figure E3-4

3.5 Using the NAND/NOR Circuit as an AND or an OR Gate

Given a gate that implements the NAND function, coupling a NOT gate to
its output would give a logical combination that would implement
NOT(NAND), which is simply the AND function. Section 3.4 shows that
logical inversion can be obtained without actually using an inverter if the
voltage symbolism associated with the Boolean function is also changed.
Thus, the RTL gate can be redefined as shown in Fig. 3-8. Since $F = \bar{D}$ and

Figure 3-8. The RTL Gate When Used as an AND Gate

$\bar{D} = \overline{A \uparrow B \uparrow C} = A \cdot B \cdot C$, the RTL gate will perform the AND function if
its inputs are interpreted as TRUE when negative and its output is interpreted
as TRUE when positive. The symbol for the RTL gate when it is used to
perform the AND function is shown on the right of Fig. 3-8.

   To further illustrate this conversion from NAND to AND, the RTL
voltage truth table given in Sec. 3.2 is repeated below:

A (V)	B (V)	C (V)
0	0	3.6
0	3.6	0
3.6	0	0
3.6	3.6	0

If the inputs, A and B, are considered as being negative-TRUE and the output, C, is considered as being positive-TRUE, the RTL voltage truth table
can be converted into a logical truth table:

A	B	C
1	1	1
1	0	0
0	1	0
0	0	0

Note that the input is TRUE when 0 V is applied, while the output is TRUE
when 3.6 V is present. The above table defines the AND operation, since
the output is TRUE only when both the inputs are TRUE.

   Consideration of the symbolism associated with the use of the RTL
circuit as a NOR gate shows that the gate can perform as an OR if the sym-

bolism at the output of the NOR is reversed, equivalent to NOT (NOR) or simply OR. The symbol for the RTL gate when used to perform the OR function is shown in Fig. 3-9.

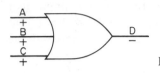

Figure 3-9. The RTL Gate When Used as an OR Gate

At first glance, the change in voltage symbolism between the inputs and the output of an RTL gate that is used to perform the AND or OR functions seems to be an inconvenience. In the case of AND/OR Boolean equations, however, the positive symbolism at the output of the RTL circuit when it is used as an AND gate is correct for input to the same circuit when it is used as an OR gate. A similar agreement applies to OR/AND gating. After passing through two levels of gating (either AND followed by OR or OR followed by AND) the outputs will have the same symbolism as the inputs.*

Comparison of the symbolism for the RTL circuit when used as an AND and when used as an OR shows that the symbolism at corresponding inputs and at the outputs is reversed. The AND/OR functions form a reversed-symbolism pair, just as the NAND/NOR functions did. The rules for assigning the symbolism so that the same circuit can be used to perform these four logical functions are summarized in Fig. 3-10.

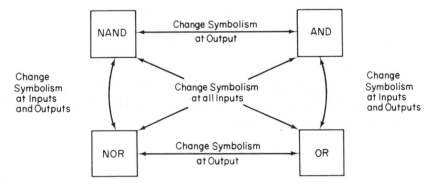

Figure 3-10. Voltage Symbolism and the Relationships between the Logical Operations

The diagonal paths in Fig. 3-10 may be explained with De Morgan's law. Consider the AND to NOR path. The AND function is

---

*It should be noted that the voltage symbol becomes an integral part of the logical diagram. When troubleshooting a logical system, it is necessary to know the voltage symbolism at every point in the system so that a measured voltage can be related to its logical meaning.

$$F = A \cdot B \cdot C \qquad (3\text{-}0)$$

Changing the voltage symbolism at the input of the AND gate requires a corresponding change in the logical value of the inputs, making the output of the AND gate the ANDing of the complements of the input variables:

$$F = \bar{A} \cdot \bar{B} \cdot \bar{C} \qquad (3\text{-}1)$$

or, by using De Morgan's theorem,

$$F = \overline{A + B + C} \qquad (3\text{-}2)$$

which is the NOR function. The proof of the NAND/OR translation is similar.

Figure 3-10 gives the conversions whereby a gating circuit that can perform any one of the four logical functions, AND, OR, NAND, or NOR, can be redefined so as to perform the other three. The two forms that are desired are the AND and the OR form, since these are used in implementing A-O-N Boolean expressions.

EXAMPLE 3-5:   The basic TTL gating circuit acts as a positive-TRUE NAND gate. Show the symbolism for this circuit when it is used to perform the NOR, AND, and OR functions.

Answer: See Fig. E3-5. Note that, as in the RTL case, two-level AND/ OR or OR/AND interconnections agree in the voltage symbolism that is required between outputs and inputs.

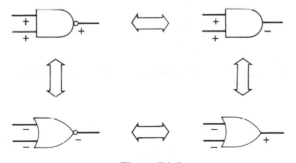

Figure E3-5

3.6 Combinational Design with Voltage Symbolism

The use of voltage symbolism separates the design process for implementing a Boolean expression into three parts. First, the *logical* layout is made, showing the required AND and OR gates without interconnection. The Boolean inputs and outputs of all gates are shown. Next, the *electrical* information, in the form of the voltage symbolism necessary for AND and OR gating, is added below the lines that represent all inputs and outputs. Finally, the *interconnections* are made, with NOT gates inserted where necessary to change either logical meaning or voltage symbolism. The AND-OR-NOT form of the expression is unchanged during this process, making the relationship between the combinational system and the Boolean expression that it implements easy to understand.

EXAMPLE 3-6:   Set up the gating circuitry necessary to implement $F = \overline{(AB)} + \overline{A}BC$. Use RTL gates (negative-TRUE NAND). Assume that A, B, and C are available with negative symbolism. The output function, F, must have positive symbolism.

Answer: The circuitry is shown in Figs. E3-6a–c.

The Logical Layout:

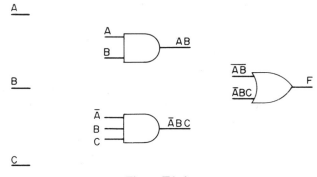

Figure E3-6a

The Electrical Layout:

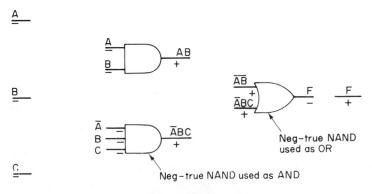

Figure E3-6b

The Interconnection:

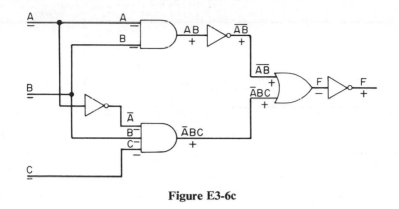

Figure E3-6c

EXAMPLE 3-7: Construct a two-input EXCLUSIVE-OR gate from TTL gates. The inputs and the output are to have positive symbolism.

Answer: The construction is shown in Figs. E3-7a–c.

The Logical Layout:

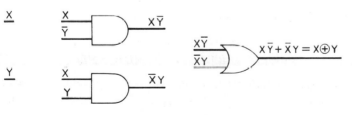

Figure E3-7a

The Electrical Layout:

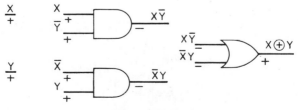

Figure E3-7b

The Interconnection:

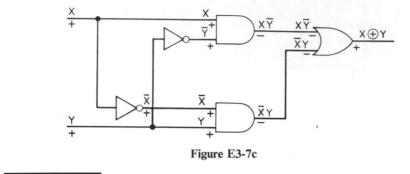

Figure E3-7c

As an interesting sidelight, voltage symbolism and the relationships given in Figs. 3-7 and 3-10 can be used to perform the conversions between standard forms that are discussed in an algebraic sense in Sec. 2-10. An AND/OR expression can easily be converted to NAND/NAND, OR/NAND, or NOR/OR form (the other forms in the AND/OR group) by reinterpreting the voltage symbolism that is used in its gating implementation. Similarly, an OR/AND expression can be converted into NAND/AND, AND/NOR, and NOR/NOR form by the reinterpretation process. The procedure is illustrated by the following example, a reexamination of Ex. 2-15.

EXAMPLE 3-8:   Example 2-15 shows that the EXCLUSIVE OR of two Boolean variables can be expressed as

$$F = A\bar{B} + \bar{A}B \qquad \text{(AND/OR)}$$
$$F = (A + B)(\bar{A} + \bar{B}) \qquad \text{(OR/AND)}$$

By using voltage symbolism, reexpress these two expressions in order to obtain the other six standard forms for the EXCLUSIVE-OR operation.

Solution: Either RTL or TTL gating can be assumed. Beginning with

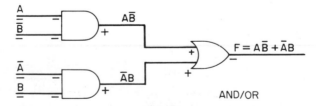

Figure E3-8a

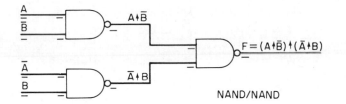

Figure E3-8b

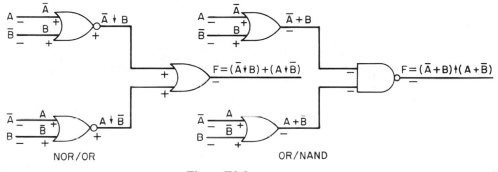

Figure E3-8c

the AND/OR expression, if RTL gating is used for its implementation, the logic diagram is as shown in Fig. E3-8a. Now, if the voltage symbolism assigned to the input and output signals is kept constant, the gating operation can be reinterpreted by changing the voltage symbolism at the inputs and outputs to each gate. Considering both levels of gating as NAND gates gives the logic diagram of Fig. E3-8b. The output is the NAND/NAND equivalent to the original AND/OR expression. Consideration of the circuitry in NOR/OR and OR/NAND forms gives the logic diagrams of Fig. E3-8c. The four equivalent expressions are

$$
\begin{aligned}
F &= A\bar{B} + \bar{A}B && \text{(AND/OR)} \\
&= (A \uparrow \bar{B}) \uparrow (\bar{A} \uparrow B) && \text{(NAND/NAND)} \\
&= (\bar{A} \downarrow B) + (A \downarrow \bar{B}) && \text{(NOR/OR)} \\
&= (\bar{A} + B) \uparrow (A + \bar{B}) && \text{(OR/NAND)}
\end{aligned}
$$

These agree with the results given in Ex. 2-15.

Beginning with the OR/AND expression (and using TTL gates, for contrast) gives the other four standard forms, as shown in Fig. E3-8d. The four expressions found here agree with the results found in Ex. 2-15.

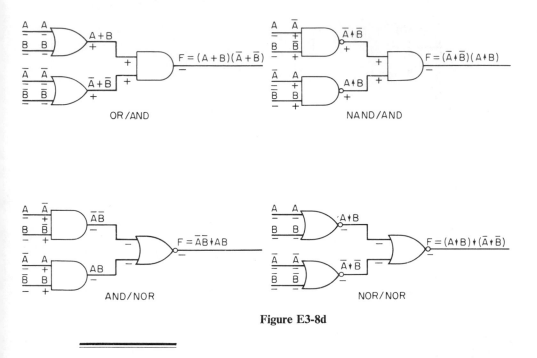

Figure E3-8d

3.7 Using the NAND/NOR Gate as an Inverter

Examples 3-6 and 3-7 show that the necessity for complementation often arises during the implementation of Boolean equations. Most digital circuit manufacturers include a NOT gate (one input, and one output that is always the complement of the input) in their product line. In addition, the NAND/ NOR gate can be used as an inverter itself, due to the voltage inversion inherent in its electronic operation.

Section 2-7 shows that the NAND and NOR functions can be used to produce the NOT function in either of two ways. Consider a three-input RTL gate that is to be used as an inverter. The RTL gate acts as a positive-TRUE NOR. Section 2.7.1 shows that the complement of a variable can be obtained by NORing it with itself:

$$\bar{A} = A \downarrow A \downarrow A \downarrow \cdots \qquad (2\text{-}22)$$

Thus, if an RTL gate is connected as shown in Fig. 3-11, it will act as an inverter. The input symbolism is not shown, since the circuit will complement an input having either symbolism.

The disadvantage of this type of connection is that the power required

Figure 3-11. The RTL Gate When Used as an Inverter, Type I

by the gate is increased as its input connections are paralleled. Thus, a three-input gate connected as shown in Fig. 3-11 will draw about three times as much power from the signal source as would a NOT gate having only one input. This makes a second method for obtaining the NOT function from the NOR more preferable. This method uses the expression

$$\bar{A} = A \downarrow 0 \downarrow 0 \downarrow \cdots \tag{2-23}$$

which is derived in Sec. 2.7.1.

If a gate is connected in accordance with Eq. (2-23), all but one of its inputs must be tied to the voltage that represents the FALSE condition at the input to a NOR gate. Since the RTL NOR gate requires positive-TRUE inputs, the FALSE condition will be 0 V, the more negative of the choices. This method for obtaining the NOT function from an RTL gate is shown in Fig. 3-12. The second method does not increase the load seen by the source of the input signal.

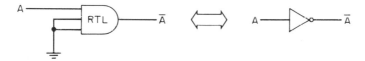

Figure 3-12. The RTL Gate When Used as an Inverter, Type II

A similar reasoning process involving the RTL NAND gate and the expression

$$\bar{A} = A \uparrow 1 \uparrow 1 \uparrow \cdots \tag{2-25}$$

shows, again, that the unused inputs should be connected to 0 V. When used as an inverter, a TTL gate requires that its unused inputs be connected to 5 V, since the TTL gate acts as a positive-TRUE NAND gate.

3.8   Open Inputs and the Use of Switches

In some situations it is convenient to use only part of a gate's input capability. For example, a three-input gate may be used to combine only two logical input signals. Of concern is the effect that the extra input has upon the operation of the gate. The logical and electrical effect that the unused gate input will have can be determined only by analysis of the gate circuit or by experimentation. Normally, if an input is left open it will act as if it were connected to one of the two voltage levels used by the circuit. In the specific case of RTL gate circuits, open inputs have the same effect as inputs that are connected to 0 V. Connection to 0 V acts as a TRUE condition to an RTL AND and an RTL NAND gate, since they both require negative-TRUE inputs. This has the logical effect of

$$A \cdot B \cdot 1 = AB \qquad (3\text{-}3)$$

or

$$A \uparrow B \uparrow 1 = A \uparrow B \qquad (3\text{-}4)$$

showing that the presence of unused, open inputs will not affect the AND or NAND operation of the gate upon those inputs that *are* connected. Similarly, the 0 V effect of an open input will appear as a FALSE input to either an RTL OR gate or an RTL NOR gate. Then, since

$$A + B + 0 = A + B \qquad (3\text{-}5)$$

and

$$A \downarrow B \downarrow 0 = A \downarrow B \qquad (3\text{-}6)$$

the open inputs are again of no logical concern.

If an open input to an RTL gate acted as if it were connected to 3.6 V, however, it would invalidate the action of the other inputs. For example, connection to 3.6 V would appear as a TRUE input to an RTL OR gate, and, since

$$A + B + 1 = 1 \qquad (3\text{-}7)$$

the gate would always indicate a TRUE condition at its output. Similar logical disasters would occur if the gate were used for the NAND, NOR, AND, or NOT operations.

Luckily, this effect does not occur with open inputs to either RTL or TTL gates, or most other types of gating circuits, but open inputs *are* more sensitive to electrical noise. Thus, it is always a good idea to "strap" unused gate inputs to a fixed voltage. The fixed voltage should be the level that acts as a TRUE input condition to the gate when it is used as an AND gate, i.e., 0 V for RTL gates and +5 V for TTL gates. Similar considerations apply to other logical gating types.

EXAMPLE 3-9: Show the connections necessary to use a three-input TTL gate as an inverter and as a two-input AND gate.

Answer: See Fig. E3-9.

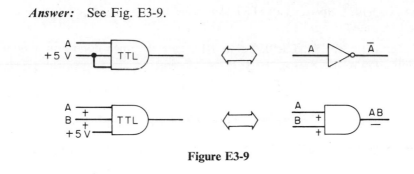

Figure E3-9

The open-input effect can be used in generating both TRUE and FALSE values for signals by using toggle switches. As an example, consider the use of a single-pole, single-throw (SPST) switch to provide a logical input, say *A*, to an RTL gate. If the switch is connected as shown in Fig. 3-13, the variable *A* will never change its logical value, since the open connection will have the same logical effect as the switched connection to 0 V. The correct connection is shown in Fig. 3-14. The connection shown in Fig. 3-14 will

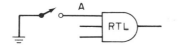   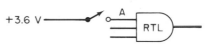

Figure 3-13. Improper Use of a Switch to Generate a Boolean Signal

Figure 3-14. The SPST Toggle Switch Used to Generate a Boolean Signal

cause the variable *A* to be positive when the switch is closed and to appear to be negative (since an open input acts as 0 V) when the switch is opened. Thus, if the variable is considered as being TRUE when the switch is closed, the variable is generated with positive-TRUE symbolism. If it is considered as TRUE when the switch is open, it is generated with negative-TRUE symbolism.

A better connection uses a single-pole, double-throw (SPDT) switch to generate the variable, as shown in Fig. 3-15. The connection shown in Fig. 3-15 eliminates the open condition, thereby reducing the effect that electrical noise can have on the gate.

Figure 3-15. The Use of an SPDT Switch to Generate a Boolean Variable

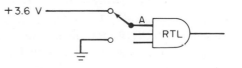

EXAMPLE 3-10: How can an SPST switch be used to provide a logical input to a TTL gate?

Answer: See Fig. E3-10. For a TTL gate the open condition acts as a connection to $+5$ V.

Figure E3-10

3.9 Special-Function Gating Circuits

With the development of integrated-circuit technology, it has become economically feasible to manufacture gating circuits that are more complex than the basic NAND/NOR gate. The EXCLUSIVE-OR gate is an example of this. The symbol and voltage truth table for the two-input *RTL* EXCLUSIVE-OR gate are shown in Fig. 3-16. The gate implements the EXCLUSIVE

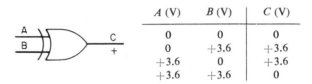

A (V)	B (V)	C (V)
0	0	0
0	$+3.6$	$+3.6$
$+3.6$	0	$+3.6$
$+3.6$	$+3.6$	0

Figure 3-16. The RTL EXCLUSIVE-OR Gate

OR of its inputs if the output is interpreted as TRUE when positive. The input symbolism for this (or any) EXCLUSIVE-OR gate can be *either* positive-TRUE or negative-TRUE. This is a consequence of the identity

$$A \oplus B = \bar{A} \oplus \bar{B} \tag{3-8}$$

If the output from the RTL EXCLUSIVE-OR gate is interpreted as TRUE when negative, the gate will perform the COINCIDENCE or EQUIVALENCE function. This function, the complement of the EXCLUSIVE OR, is TRUE whenever its inputs have the *same* logical value. For more than two arguments, the COINCIDENCE function is TRUE whenever all the inputs are FALSE or an even number of the inputs are TRUE. The COINCIDENCE function is given the ring-dot symbol:

$$F(A,B) = A \odot B = \bar{A}\bar{B} + AB = \overline{A \oplus B} \tag{3-9}$$

Thus, the positive-TRUE RTL EXCLUSIVE-OR gate is also a negative-TRUE COINCIDENCE gate. The symbol shown in Fig. 3-17 has been defined for use as a COINCIDENCE gate.

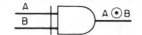

Figure 3-17. The COINCIDENCE Gate

EXAMPLE 3-11: Show how the two-input RTL EXCLUSIVE-OR gate can be used to implement

$$F = A \oplus B \oplus C$$

Consider both positive-TRUE and negative-TRUE inputs.

Solution: See Figs. E3-11a and b.

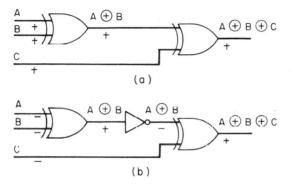

Figure E3-11

EXCLUSIVE-OR gates are also available in most other families of gating circuits. The voltage symbolism at the output of the gate varies, however. The TTL EXCLUSIVE OR gate has positive-TRUE output symbolism.

One of the real advantages of the voltage symbolism technique is the ease with which it permits special-function gating circuits to be incorporated into A-O-N combinational systems. The following examples illustrate this point.

EXAMPLE 3-12: Show an RTL implementation for

$$F = \bar{A}B + A\bar{B} + B\bar{C}$$

Assume negative-TRUE inputs and positive-TRUE output.

Solution: The output function may be reexpressed as

$$F = (A \oplus B) + B\bar{C}$$

This can be implemented as shown in Fig. E3-12.

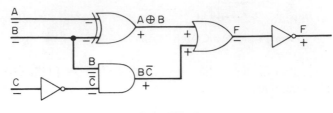

Figure E3-12

EXAMPLE 3-13: Repeat Ex. 3-12 assuming positive-TRUE input symbolism.

Answer: See Fig. E3-13.

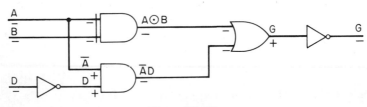

Figure E3-13

EXAMPLE 3-14: Implement the following expression with TTL gating. Assume negative-TRUE inputs and output.

$$G = AB + \bar{A}\bar{B} + \bar{A}D$$

Solution: G may be reexpressed as

$$G = (A \odot B) + \bar{A}D$$

This can be implemented as shown in Fig. E3-14.

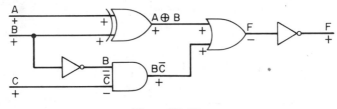

Figure E3-14

     The voltage symbolism relationships that arise when the EXCLUSIVE-OR gate is used can have some unusual characteristics. Consider the following logical expression:

$$F = (\bar{X} \oplus Y) \cdot Z \qquad (3\text{-}10)$$

The RTL gating for implementing Eq. (3-10) with negative-TRUE input symbolism is shown in Fig. 3-18. Two inverters are required—one at the input to the EXCLUSIVE-OR gate and one between its output and the next level of gating.

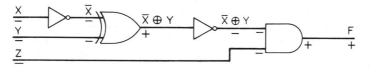

Figure 3-18. RTL Implementation of Eq. (3-10)

     However, Eq. (3-10) can be reexpressed as

$$F = \overline{(X \oplus Y)} \cdot Z \qquad (3\text{-}11)$$

by using the relationship

$$\bar{X} \oplus Y = \overline{X \oplus Y} \qquad (3\text{-}12)$$

which was proved as part of Prob. 2-2. The RTL implementation for Eq. (3-11) is given in Fig. 3-19. This modification eliminates the need for the two inverters that are shown in Fig. 3-18. In general, whenever a logical system contains an EXCLUSIVE-OR gate that has inverters in series with any two

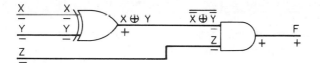

Figure 3-19. Modified RTL Implementation of Eq. (3-10)

of its inputs or output, the logical system can be modified so as to eliminate the need for both inverters. This property is explored further in Prob. 3-15.

     Another special-function gate is the AND-OR-INVERT (A-O-I) gate found in TTL families of digital circuits. This type of gate performs the A-O-I sequence of operations with positive-TRUE symbolism, as shown in Fig. 3-20. By redefining the input and output symbolism, the A-O-I gate

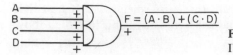

$F = \overline{(A \cdot B) + (C \cdot D)}$

Figure 3-20. The TTL AND-OR-INVERT Gate

can be interpreted in *either* AND/OR or OR/AND form, with the inversion of the output incorporated into the output symbolism. Figure 3-21 shows the symbolism applicable to these two cases. The extra inputs available at the second level of the A-O-I gate provide for expansion of the first level of logical operation.

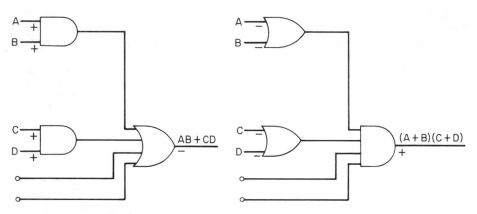

Figure 3-21. The TTL A-O-I Gate Used to Perform AND/OR or OR/AND Logic

Special-function gates such as the A-O-I and EXCLUSIVE-OR gates are useful in many combinational design applications. The use of voltage symbolism makes it quite easy for the designer to incorporate these (and other) combinational devices into the combinational systems that he designs. It should be remembered, however, that the basic NAND/NOR gate is sufficient for all combinational needs.

3.10 Other Considerations

When assembling gating circuits into a combinational system and when comparing different approaches to the same design problem, concepts of cost, speed, and loading are of concern. This section summarizes some general information on several such *practical* aspects of combinational design. Information on some of these topics is also available from digital circuit manufacturers in the form of device specifications or applications notes. (See Refs. [2] and [13], for example.)

3.10.1. Gates in Parallel.

If the output terminals of two NAND/NOR gates are tied together, the resultant logical and electrical effects will vary depending on the type of circuitry used in manufacturing the gate. Parallel RTL gates will exhibit *parallel expansion,* in that two parallel two-input gates will look like a single four-input gate with the same logical function. This is illustrated in Fig. 3-22. Parallel expansion can be continued up to

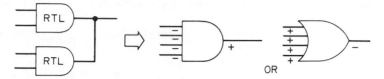

Figure 3-22. Parallel Expansion of RTL Gates

the limit specified by the manufacturer of the gate [13], usually about 10 to 12 inputs. This limit is often referred to as the allowable *fan-in* for the particular gating circuit.

The paralleling of gates of the DTL variety has a markedly different effect. In A-O-N terms, the effect is shown in Fig. 3-23 (negative-TRUE NAND gates shown). The paralleled outputs act as an OR following an AND operation or as an AND following an OR operation, although there is no change in voltage symbolism as there would be if another DTL gate were used to combine the outputs from the first level of gating. The net logical function is the same as the AND-OR-INVERT gate described in Sec. 3-8. This logical behavior is commonly called the *wired-OR* property, but *wired-AND* applies when parallel connection follows first-level OR operations. (For another interpretation of this effect, see Ref. [3].) A common symbolism for the "wired" operations is shown at the right of Fig. 3-23. The use of this type of symbol makes it clear that the wired logic property, rather than an actual gate circuit, is used to implement the second level of gating.

TTL gates also exhibit the wired-OR property, although the paralleled outputs may cause electrical damage to the gating circuits during changes in the input signals. For this reason the wired-OR property is not often used with normal TTL gates. An alternative type of TTL gating circuit, the *three-state TTL gate*, can implement the wired logic operations without electrical damage (Refs. [5] and [15]). The wired-OR is not available from RTL gates, since they exhibit parallel expansion.

The behavior of any type of gating circuit when connected in parallel will always fall into one of these three classes: parallel expansion, wired logic, or electrical damage. The action of a particular circuit can be determined by experimentation, by circuit analysis, or from the manufacturer's specifications.

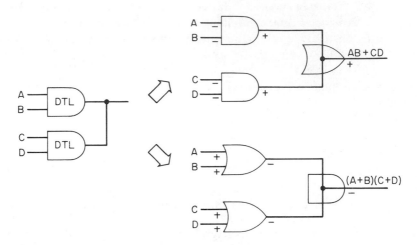

Figure 3-23. The Parallel Connection of DTL Gates

Parallel expansion is useful when gates having several inputs are needed. The basic RTL gate comes in two-, three-, and four-input forms. By paralleling these input forms, gates having more inputs may be obtained.

EXAMPLE 3-15:   Implement $F = ABCDE$ with RTL gates.

Solution: See Fig. E3-15.

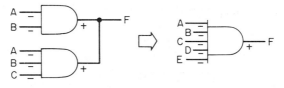

Figure E3-15

With DTL gates, input expansibility is generally obtained by adding additional diodes in parallel with the input connections. Groups of diodes, commonly called *clusters*, are manufactured for this purpose.

Expanding the number of inputs to a TTL gate is more difficult. The basic TTL gate types include circuits with two, three, four, and eight inputs. The AND-OR-INVERT gates offer some further expansibility, but obtaining a NAND/NOR gate with more than eight inputs involves some complex logical manipulation.

EXAMPLE 3-16: How may a ten-input TTL AND gate be implemented?

Solution: Start with an eight-input gate and a two-input gate (see Fig. E3-16a). Invert their outputs and apply them to another two-input gate, as shown in Fig. E3-16b. The output from this combination is the ANDing of the ten input signals.

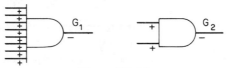

Figure E3-16a

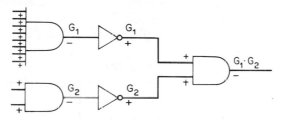

Figure E3-16b

3.10.2. Cost Measurements. When comparing two different gating systems for implementing the same Boolean function, some simple cost criterion proves to be useful. The discrete-component designer who uses DTL gates commonly counts the number of active elements used in the discrete-component gates and expresses the system cost in terms of the total number of diodes and transistors used to implement the Boolean function. Each DTL gate uses one transistor plus a diode for each input to the gate. Thus, the implementation of

$$F = ABC + DE \qquad (3\text{-}13)$$

would take seven diodes (a three-input AND, a two-input AND, and a two-input OR) and three transistors.

A modification of this cost criterion counts only diodes and is commonly called *gate-leg* cost in that each diode represents an input "leg" to a DTL NAND/NOR gate. For further comments on this approach to cost comparison, see Ref. [8].

EXAMPLE 3-17: Find the gate-leg cost for

$$F = AB + AC + DE$$

Answer: Cost $= 2 + 2 + 2 + 3 = 9$ gate legs (diodes). Note that by factoring F as

$$F = A(B + C) + DE$$

the cost is reduced to 8 gate legs.

When using digital circuits that are constructed with integrated circuit technology, counting the number of active elements in order to estimate cost is not as valid as it is when comparing discrete-component gating systems, since the cost of manufacturing an integrated-circuit gate is not directly related to the complexity of the gating circuit. The most direct method for comparing cost between two different integrated-circuit gating arrangements would be to use actual package count and actual cost.

However, consideration of most integrated-circuit gating circuits shows that there is a loosely linear relationship between cost per package and the number of input connections to the gates in the package. This points to the fact that the cost of an integrated-circuit device is somewhat independent of the complexity of the electronics *within* the package but is determined, in a large part, by the number of input and output connections required by the device. Thus, as a simple cost comparison, counting the number of gate inputs is a valid technique for both discrete-component and integrated-circuit gating. The cost of the gating can then be expressed in equivalent *cost units* (c.u.), which correspond to the number of input connections.

EXAMPLE 3-18: Equation (2-18) gives the following equality:

$$(A + B) \cdot (A + C) = A + BC$$

Compare the two expressions in terms of the cost units necessary to implement them.

Answer: $(A + B)(A + C)$ uses two two-input OR gates and a single two-input AND gate, a total of six inputs, or a cost of 6 c.u. $A + BC$ uses only four gate inputs, or 4 c.u., a savings of 2 c.u. by using the latter expression.

The use (when possible) of special-function gates such as the EXCLU-SIVE-OR gate will generally save gating cost relative to implementation with NAND/NOR gates alone. The use of gate-input cost units with EXCLU-SIVE-OR gates is not as straightforward as it is with the NAND/NOR gate, however. As an estimate, a two-input EXCLUSIVE-OR gate costs approximately 3 c.u., making each of its inputs have a cost of 1.5 c.u. This estimate can be used to describe the relative cost of implementing logical expressions such as Eq. (3-10) in an EXCLUSIVE-OR form, as shown in Fig. 3-19. The gating shown in Fig. 3-19 would cost approximately 5 c.u. Without the use of the EXCLUSIVE-OR gate, the implementation of Eq. (3-10), with the inputs having only negative-TRUE voltage symbolism, would require 10 c.u.

The process of minimizing and manipulating a Boolean expression to save cost is the subject of Chapter 4. In all design examples in the remainder of this text, cost will be expressed in terms of the gate-input cost units defined above.

3.10.3. Delay.

When the input to an electronic gate circuit changes, there is a finite time delay before the change in input is seen at the output terminal. This delay is referred to as the *propagation delay* of the gate (t_g) and is about 12 nsec for RTL gates and about 5 nsec for TTL gates. The delay through a gating system becomes critical if the states of the inputs are expected to change rapidly. If so, the input may change before the output has settled to its proper value, leading to erroneous output information.

Of primary concern is the path from input to output with the most propagational delay, the *worst-case delay path*. Since all combinational functions can be expressed in the AND/OR or OR/AND form, the worst-case delay need not exceed twice the propagational delay for a single gate, plus the delay in any inverters needed at the inputs or output of the gating system. Often, however, Boolean manipulation used to save cost will result in added delay.

EXAMPLE 3-19: Given $F_1 = AB + AC + DE$, compare the cost and worst-case delay for generating F in this form and in the equivalent form, $F_2 = A(B + C) + DE$.

Answer: F_1 costs 9 c.u. and two delays. F_2 costs 8 c.u. and three delays, as shown in Fig. E3-19. The choice as to which form to use is dependent on

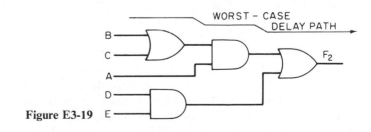

Figure E3-19

the application. If speed is critical, the first expression is preferable even though it costs more to implement.

3.10.4. Loading. Electronic gate circuits consume power. A gate must be capable of supplying the power requirements of all other gates that are connected to its output. Loading requirements for digital circuits are generally expressed in terms of current requirements, with the current drawn by a single gate input referred to as a *unit load*. The driving capability of a gate output is then expressed in terms of the number of unit loads (other gate inputs) that it can supply and still operate correctly. Knowledge of the actual value of the unit-load current is therefore unnecessary in calculating load requirements. The output drive capability of a gate, expressed in unit loads, is commonly called the *fan-out* of the gate.

For a typical RTL gate the fan-out is 5 unit loads, meaning that the output from a gate can be connected only to five other gate inputs. To provide more output power, a power amplifier (*buffer*) is included in most digital circuit families. An RTL buffer draws two unit loads at its input and can drive 25 unit loads. The RTL buffer acts as an inverter as well as a power amplifier. The fan-out of a normal TTL gate is 10 unit loads. A TTL buffer (inverting) will drive 40 unit loads. Buffers are available in almost all families of logic circuits.

The consideration of loading factors as a part of digital design is essential. Whenever load limits are exceeded, a buffer should be used to amplify the signal that is overloaded.

3.11 Combinational Design with MSI Devices

The state-of-the-art in digital circuit design is rapidly pushing to higher and higher levels within the component/functional/systems design hierarchy. Only a few years ago the digital design process began with discrete electronic components (transistors, resistors, etc.); designers had to construct the required gating circuits. The progress in integrated-circuit technology has made it possible to produce the inexpensive, ready-made gating circuits that have now become the basic "component" of the design process. The development of this process has accelerated rapidly; medium- and large-scale integration (MSI and LSI) techniques promise to make today's functional-level devices tomorrow's (or even today's) basic components.

As new functional-level devices become available in prepackaged form, designers should seek to use them in their systems. The voltage symbolism technique is a powerful tool for this. The input and output signals to these higher-level devices do not always use the same voltage level for the TRUE condition, however, and the use of voltage symbolism allows these input/output conditions to be clearly specified.

The three-step design process described in Sec. 3.6 is still maintained. Voltage symbolism allows the interconnections between various combinational MSI devices and other components of a digital system to be made with assurance. The need for inversion, where necessary, is clearly identified. The following examples show the use of voltage symbolism with several combinational TTL and MSI devices.

EXAMPLE 3-20:   The use of various Boolean codes to represent the ten decimal digits is covered in depth in Chapter 10. MSI devices for decoding some common codes into the ten different decimal values are available. The voltage symbolism that is applicable to one such device is shown in Fig. E3-20. This

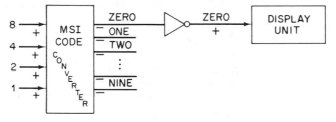

Figure E3-20

combinational circuit accepts a four-bit binary number ranging from zero to 9 and converts that input code into one of ten outputs (8421 BCD-to-decimal conversion). The input signals must be provided in single-rail, positive-TRUE form. The outputs from this device are generated with negative-TRUE voltage symbolism. Thus, if this device is used to drive a decimal display unit that requires positive-TRUE inputs, the voltage symbolism indicates the need for inverters between the outputs from the converter and the inputs to the display unit.

Combinational devices such as the one shown above are available for converting a variety of codes (see Chapter 10) into their decimal equivalents.

EXAMPLE 3-21:   The 74156 is a TTL MSI device that contains a pair of 2-line-to-4-line decoders. These decoders are combinational devices that accept a pair of single-rail Boolean inputs, say A and B, and produce four output signals, say m_0, m_1, m_2, and m_3, corresponding to the four possible combinations of the two-valued inputs. The two decoders can be represented by the symbols shown in Fig. E3-21a.

The G and C signals act as enabling signals for the decoder outputs. Note that the C input is positive-TRUE to one decoder and is negative-TRUE to the other. Thus, if the A, B, C, and G inputs to the two decoders are paralleled, the devices can be used as a three-to-eight decoder. A, B, and C then serve as three single-rail Boolean inputs and G serves as an enabling signal (GATE). The decoder whose C input is positive-TRUE produces m_4 through m_7, since it is enabled when the C input is HIGH. The composite diagram for the three-to-eight decoder is shown in Fig. E3-21b.

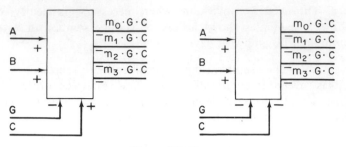

Figure E3-21a

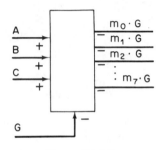

Figure E3-21b

EXAMPLE 3-22:   Show how the three-to-eight decoder described in Ex. 3-21 can be used to implement:

$$F(A,B,C) = \sum 0,2,3,4,7$$

in both AND/OR and OR/AND form.

Solution: The decoder produces eight signals that correspond to the eight minterms. ORing the m_0, m_2, m_3, m_4, and m_7 outputs will produce F in sum-of-minterms form, as shown in Fig. E3-22a. The G input is strapped to the TRUE voltage condition.

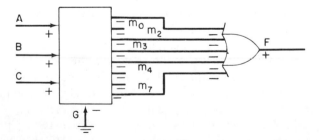

Figure E3-22a

Similarly, ORing the m_1, m_5, and m_6 outputs will produce $\bar{F}$ in AND/ OR form. Changing the voltage symbolism at the $\bar{F}$ output will produce the desired output signal, as shown in Fig. E3-22b.

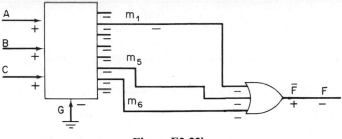

Figure E3-22b

EXAMPLE 3-23: The 54H87 is a three-input combinational device that uses two of the inputs to modify the logical state of the third. A symbol for the device is shown in Fig. E3-23a. When B and C are both TRUE, the $f(A)$

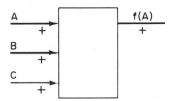

Figure E3-23a

output is forced to be FALSE. It is forced TRUE for $B\bar{C}$, equated to A for $\bar{B}C$, and equated to $\bar{A}$ for $\bar{B}\bar{C}$. These conditions are tabulated below.

B	C	$f(A)$	*Output*
0	0	$\bar{A}$	Complement of A
0	1	A	True Value of A
1	0	1	One (Forced)
1	1	0	Zero (Forced)

This TRUE/COMPLEMENT/ZERO/ONE function is useful in arithmetic processes.

As an example, consider the design of a combinational controller that accepts four signals: T(True), CP(ComPlement), Z(Zero), and N(oNe), and produces the proper values of B and C for controlling the arithmetic network.

Solution: Using the information given above, the following table can be generated:

T	CP	Z	N	B	C
0	0	0	1	1	0
0	0	1	0	1	1
0	1	0	0	0	0
1	0	0	0	0	1

Inspection of the above truth table shows that:

$$B = Z + N$$
$$C = T + CP$$

Thus, the element and its control logic can be assembled as shown in Fig. E3-23b.

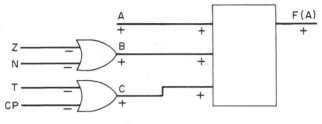

Figure E3-23b

Other examples of the use of combinational MSI devices are given in Chapters 4, 8, 9, and 10. Other types of combinational MSI devices include arithmetic comparators, multiple-bit adders, parity generators (see Chapter 10), and various code-to-code converters. Keeping up with the availability of these types of devices is a necessity for the practicing digital designer. Fortunately, the use of voltage symbolism simplifies the incorporation of these new building blocks into digital systems as soon as they become available.

3.12   Summary

This chapter is concerned with the implementation of Boolean expressions by using electronic gating circuits. Voltage symbolism, a technique whereby NAND/NOR gates can be used to implement A-O-N expressions, is introduced. Several examples of the implementation process are given. The use of special-function gates is also discussed. Practical considerations such as cost, delay, and loading are considered as well. The voltage symbolism approach is used throughout the remainder of this text.

3.13 Did You Learn?

1. How to convert a voltage truth table to a logical truth table by using voltage symbolism?

2. The voltage truth tables for the RTL and TTL integrated-circuit gates?

3. How an inverter can change either logical meaning *or* voltage symbolism? How a piece of wire can change *both*?

4. What single-rail and double-rail input means?

5. The AND-OR-NAND-NOR relationships given in Fig. 3-10?

6. How voltage symbolism separates the logical and electrical aspects of combinational design?

7. How to use a NAND/NOR gate as an inverter?

8. How to use a single-pole, single-throw switch as a logical signal source? An SPDT switch?

9. What effect open (unused) gate inputs have? How to strap them down to reduce the effects of electrical noise?

10. How an EXCLUSIVE-OR gate can be used in conjunction with NAND/NOR gates? How it can act as a COINCIDENCE gate? Why its input symbolism can be *either* positive-TRUE or negative-TRUE?

11. The output voltage symbolism for the RTL and TTL EXCLUSIVE-OR gates? The RTL and TTL COINCIDENCE gates?

12. The logical operation of the TTL AND-OR-INVERT gate?

13. How parallel expansion operates? How to combine RTL gates in parallel?

14. What use Wired-OR and Wired-AND have?

15. How to estimate the cost of implementing a given logical expression? How cost and worst-case delay time can be traded off?

16. How to measure loading in a logical circuit? How and when to use a buffer?

17. How to incorporate MSI devices into combinational systems?

3.14 References

[1] Asija, S. P., "Instant Logic Conversions," *IEEE Spectrum*, 5, No. 12 (December 1968), 77–80.

[2] Crowe, Ronald, and Louis Delhorn, "Logic Design with Series 54/74 Gates," *Texas Instruments Application Report No. CA-112.* Dallas: Texas Instruments, Inc., October 1968. See also R. L. Morris and J. R. Miller, editors, *Designing with TTL Integrated Circuits.* New York: McGraw-Hill Book Company, 1971.

[3] Danielsson, Per. E., "A Note on Wired-OR Gates," *IEEE Trans. on Computers*, C-19, No. 9 (September 1970), 849–850.

[4] Ennis, Victor, "Positive and Negative Logic," *Computer Design*, 9, No. 9 (September 1970), 79–87.

[5] Fleming, Don, "Enhancement of Modular Design Capability by Use of Tri-State Logic," *Computer Design*, 10, No. 6 (June 1971), 59–64.

[6] Garrett, Lane S., "Integrated-circuit Digital Logic Families," *IEEE Spectrum*, 7, No. 10, Part I (October 1970), 46–58; 7, No. 11, Part II (November 1970), 63–72; 7, No. 12, Part III (December 1970), 30–42.

[7] Hill, Frederick J., and Gerald R. Peterson, *Introduction to Switching Theory and Logical Design*, New York: John Wiley & Sons, Inc., 1968.

[8] Hurley, R. B., *Transistor Logic Circuits.* New York: John Wiley & Sons, Inc., 1961.

[9] Kintner, P. M., "A Simple Method of Designing NOR Logic," *Control Engineering* (February 1963), 77–79. See also P. M. Kintner, *Electronic Digital Techniques.* New York: McGraw-Hill Book Company, 1968.

[10] Maley, Gerald A., and J. Earle, *The Logic Design of Transistor Digital Computers.* Englewood Cliffs, N.J.: Prentice-Hall, Inc., 1963.

[11] Malmstadt, H. V., and C. G. Enke, *Digital Electronics for Scientists.* Reading, Mass.: W. A. Benjamin, Inc., 1969.

[12] Marcus, Mitchell P., *Switching Circuits for Engineers.* Englewood Cliffs, N.J.: Prentice-Hall, Inc., 1962.

[13] Maul, Lloyd, "Loading Factors and Paralleling Rules for MRTL Integrated Circuits," *Motorola Integrated Circuits Application Note AN-285.* Motorola Semiconductor Products, Inc., Phoenix, 1967.

[14] Millman, Jacob, and Herbert Taub, *Pulse, Digital, and Switching Waveforms.* New York: McGraw-Hill Book Company, 1965.

[15] Sheets, John, "Three-state Switching Brings Wired-OR to TTL," *Electronics*, 43, No. 19 (September 14, 1970), 78–84.

[16] Washburn, S. H., "An Application of Boolean Algebra to the Design of Electronic Switching Circuits," *Trans. of AIEE* (September 1953), 380–388.

[17] Wickes, W. E., *Logic Design with Integrated Circuits.* New York: John Wiley & Sons, Inc., 1968.

3.15 Problems

3-1. Considering the transistors shown in the circuits in Fig. P3-1 as ideal switching elements, develop the voltage truth tables for the two gating circuits. The input voltage levels are 0 and +5 V.

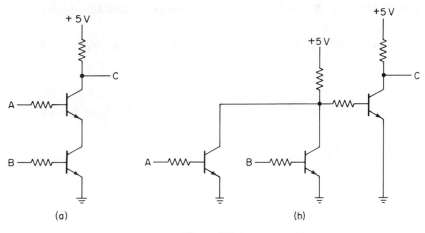

Figure P3-1

3-2. The voltage truth tables that describe several gating circuits are given below. Determine the voltage symbolism assignments that must be made in order to use each circuit as a NAND, NOR, AND, and OR (if possible). Show the four gate symbols for each case, with the required input and output symbolism included on the symbol.

a.

A (V)	B(V)	C (V)
−3	−3	+3
−3	+3	+3
+3	−3	−3
+3	+3	+3

c.

A (V)	B (V)	C (V)
0	0	0
0	−10	−10
−10	0	0
−10	−10	0

b.

W (V)	X (V)	Y (V)
0	0	0
0	+5	+5
+5	0	+5
+5	+5	+5

d.

A (V)	B (V)	F (V)
0	0	+3
0	+3	0
+3	0	+3
+3	+3	0

3-3. Using the voltage truth tables that are found in Prob. 3-1, show the input and output voltage symbolism required to use each of the two circuits as a NAND, NOR, AND, and OR.

3-4. What is the logical effect of leaving the B input open on the circuits given in Prob. 3-1a and b? To what voltage level should unused inputs for those two circuits be strapped in order to assure correct logical operation?

3-5. Show a complete logic diagram for implementing each of the following logical expressions. Assume positive-TRUE inputs and provide positive-TRUE outputs. Use only two- and three-input RTL gates. Give the cost (in c.u.) for each circuit. Estimate the worst-case propagational delay.

a. $F = A\bar{B} + \bar{C}D$

b. $G = (A + \bar{B})(C + B\bar{D})$

c. $F = \overline{(A + B)} + \bar{B}$

d. $g = \bar{a} + b\bar{c} + \overline{def}$

e. $G = WX(\overline{ABCD})$

f. $g = w(x\bar{y} + yz)$

g. $F = (A \downarrow B) \downarrow (\bar{C} \downarrow D)$

h. $G = W \downarrow \overline{(X \downarrow \bar{Y})}$

i. $F = (W \uparrow \bar{X}) \downarrow (A + \bar{B}C)$

j. $F = A\bar{B}C\bar{D}E + \bar{A}B\bar{C}D\bar{E}$

3-6. Repeat Prob. 3-5 under each of the following conditions.

a. RTL gates, negative-TRUE inputs, positive-TRUE output.

b. TTL gates, negative-TRUE inputs and output.

c. TTL gates, positive-TRUE inputs and output.

d. TTL gates, double-rail input, unspecified output symbolism.

3-7. Show a complete logic diagram for implementing each of the following AND/OR equations. Assume RTL gates and negative-TRUE input and output symbolism. Then, by using the techniques given in Sec. 2-10 and Prob. 2-21, convert each expression to NAND/NAND form. Implement each NAND/NAND expression with RTL gates and negative-TRUE input and output symbolism. Compare the gating arrangements that result from the implementation of each expression in its equivalent forms.

a. $F = A\bar{B} + \bar{C}D$

b. $F = \bar{A} + B\bar{C} + CD$

c. $G = WX\bar{Y} + W\bar{X}Y + \bar{W}XY$

d. $G = wx + \bar{w}y$

3-8. Repeat the comparison procedure of Prob. 3-7 for the following OR/AND expressions and their equivalent NOR/NOR forms. Use TTL gates and positive-TRUE input and output symbolism.

a. $F = (A + \bar{B})(\bar{C} + D)$

b. $G = w(x + \bar{y})(\bar{w} + y)$

c. $G = (A + \bar{B} + C)(A + B + \bar{C})(\bar{A} + B + C)$

d. $F = (\bar{w} + x)(\bar{y} + \bar{z})x$

3-9. Using the AND/OR gating found for each expression given in Prob. 3-7, reinterpret the gating functions and find the NAND/NAND, OR/

NAND, and NOR/OR expressions that are equivalent to the original AND/
OR expressions.

3-10. Using the OR/AND gating found for each expression given in Prob.
3-8, reinterpret the gating functions and find the AND/NOR, NAND/AND,
and NOR/NOR expressions that are equivalent to the original OR/AND
expressions.

*3-11. The logic diagrams in Fig. P3-11 show various gating arrangements.
The input and output voltage symbolism is indicated, but the *function* of each
gate is not. Interpret the logical operation of each gating arrangement assuming that RTL gates are used, that the first-level gating acts as an AND, and
that the second-level gating acts as an OR. Repeat the analysis for OR/AND,
NAND/NAND, and NOR/NOR interpretations. Prove that the logical
expressions that result from each interpretation are equivalent.

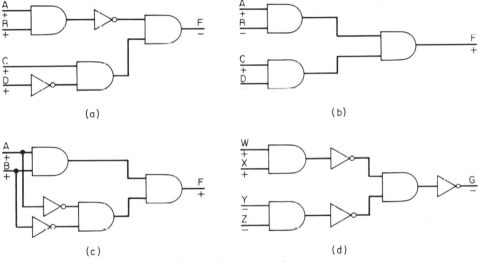

Figure P3-11a, b, c, d

*3-12. Repeat Prob. 3-11 assuming that TTL gates are used instead of RTL
gates.

3-13. Interpret the gating arrangements given in Prob. 3-11 as if RTL
gates are used and the two levels of gating are *both* AND gates. Repeat for
OR/OR gating. Repeat for TTL in both the AND/AND and OR/OR forms.
Prove that these forms are equivalent to the AND/OR expressions found in
Probs. 3-11 (RTL) and 3-12 (TTL).

*3-14. Give an A-O-N form for the logical operation of each of the gating
circuits in Fig. P3-14. Consider both RTL and TTL gating.

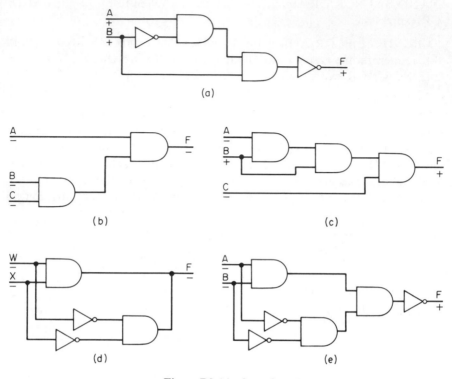

Figure P3-14a, b, c, d, e

*3-15. Prove that each of the pairs of (or groups of three) gating circuits shown in Fig. P3-15 implement the same logical expression. Assume RTL gates.

*3-16. Implement each of the following expressions with only RTL gates and inverters. Show a complete logic diagram for each implementation. Repeat for TTL. Repeat for RTL assuming that RTL EXCLUSIVE-OR gates are available. Repeat for TTL, with TTL EXCLUSIVE-OR gates available.

  a. $F = (A\bar{B} + \bar{A}B)C$ (positive-TRUE inputs and output)
  b. $G = A\bar{B} + AB + \bar{A}\bar{B}$ (negative-TRUE inputs and output)
  c. $F = \bar{A}B\bar{C} + A(\bar{B} + C)$ (double-rail inputs, positive-TRUE output)
  d. $F = (w + x)(y\bar{z} + \bar{y}z)$
                    (negative-TRUE inputs, positive-TRUE output)

3-17. Show an implementation for each of the following expressions assuming that TTL gating, including the TTL A-O-I gate, is available.

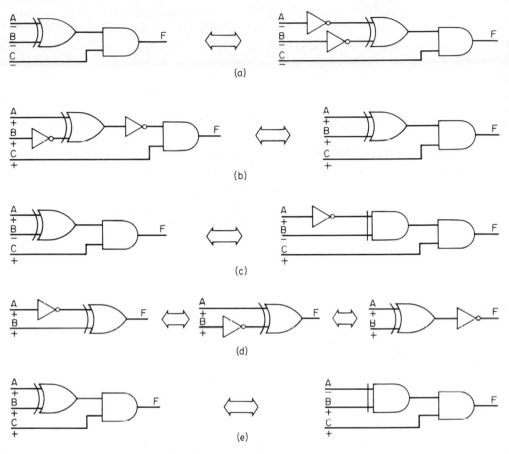

Figure P3-15a, b, c, d, e

a. $F = A\bar{B} + A\bar{C}$ (negative-TRUE inputs and output)

b. $F = (A + B)(A + C)$
   (negative-TRUE inputs and positive-TRUE output)

c. $F = (A\bar{B} + A\bar{C})(D\bar{E})$ (any input and output symbolism)

*3-18. Assume that only two-input RTL gates are available. Show a logic diagram for implementing each of the following expressions.

a. $F = AB\bar{C} + \bar{A}BC + A\bar{B}C$ (negative-TRUE inputs and output)

b. $F = ABCD + \bar{A}\bar{B}\bar{C}\bar{D}$ (positive-TRUE inputs and output)

c. $G = (A + \bar{B} + C + \bar{D} + E)(\bar{A} + \bar{B})$
   (negative-TRUE inputs, positive-TRUE output)

*3-19. Assume that only two-input TTL gates are available. Show the gating interconnections, input and output voltage symbolism, approximate worst-case delay, and total cost (c.u.) for implementing each of the following composite gate types. Include a complete diagram of the gating configuration.

    a. A three-input AND gate
    b. A four-input OR gate
    c. A five-input NAND gate
    d. A six-input NOR gate
    e. A ten-input AND gate

3-20. Give the gating cost for implementing each of the following logical expressions. Include the cost of inverters. Consider both RTL and TTL gates.

    a. $F = (A + \bar{B})(C + \bar{D})$ (positive-TRUE inputs and output)
    b. $G = (ABC + \bar{B}\bar{C} + A\bar{C})(DE\bar{F})$
                    (double-rail inputs, negative-TRUE outputs)
    c. $G = \overline{(A + B)}(\bar{C} + \bar{D})(C + D)$
                    (any convenient input and output symbolism)

    d–g. All the expressions given in Prob. 3-7
    h–k. All the expressions given in Prob. 3-8

*3-21. Show a complete logic diagram for an RTL half-adder:

    a. With positive-TRUE inputs and output
    b. With negative-TRUE inputs and output
    c. With an RTL EXCLUSIVE-OR gate available and positive-TRUE inputs and output.
    d. With an RTL EXCLUSIVE-OR gate and negative-TRUE inputs and output.

3-22. Repeat Prob. 3-21 for TTL gates.

*3-23. Repeat the RTL and TTL analyses of Probs. 3-21 and 3-22 for

    a. The full adder
    b. The half-subtractor
    c. The full subtractor
    d. The three-input majority gate (Prob. 2-11)
    e. The four-input EXCLUSIVE-OR gate (Prob. 2-12)

*3-24. The circuit shown in Fig. P3-24 will generate the EXCLUSIVE OR of its inputs when implemented with either RTL or TTL gates. Analyze the circuit in AND/OR and OR/AND terms for both the RTL and the TTL cases. Determine the output symbolism that produces the EXCLUSIVE-OR function (as opposed to the COINCIDENCE function). Consider both positive-TRUE and negative-TRUE input symbolism.

3-25. Prove that if a combinational system acts as a full adder when its inputs and outputs are interpreted as being TRUE when positive, then the device will also act as a *negative*-TRUE full adder. Repeat for a half-adder, a full subtractor, and a half-subtractor.

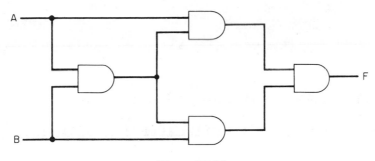

Figure P3-24

3-26. Does the reversed-symbolism independence described in Prob. 3-25 hold for a four-input majority gate? Any majority gate? A three-input EXCLUSIVE-OR gate? A four-input EXCLUSIVE-OR gate? An *n*-input EXCLUSIVE-OR gate?

3-27. The logic diagram of Fig. P3-27 is offered by a manufacturer for describing the logical operation of a complex ECL gate. Positive-TRUE symbolism is assumed. Give the AND/OR expressions for each output function. Remember that the circle at an output indicates complementation.

3-28. Repeat the analysis of the gating circuit in Prob. 3-27 with negative-TRUE voltage symbolism assumed.

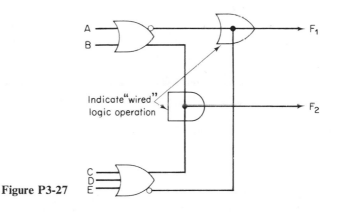

Figure P3-27

4 Minimization of Combinational Functions

4.0 Introduction

Chapters 2 and 3 introduce techniques for converting a combinational design problem into a set of Boolean equations and for implementing these expressions with modern gating circuits. As explained in Sec. 2.10, the two A-O-N forms for these expressions, the sum of minterms and the product of maxterms, are the most useful ways to describe the combinational activity of the system being designed. Unfortunately, these forms are not always the simplest ways to express a given Boolean equation, since they may contain logical redundancies, as shown in Sec. 2-8.

EXAMPLE 4-0: Given $F(A,B) = \sum 0, 1$. Is it in minimal form?

Answer: No, since

$$F = \bar{A}\bar{B} + \bar{A}B$$
$$= \bar{A}(\bar{B} + B)$$
$$= \bar{A}$$

This chapter discusses techniques whereby Boolean expressions can be simplified, or *minimized*, so as to reduce the cost of the gating systems that implement them. Two types of minimization techniques are discussed: graphical techniques and tabular techniques. A computer-aided version of the tabular technique is also presented.

4.1 Basic Minimization Concepts

The key to all minimization techniques involving A-O-N Boolean expressions is the simple Boolean identity

$$A(B + \bar{B}) = A \qquad (4\text{-}0)$$

or its equivalent form,

$$(A + B)(A + \bar{B}) = A \qquad (4\text{-}1)$$

If a Boolean expression in sum-of-minterms form contains two minterms that differ in only one variable, $m_0 = \bar{A}\bar{B}\bar{C}\bar{D}$ and $m_1 = \bar{A}\bar{B}\bar{C}D$, for example, then the terms that are common to the two minterms can be factored out, leaving an expression in the form of Eq. (4-0). The two minterms can therefore be combined into a single term that is cheaper to implement than the original pair of terms. A similar simplification applies to pairs of maxterms which differ in only one variable. That reduction follows from Eq. (4-1).

EXAMPLE 4-1:   Can $F(A,B,C) = \sum 4, 6$ be reduced?

Answer: Yes, as

$$\begin{aligned} F &= A\bar{B}\bar{C} + AB\bar{C} \\ &= A\bar{C}(\bar{B} \mid B) \\ &= A\bar{C} \end{aligned}$$

The sum of minterms requires 8 c.u. for implementation; the reduced form requires only 2 c.u.

EXAMPLE 4-2:   Reduce $F(A,B,C) = \prod 3, 7$.

Answer:

$$\begin{aligned} F &= (\bar{A} + B + C)(A + B + C) \\ &= [(B + C) + \bar{A}][(B + C) + A] \\ &= (B + C) \qquad \text{from Eq. (4-1)} \end{aligned}$$

Again, a reduction from 8 to 2 c.u.

A second property of Boolean algebra that is of use in the minimization process is

$$A = A + A + A + \cdots \qquad \text{from Eq. (2-8)}$$

or, equivalently,

$$A = A \cdot A \cdot A \cdot \cdots \qquad \text{from Eq. (2-9)}$$

These identities show that a term present in a sum of minterms or product of maxterms can be duplicated as many times as is useful for combination with other terms.

EXAMPLE 4-3: Reduce $F(A,B,C) = \sum 0, 1, 2$.

  Solution: $F = \bar{A}\bar{B}\bar{C} + \bar{A}\bar{B}C + \bar{A}B\bar{C}$. Duplicating m_0 gives

$$F = \bar{A}\bar{B}\bar{C} + \bar{A}\bar{B}C + \bar{A}\bar{B}\bar{C} + \bar{A}B\bar{C}$$
$$= (\bar{A}\bar{B})(\bar{C} + C) + (\bar{A}\bar{C})(\bar{B} + B)$$
$$= \bar{A}\bar{B} + \bar{A}\bar{C}$$

The simplification reduces the cost of implementing F from 12 to 6 c.u.

These two simple concepts, the pairing of minterms that differ in only one variable and the repetition of terms that can combine with more than one other term, are the bases for combinational minimization. Combinations that result from a sum-of-minterms beginning will generally maintain the AND/OR form and are referred to as simplified sum-of-products expressions, or **SOP** forms. Similarly, reductions of product-of-maxterm expressions are called simplified product-of-sums (**POS**) expressions. In some cases the SOP and POS forms for a function may be identical.

EXAMPLE 4-4: Express the function given below in both SOP and POS form.

A	B	F
0	0	1
0	1	1
1	0	0
1	1	0

  Solution: The minterm forms for F and $\bar{F}$ can be found from the truth table:

$$F = \sum 0, 1 = \bar{A}\bar{B} + \bar{A}B$$
$$\bar{F} = \sum 2, 3 = A\bar{B} + AB$$

Using De Morgan's theorem on $\bar{F}$ gives

$$F = (\bar{A} + B)(\bar{A} + \bar{B})$$

Both forms for F can be simplified:

$$F = \bar{A}(\bar{B} + B) \qquad F = (\bar{A} + B)(\bar{A} + \bar{B})$$
$$- \bar{A} \qquad\qquad = \bar{A}$$

The result, $F = \bar{A}$, can be called either an SOP or a POS form.

4.2 Expansion to Standard Form

The minimization process expects a function to be in either the sum-of-minterms or the product-of-maxterms form. Not all A-O-N expressions are in these expanded forms, however. Consider

$$F(A,B,C) = AB + \bar{A}BC \tag{4-2}$$

This is not a standard form, since the AB term is not a minterm. The expression does have AND/OR form, however, and is related to the sum-of-minterms form. Similarly, the expression

$$G(A,B,C) = (\bar{A} + \bar{B})(B + \bar{C}) \tag{4-3}$$

is not a product of maxterms, despite its OR/AND structure.

In a function such as $F = AB + \bar{A}BC$, the separate terms are commonly called *implicants*, since whenever any one of the terms is TRUE it is *implied* that the function is TRUE. The implicants may or may not be minterms. If all the implicants are minterms, then an expression is in standard sum-of-minterms form. If not, then each implicant is the result of a combination of minterms that simplifies the expression below its sum-of-minterms form.

EXAMPLE 4-5:   Simplify $F(A,B,C) = \sum 0, 1, 7$.

> *Solution:*

$$F - \bar{A}\bar{B}\bar{C} + \bar{A}\bar{B}C + ABC$$
$$= \bar{A}\bar{B}(\bar{C} + C) + ABC$$
$$= \bar{A}\bar{B} + ABC$$

The implicants are $\bar{A}\bar{B}$ and ABC. $\bar{A}\bar{B}$ results from the combination of m_0 and m_1.

Just as reduced SOP forms can be derived from the sum of minterms, combinational expressions in reduced POS forms may result from combinations within a product-of-maxterms expression.

EXAMPLE 4-6:   Simplify $F(A,B,C) = \prod 0, 1, 7$.

> *Solution:*

$$F = (\bar{A} + \bar{B} + \bar{C})(\bar{A} + \bar{B} + C)(A + B + C)$$
$$= (\bar{A} + \bar{B})(A + B + C)$$

The reduced form costs 7 c.u., 5 c.u. less than the original expression.

The terms in a reduced POS expression are called *implicates* and imply that the overall function is FALSE whenever any one of the implicates is FALSE.

Reduced expressions can appear in any of the other six standard forms as well, but only reduced AND/OR and OR/AND expressions are considered herein. Other forms can be converted into either of these by following the procedures given in Sec. 2-10.

Of concern when dealing with reduced expressions is the presence of further redundancy; i.e., can the expression be simplified even more? In the expression $F = AB + \bar{A}BC$, for example, the first implicant can be expanded by using Eqs. (2-4) and (2-7):

$$F = AB(C + \bar{C}) + \bar{A}BC \tag{4-4}$$

yielding

$$F = ABC + AB\bar{C} + \bar{A}BC \tag{4-5}$$

which is in minterm form. Equation (4-5) can be simplified:

$$\begin{aligned} F &= (ABC + AB\bar{C}) + (\bar{A}BC + ABC) \\ &= AB(C + \bar{C}) + BC(\bar{A} + A) \\ &= AB + BC \end{aligned} \tag{4-6}$$

The simplified expression is 1 c.u. cheaper than the original expression.

To begin the minimization process, a Boolean expression that is in partially reduced form must be expanded into minterm or maxterm form. It seems strange, but to simplify a Boolean expression, it is first necessary to express it in its most expanded form. The expansion process can be done either algebraically or by using a truth table.* Algebraic expansion uses Eq. (4-0) in reverse. Each implicant is expanded into an equivalent set of minterms, with the number of minterms in the set being 2^M, where M is the number of variables that were missing in the original implicant.

EXAMPLE 4-7:   Expand $F(A,B,C,D) = AB$.

Solution:

$$\begin{aligned} F &= AB(C + \bar{C})(D + \bar{D}) \\ &= ABCD + ABC\bar{D} + AB\bar{C}D + AB\bar{C}\bar{D} \\ &= \sum 12, 13, 14, 15 \end{aligned}$$

EXAMPLE 4-8:   Expand $G(A,B,C,D,E) = A\bar{D}$.

*A third technique, the use of the Karnaugh map as an expansion tool, is covered in Sec. 4.3.2.

Solution:

$$G = A\bar{D}(B + \bar{B})(C + \bar{C})(E + \bar{E})$$
$$= ABC\bar{D}E + ABC\bar{D}\bar{E} + AB\bar{C}\bar{D}E + AB\bar{C}\bar{D}\bar{E}$$
$$+ A\bar{B}C\bar{D}E + A\bar{B}C\bar{D}\bar{E} + A\bar{B}\bar{C}\bar{D}E + A\bar{B}\bar{C}\bar{D}\bar{E}$$
$$= \sum 16, 17, 20, 21, 24, 25, 28, 29$$

Note that the implicant $A\bar{D}$ has three variables (B, C, and E) missing and expands into $2^3 = 8$ minterms.

<hr>

A similar expansion process uses Eq. (4-1) to expand reduced POS expressions into maxterm form. The implicates are each expanded by including their missing variables.

EXAMPLE 4-9: Expand $F(A,B,C,D) = (A + B)$.

Solution:

$$F = (A + B + C)(A + B + \bar{C})$$
$$= (A + B + C + D)(A + B + C + \bar{D})(A + B + \bar{C} + D) \cdot$$
$$(A + B + \bar{C} + \bar{D})$$
$$= \prod 12, 13, 14, 15$$

<hr>

To use the truth table as a means for expansion of an AND/OR expression, each implicant of a reduced SOP expression is tabulated separately. The separate columns (one per implicant) are then merged to give the output column for the overall function. That column is used to obtain the minterm and maxterm forms for the expression.

EXAMPLE 4-10: Expand $F = AB + \bar{A}BC$ by tabulation.

Solution:

A	B	C	AB	$\bar{A}BC$	F
0	0	0	0	0	0
0	0	1	0	0	0
0	1	0	0	0	0
0	1	1	0	1	1
1	0	0	0	0	0
1	0	1	0	0	0
1	1	0	1	0	1
1	1	1	1	0	1

The last column shows that $F = \sum 3, 6, 7$.

<hr>

The tabular process for expanding a reduced POS expression is similar to that of Ex. 4-10, except that the output column for each *implicate* is determined and the *zeros* in the separate columns are merged to give the output column for the function.

For combinational problems that involve many variables, the expansion process and the recognition and elimination of redundancy become more and more difficult. The algebraic expansion and reduction methods of Secs. 4.1 and 4.2 are not very satisfactory for use with such problems. As a result, the nonalgebraic minimization techniques given in the rest of this chapter have been developed.

4.3   Karnaugh Mapping (Refs. [6] and [13])

A Karnaugh map, hereafter called a K-map or simply a map, is a graphical method for representing a Boolean function. It is similar to a truth table in that the map supplies the TRUE or FALSE value of a Boolean function for all possible combinations of its logical arguments. There are many ways in which a K-map can be arranged. The most important considerations of the arrangement are

1. There must be a unique location on the K-map for entering the TRUE/FALSE value of the function that corresponds to each combination of input variables.

2. The locations should be arranged so that reductions of the form shown in Sec. 4.1 are readily apparent to the trained observer.

The second consideration implies that a successful K-mapping arrangement should point to groups of minterms or maxterms that can be combined into reduced forms. K-maps are also useful in expanding partially reduced expressions into standard form prior to the minimization process.

K-maps can be used to represent functions of any number of variables, but they are most useful with functions having fewer than six arguments. Consider first a function involving only one variable, say $F(Z)$. The input variable can take on only two values, so the K-map for this function needs only two locations, as shown in Fig. 4-0. The circle identifies the map as being representative of the function F. The $\bar{Z}$ region (wherein $Z = 0$) is the left half of the map, and the value of F when Z is FALSE should be entered into that square. The right half of the map corresponds to the region wherein Z is TRUE and is the location at which the value of the function when Z is TRUE should be entered. Only the Z region need be labeled, since the remainder of the map must be the $\bar{Z}$ region.

Data concerning the behavior of F can be entered into the map directly from a truth table, as shown in Fig. 4-1. If F is expressed in a sum-of-min-terms form, the minterms present in the summation can be entered as 1s on the map, with the remaining locations entered as zeros.

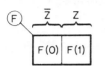

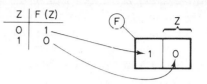
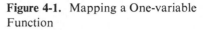

Figure 4-0. The One-variable K-Map   **Figure 4-1.** Mapping a One-variable Function

EXAMPLE 4-11: Given $G(Z) = m_1$, plot this function on a K-map.

Answer: See Fig. E4-11. The right square contains a 1 since m_1 is present in the sum-of-minterms expression for G.

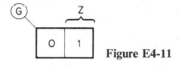

Figure E4-11

The one-variable K-map can be represented in general by Fig. 4-2, showing that m_0, if present in the minterm expression for F, will enter as a 1 in the left square. The only other minterm, m_1, will enter into the right square.

Higher-ordered K-maps can be obtained by successively doubling the size of the single-variable K-map. One technique for this uses the concept of "folding out" the larger map from the smaller one. Consider a two-argument function, say $F(Y,Z)$. Its K-map is obtained as shown in Fig. 4-3. Reading from left to right, the squares on the two-variable map correspond to $\bar{Y}\bar{Z} - m_0$, $\bar{Y}Z - m_1$, $YZ - m_3$, and $Y\bar{Z} - m_2$. The unfolded squares represent minterms whose subscripts are 2^1 greater than the subscripts of the minterms whose squares previously covered them. The nonordered

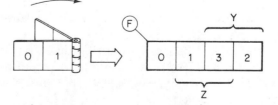

Figure 4-2. The General-ized One-variable K-map   **Figure 4-3.** Folding Out the Two-variable K-map

sequence is a consequence of the second consideration mentioned earlier and is of use in simplifying the function. The behavior of F for all four possible combinations of Y and Z can be plotted on the map given in Fig. 4-3. The numbers show the locations of the minterms. Thus, given F in sum-of-minterms form, 1s are entered in the locations corresponding to the minterms in the summation, while zeros are entered elsewhere.

The three-variable K-map is produced as shown in Fig. 4-4. Note that each square in the new row has a minterm number that is 2^2 greater than the square from which it unfolded.

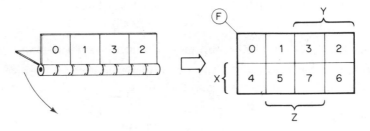

Figure 4-4. The Three-variable K-map

Continuing to the four-variable map gives the result shown in Fig. 4-5. The new squares represent minterms that are 2^3 greater than their unfolded counterparts.

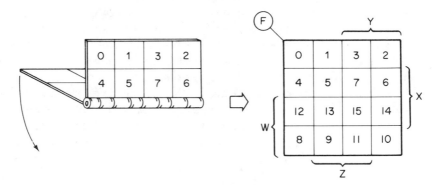

Figure 4-5. The Four-variable K-map

To verify the locations of the minterm numbers, consider location 13, as in Fig. 4-6. This square is in the *lower* half of the map, so it is in the W region. Similarly, it lies in the X, $\bar{Y}$, and Z regions. Since $m_{13} = WX\bar{Y}Z$, the location is correct.

The five-variable map is shown in Fig. 4-7. The new locations are 2^4 greater than those from which they unfolded. Note that the W region of this map is in two disjoint sections. This makes the Boolean simplification process

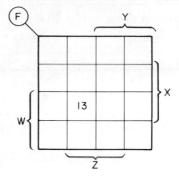

Figure 4-6. Locating m_{13} on the K-map

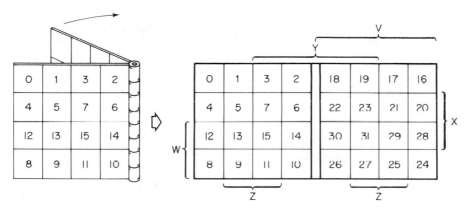

Figure 4-7. The Five-variable K-map

more difficult to perform on the five-variable map than on lower-ordered maps. The compounding of this problem, coupled with the increase in the complexity of the map itself, limits the effectiveness of K-map techniques for higher-ordered functions.

4.3.1. Entering Data into the Map.   Plotting functional behavior on the K-map is very easy if the function is given in the sum-of-minterms form.

EXAMPLE 4-12: $F(A,B,C) = \sum 0, 1,$ 3, 5. Plot F on a K-map.

Solution: See Fig. E4-12. Note that 1s are entered at the locations corresponding to $m_0, m_1, m_3,$ and m_5, with 0s entered elsewhere.

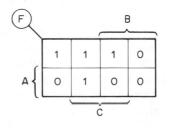

Figure E4-12

EXAMPLE 4-13: Plot $G(W,X,Y,Z) = \sum 0, 2, 4, 10, 12, 15$ on a K-map.

Solution: See Fig. E4-13.

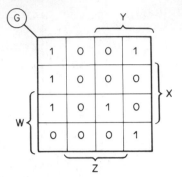

Figure E4-13

EXAMPLE 4-14: Given $F(A,B,C,D,E) = \sum 1, 9, 15, 17, 23, 28, 31$. Plot F on a K-map.

Solution: See Fig. E4-14.

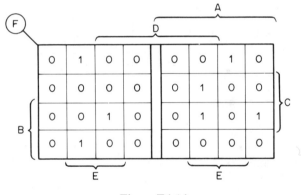

Figure E4-14

Learning the numbered locations of each minterm on the K-map permits a sum-of-minterms expression to be plotted directly, with no involved conceptual processes. Plotting the minterms by numbered location is the recommended technique, since it offers little chance of error in locating the minterm positions.

If a function is given in the product-of-maxterms form, De Morgan's theorem or the procedure given in Prob. 2-8 can be used to express the complement of the function as a sum of minterms. The 0 symbol can then be entered at the locations corresponding to those minterms, since a function is FALSE whenever its complement is TRUE. The remaining locations are filled with 1s.

EXAMPLE 4-15:   Given $F(A,B,C,D) = \prod 0, 2, 5, 9, 14$. Plot F on a K-map.

Solution:

$$F = (\bar{A} + \bar{B} + \bar{C} + \bar{D})(\bar{A} + \bar{B} + C + \bar{D})(\bar{A} + B + \bar{C} + D) \cdot$$
$$(A + \bar{B} + \bar{C} + D)(A + B + C + \bar{D})$$

Hence, using De Morgan's theorem,

$$\bar{F} = ABCD + AB\bar{C}D + A\bar{B}C\bar{D} + \bar{A}BC\bar{D} + \bar{A}\bar{B}\bar{C}D$$
$$= \sum 15, 13, 10, 6, 1$$

Entering 0 at those locations on the K-map gives the map shown in Fig. E4-15a. The map is completed by placing a 1 in each of the remaining squares, as shown in Fig. E4-15b.

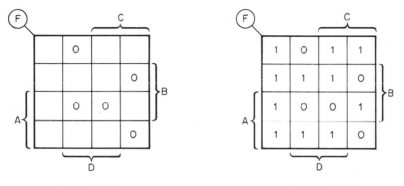

Figure E4-15a   **Figure E4-15b**

The results of Prob. 2-8 can be used to convert from the maxterm subscripts to the minterm subscripts without using De Morgan's theorem. For the four-variable case, each maxterm subscript is subtracted from $2^4 - 1 = 15$. Thus, given

$$F(A,B,C,D) = \prod 0, 2, 5, 9, 14$$

the equivalent expression is

$$\bar{F}(A,B,C,D) = \sum 15, 13, 10, 6, 1$$

These minterms are entered as 0s, as shown in Figs. E4-15a and b.

4.3.2. Plotting Partially Reduced Functions.   Section 4.1 presents both algebraic and tabular techniques for converting a partially reduced Boolean expression to minterm or maxterm form. Either of these approaches can be used to plot a reduced expression. However, a more direct approach is to enter each implicant or implicate directly into the map without first expanding it into its minterm or maxterm components.

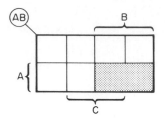

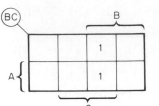

Figure 4-8. The *AB* Region on a Three-variable K-map

Figure 4-9. The *BC* Implicant Plotted on a Three-variable K-map

Consider $F = AB + BC$. The first implicant is TRUE whenever both *A* and *B* are TRUE, independent of the value of *C*. On a three-variable K-map the region wherein both *A* and *B* are TRUE covers two squares, as shown in Fig. 4-8. This region corresponds to m_6 and m_7. Thus, the *AB* implicant can be plotted by filling the *AB* region of the map with 1s. The *BC* implicant is plotted in Fig. 4-9. Clearly, $BC = m_3 + m_7$. Combining the maps for *AB* and *BC* gives the K-map that is shown in Fig. 4-10. Note from Fig. 4-10 that $F = \sum 3, 6, 7$.

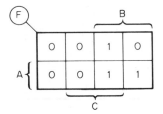

Figure 4-10. The K-map for $F(A,B,C)$ $= AB + BC$

EXAMPLE 4-16: Plot $F_1(A,B,C,D) = A\bar{B}$.

Answer: See Fig. E4-16. The 1s fill the region of the map wherein *A* is TRUE and *B* is FALSE. The minterm expansion for F_1 is

$$F_1(A,B,C,D) = A\bar{B} = \sum 8, 9, 10, 11$$

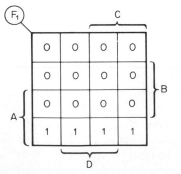

Figure E4-16

EXAMPLE 4-17:   Show $F_2 = ABC + \bar{A}C + A\bar{B}$ on a three-variable K-map.

Answer: See Fig. E4-17. Note that $F_2(A,B,C) = \sum 1, 3, 4, 5, 7$.

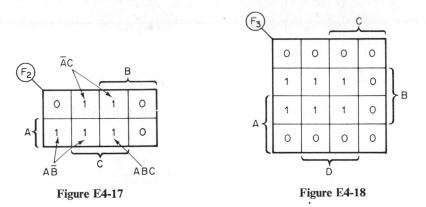

Figure E4-17   Figure E4-18

EXAMPLE 4-18:   Plot $F_3(A,B,C,D) = AB\bar{C}D + BD + B\bar{C}\bar{D}$.

Answer: See Fig. E4-18. Note that m_{13} is covered twice and that

$$F_3(A,B,C,D) = \sum 4, 5, 7, 12, 13, 15$$

The K-map procedure for expanding a partially reduced POS (OR/AND) expression is similar to the SOP expansion process. De Morgan's theorem is used to convert the OR/AND expression into its complemented AND/OR form. Then the implicants that make up the AND/OR expression are plotted as 0s on the K-map, since implicants that make the complement of the function be TRUE will make the function itself become FALSE. The remainder of the K-map is filled with 1s.

EXAMPLE 4-19:   Plot $G(A,B,C,D) = (A + B)(B + \bar{C})(A + D)$.

Solution: By De Morgan's theorem,

$$\bar{G} = \bar{A}\bar{B} + \bar{B}C + \bar{A}\bar{D}$$

Thus, plotting the implicants as 0s and filling the remaining squares with 1, we obtain the map shown in Fig. E4-19. From inspection of the map,

$$G(A,B,C,D) = \sum 5, 7, 8, 9, 12, 13, 14, 15$$

4.3.3. Groupings of Minterms on the K-map.   Once a Boolean expression has been plotted on a K-map, the identification of useful combinations of minterms of the type described in Sec. 4.1 is facilitated by the geometrical arrangement of the map. The simplest case of Boolean simplification is the combination of two minterms that are similar in all but one of their variables.

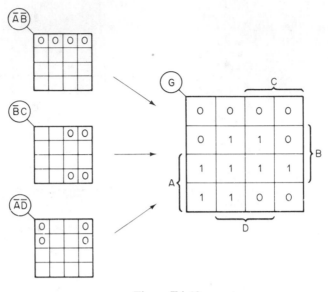

Figure E4-19

Section 4.1 shows that $F(A,B,C,D) = m_0 + m_1$ can be reduced to $F = \bar{A}\bar{B}\bar{C}$ by using purely algebraic methods. Figure 4-11 shows the positions of these two minterms on a K-map.* Note that m_0 and m_1 together completely fill the area on the K-map that corresponds to $\bar{A}\bar{B}\bar{C}$ and that the variable whose TRUE/FALSE boundary lies between m_0 and m_1 is D. The map shows that the two minterms have $\bar{A}\bar{B}\bar{C}$ in common and can be combined into one reduced term that does not contain the variable D. The ring shown in Fig. 4-11 indicates that m_0 and m_1 can be combined to give $F = \bar{A}\bar{B}\bar{C}$.

For any given four-variable minterm there are four other minterms with which it can be combined. Each pairing will form a reduced term containing only three of the four variables. Consider $F(A,B,C,D) = \sum 1, 4, 5, 7, 13$. This function is K-mapped in Fig. 4-12. The five minterms that make up F combine into four pairs of two:

$$m_1 + m_5 = \bar{A}\bar{C}D, \qquad m_4 + m_5 = \bar{A}B\bar{C},$$
$$m_7 + m_5 = \bar{A}BD, \qquad m_{13} + m_5 = B\bar{C}D \tag{4-7}$$

The sum-of-minterms expression can be reduced to

$$F = \bar{A}\bar{C}D + \bar{A}B\bar{C} + \bar{A}BD + B\bar{C}D \tag{4-8}$$

Minterm m_5 is used *four times* in combination with other minterms, a consequence of Eq. (2-8).

*In most of the K-maps given in the remainder of this text, squares that contain 0s are left blank as a simplification of the mapping process.

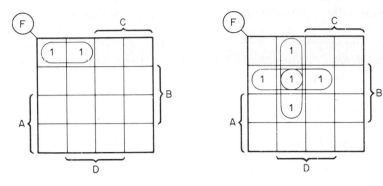

Figure 4-11. Combination of m_0 and m_1 on a K-map

Figure 4-12. The K-map for $F = \sum 1, 4, 5, 7, 13$

The expression for F that is given in Eq. (4-8) can be called the *simplified SOP form*. It requires 16 c.u. for implementation, exclusive of inversion. A modification of the expression for F is

$$F = \bar{C}D(\bar{A} + B) + \bar{A}B(\bar{C} + D) \qquad (4-9)$$

which can be implemented as shown in Fig. 4-13. The modified gating arrangement shown in Fig. 4-12 requires only 12 c.u. (plus two inverters).

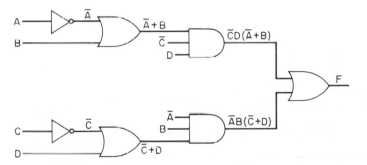

Figure 4-13. A Logical Implementation of Eq. (4-9)

The selection of the *best* gating arrangement is sometimes quite difficult. The simplified SOP expression for F given in Eq. (4-8) costs more to implement, but it preserves the two-level AND/OR logical sequence. Its worst-case delay (including inversion) is only three gate delays. The modified expression, Eq. (4-9), is 4 c.u. cheaper, but its worst-case delay is four gate delays.

With modern gating circuits the cost per gate has become low enough so that the minimization of cost is no longer the primary concern in evaluating alternative implementations of the same expression. Rather, logical simplicity and ease of troubleshooting are often more desirable features, even at some increase in gating costs. For these reasons the procedures given in this text aim only toward finding the simplified SOP and POS forms for a given combinational expression. These two forms maintain the two-level A-O-N logical sequence. Simplification beyond these forms is often possible and can be made if desirable, but the development of two-level simplified SOP and POS expressions can be considered as a reasonable stopping point that combines reasonable amounts of cost reduction, design effort, and logical simplicity.*

The K-maps given in Fig. 4-14 show several other two-minterm combinations. Examination of Fig. 4-14 shows that two-minterm combinations exist whenever a K-map contains 1s which are

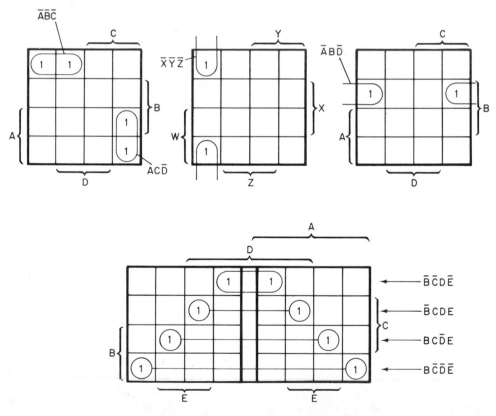

Figure 4-14. Combinations of Two Minterms

*Readers interested in further techniques for minimization beyond the simplified SOP and POS forms are referred to Ref. [1]. As an interesting sidelight to this topic, also see Refs. [8] and [9].

1. Adjacent to each other in the same row or column.

2. At opposite ends of the same row or column.

3. In "mirrored" positions in the left/right halves of a five-variable map. (In effect, they would be on top of each other if the map were folded back into a four-variable map.)

Any of these pairings yields a reduced implicant with one less than the maximum number of variables.

EXAMPLE 4-20:   Reduce $F(V,W,X,Y,Z) = \sum 0, 4, 9, 14, 25, 30$.

Answer: See Fig. E4-20.

$$F = \bar{V}\bar{W}\bar{Y}\bar{Z} + WXY\bar{Z} + W\bar{X}\bar{Y}Z$$

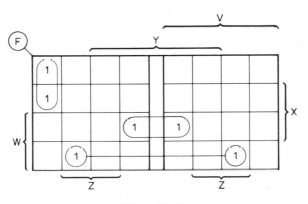

Figure E4-20

Now, consider the K-map given in Fig. 4-15. The two vertical pairs of minterms can be combined to give

$$F = \bar{A}\bar{C}\bar{D} + A\bar{C}\bar{D} \qquad (4\text{-}10)$$

This can be reduced further:

$$\begin{aligned} F &= \bar{C}\bar{D}(\bar{A} + A) \\ &= \bar{C}\bar{D} \end{aligned} \qquad (4\text{-}11)$$

showing that the four original minterms can be combined into a single term having two less than the maximum number of variables. The group of four minterms completely fills the $\bar{C}\bar{D}$ area on the K-map. Some other groups of four are shown in Fig. 4-16. The groups shown in Fig. 4-16 can be recognized as squares, complete rows or columns, and *wrapped-around* squares such as $\bar{A}\bar{D}$. The four corners are also a combinable set of minterms.

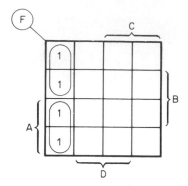

Figure 4-15. A Grouping of Four Minterms

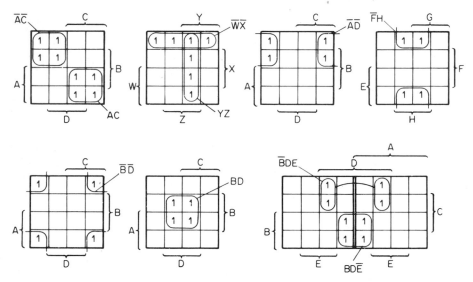

Figure 4-16. Combinations of Four Minterms

Groups of eight minterms can also be recognized by using the K-mapping technique. These groups reduce the eight minterms to a single implicant that contains three fewer variables than the minterms had to start with. Several groups of eight are shown in the K-map of Fig. 4-17. Larger groupings (always in powers of 2) can be found on K-maps involving more than four variables.

4.3.4. Selection of Implicants. Once a Boolean function has been plotted on a K-map, all possible groupings of minterms can be identified. These groupings define the set of implicants that can be ORed together to express the original function in its simplified SOP form.

EXAMPLE 4-21: Simplify $F(A,B,C,D) = \sum 0, 2, 5, 8, 10, 14$.

Solution: See Fig. E4-21.

$$F = \bar{B}\bar{D} + AC\bar{D} + \bar{A}B\bar{C}D$$

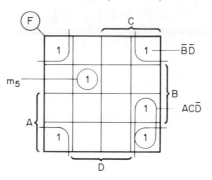

Figure E4-21

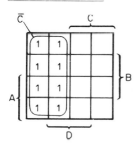
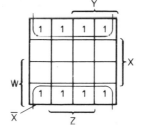
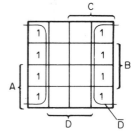
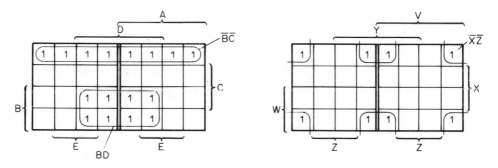

Figure 4-17. Groups of Eight Minterms

Often there are overlaps between several implicants. As a result, there may be no clear-cut way to select the best set of implicants. In fact, some of the possible implicants may not need to be used at all.

When selecting the implicants, there are two useful concepts that apply. First, if any minterm is covered by only one grouping, the implicant defined by that group is an *essential implicant,* meaning that it *must* be included in the set of implicants that is used to express the function. Omission of an essential implicant would mean that the minterm (or minterms) that make it essential would not be covered by the simplified SOP expression. Essential implicants can be identified by marking all minterms that are covered by only one group. A common marking technique uses a dot to mark these squares.

EXAMPLE 4-22: Given $F(A,B,C,D) = \sum 1, 3, 5, 6, 7, 10, 13$. Mark the singly covered minterms and list the essential implicants.

Solution: See Fig. E4-22. The singly covered minterms are m_1, m_3, m_6, m_{10}, and m_{13}. The essential implicants are $\bar{A}D$, $\bar{A}BC$, $B\bar{C}D$, and $A\bar{B}C\bar{D}$. The function *must* contain those four terms. Those implicants cover all the 1s on the K-map, so that

$$F = \bar{A}D + \bar{A}BC + B\bar{C}D + A\bar{B}C\bar{D}$$

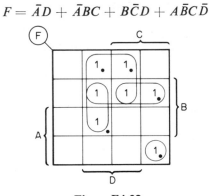

Figure E4-22

The second concept of use in implicant selection is that each minterm should always be covered with the largest group that can be found, provided that the minterm is not already covered by some other group. A group that is large enough so that it is not completely covered by any other grouping is said to define a *prime implicant*. By using the prime implicants the maximum cost reduction can be obtained. The use of large groups can lead to trouble, however, as Ex. 4-23 shows.

EXAMPLE 4-23: Simplify $F(A,B,C,D) = \sum 1, 3, 6, 7, 10, 11, 13, 15$.

Solution: See Fig. E4-23. The essential implicants are $\bar{A}\bar{B}D$, $\bar{A}BC$, ABD, and $A\bar{B}C$. They completely cover *all* the 1s, making the implicant CD unnecessary, even though it is a group of four. Thus

$$F = \bar{A}\bar{B}D + \bar{A}BC + ABD + A\bar{B}C$$

is the proper solution.

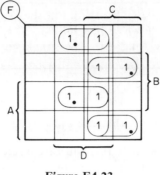

Figure E4-23

The selection of implicants is best learned by experience. Eventually a designer learns to select the proper groupings without a great deal of auxiliary effort. Until that intuitive skill is developed, however, the following "cookbook" approach can be used to help in selecting the best set of implicants:

1. Identify all possible groupings, making each group as large as possible.

2. Mark all singly covered minterms and list the essential implicants.

3. Cross out all 1s that are covered by the essential implicants.

4. Select implicants to cover the remaining 1s, using the largest groupings possible. The minterms that are crossed out can be used in these groupings.

EXAMPLE 4-24:   Simplify $F(A,B,C,D) = \sum 3, 4, 5, 7, 8, 9, 13, 14, 15$.

Solution: The K-map, with all possible groups and the singly covered minterms indicated is given in Fig. E4-24a. The essential implicants are

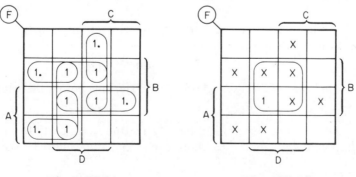

Figure E4-24a   **Figure E4-24b**

$\bar{A}CD$, $\bar{A}B\bar{C}$, ABC, and $A\bar{B}\bar{C}$. Crossing out the 1s that are covered by these groups leaves the map shown in Fig. E4-24b. The single remaining 1 is best covered by the group of four, BD. The simplified expression for F is

$$F = \bar{A}CD + \bar{A}B\bar{C} + ABC + A\bar{B}\bar{C} + BD$$

having a cost of 19 c.u.

In some cases all the minterms will be covered by two or more groups. This condition is referred to as a *cyclic map*, and the procedure given earlier is of no use. Several different sets of implicants may be used to cover the function. The choice is up to the designer. The elimination of unnecessary implicants and the use of large groups should always be the goals, however.

EXAMPLE 4-25:   Simplify $F(A,B,C,D) = \sum 0, 1, 5, 7, 8, 10, 14, 15$.

Solution: See Fig. E4-25a. Since each minterm is covered by two groups, a choice must be made. Selecting the implicant $\bar{A}\bar{B}\bar{C}$ *as if it were essential* leads to the reduced K-map given in Fig. E4-25b. Minterms m_5

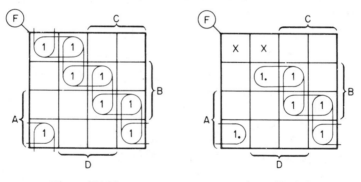

Figure E4-25a   Figure E4-25b

and m_8 are now singly covered, forcing the use of implicants $\bar{A}BD$ and $A\bar{B}\bar{D}$. Continuing the process yields the expression

$$F = \bar{A}\bar{B}\bar{C} + \bar{A}BD + ABC + A\bar{B}\bar{D}$$

Had the implicant $\bar{B}\bar{C}\bar{D}$ been chosen as a starting point, the expression

$$F = \bar{B}\bar{C}\bar{D} + \bar{A}\bar{C}D + BCD + AC\bar{D}$$

would arise. The two expressions for F have the same cost, 16 c.u. each.
Additional methods for handling cyclic maps are given in Sec. 4.6.2.

4.3.5. Simplified POS Forms. Sections 4.3–4.3.4 show how the K-mapping process can be used to find the simplified SOP form for a Boolean function. To find the simplified POS form, the following steps can be used:

1. Make a K-map of $\bar{F}$, the complement of the given function.

2. Find the simplified SOP form for $\bar{F}$.

3. Use De Morgan's theorem to convert the SOP form for $\bar{F}$ into a simplified POS form for F.

The first step can be omitted if the map for F is used and groupings of 0s, rather than 1s, are found.

EXAMPLE 4-26: Given $F(A,B,C) = \sum 0, 1, 2, 3, 5, 6$. Express F in simplified SOP and POS forms.

Solution: The K-map for F is given in Fig. E4-26a. Grouping the 1s gives the map of Fig. E4-26b, showing that

$$F = \bar{A} + B\bar{C} + \bar{B}C \qquad (7 \text{ c.u.})$$

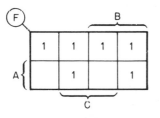
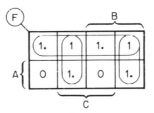

   Figure E4-26a   **Figure E4-26b**

Grouping the 0s gives the map of Fig. E4-26c, from which

$$\bar{F} = A\bar{B}\bar{C} \mid ABC$$

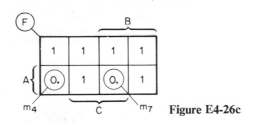

Figure E4-26c

Thus, by using De Morgan's theorem,

$$F = \overline{(A\bar{B}\bar{C}) + (ABC)}$$
$$= (\bar{A} + B + C)(\bar{A} + \bar{B} + \bar{C}) \qquad (8 \text{ c.u.})$$

The simplified SOP form is preferable since it is cheaper.

In general, a digital designer should examine both the simplified SOP and POS forms for a given Boolean expression and select the one that is cheapest to implement. Other considerations, such as constrained voltage symbolism at inputs or output, may influence this decision.

EXAMPLE 4-27:   If the function given in Ex. 4-26 is to be implemented with RTL gates and negative-TRUE inputs and positive-TRUE outputs are required, which form is cheaper?

Solution: The SOP implementation is shown in Fig. E4-27a, with a total cost of 10 c.u., including the necessary inverters. The POS implementation is shown in Fig. E4-27b, with a total cost of 10 c.u. When the input

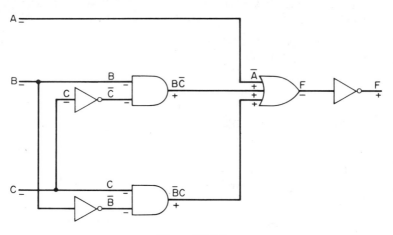

Figure E4-27a

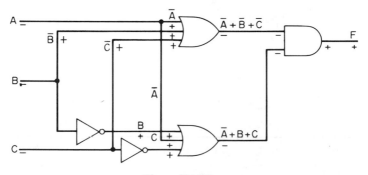

Figure E4-27b

and output constraints are considered, the net costs of implementing both expressions are the same.

In some cases, as shown in Sec. 4-1, the simplified SOP and POS forms may be identical.

EXAMPLE 4-28:   Given $G(W, X, Y) = \sum 0, 1, 2, 3, 5, 7$. Find its simplified SOP and POS forms.

Solution: See Fig. E4-28.

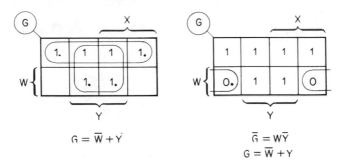

$$G = \overline{W} + \overline{Y} \qquad\qquad \overline{G} = W\overline{Y}$$
$$G = \overline{W} + Y$$

Figure E4-28

4.3.6. More Examples.

EXAMPLE 4-29:   To illustrate the complete combinational design process, consider the design of a logical device that receives a four-bit binary number, $A_8 A_4 A_2 A_1$, and is to indicate whenever the number is evenly divisible by either 4 or 5. The device can be represented by the block diagram and truth table given in Fig. 4-18. Assume that TTL logic is available and that the input signals are available with both positive-TRUE and negative-TRUE symbol-

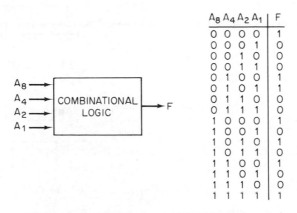

A_8	A_4	A_2	A_1	F
0	0	0	0	1
0	0	0	1	0
0	0	1	0	0
0	0	1	1	0
0	1	0	0	1
0	1	0	1	1
0	1	1	0	0
0	1	1	1	0
1	0	0	0	1
1	0	0	1	0
1	0	1	0	1
1	0	1	1	0
1	1	0	0	1
1	1	0	1	0
1	1	1	0	0
1	1	1	1	1

Figure 4-18. Description of Division Detector

ism (double-rail inputs) and that the output must have negative-TRUE symbolism. From the truth table given in Fig. 4-18 it is apparent that

$$F = \sum 0, 4, 5, 8, 10, 12, 15 \tag{4-12}$$

This function is K-mapped in Fig. 4-19. The groupings shown in Fig. 4-19 yield the simplified SOP expression for F, which is

$$F = \bar{A}_2\bar{A}_1 + \bar{A}_8 A_4 \bar{A}_2 + A_8 \bar{A}_4 \bar{A}_1 + A_8 A_4 A_2 A_1 \tag{4-13}$$

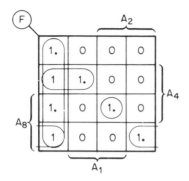

Figure 4-19. K-map Representation of the Division Detector

Implementation of Eq. (4-13) costs 16 c.u. Grouping the 0s on the K-map of Fig. 4-19 gives

$$\bar{F} = \bar{A}_8 A_2 + \bar{A}_4 A_1 + A_4 A_2 \bar{A}_1 + A_8 \bar{A}_2 A_1 \tag{4-14}$$

from which De Morgan's theorem gives

$$F = (A_8 + \bar{A}_2)(A_4 + \bar{A}_1)(\bar{A}_4 + \bar{A}_2 + A_1)(\bar{A}_8 + A_2 + \bar{A}_1) \tag{4-15}$$

The POS form given in Eq. (4-15) costs 14 units, 2 less than the SOP form given in Eq. (4-13).

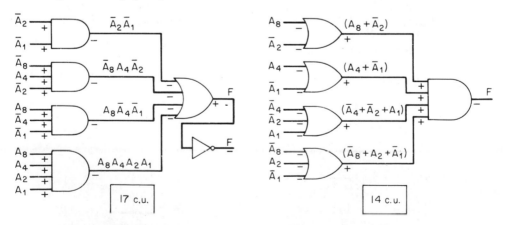

Figure 4-20. The SOP and POS Implementations for the Division Detector

The TTL implementations for both forms are shown in Fig. 4-20. The POS implementation is preferred since its overall cost is 3 c.u. cheaper.

EXAMPLE 4-30:   As a second example, consider the K-mapped design of a full adder. This device is introduced in Sec. 1.5.2 and considered algebraically in Probs. 2-14 and 3-23. The truth table for this functional device is given in Table 1-4 and is repeated here as Table 4-0. A_i and B_i are the Boolean

Table 4-0:   The Binary Full Adder

A_i	B_i	C_i	S_i	C_{i+1}
0	0	0	0	0
0	0	1	1	0
0	1	0	1	0
0	1	1	0	1
1	0	0	1	0
1	0	1	0	1
1	1	0	0	1
1	1	1	1	1

variables that represent the binary inputs to the ith full adder in a string of adders that are used to add two multibit binary numbers. C_i is the carry input to the ith stage. The outputs from the full adder are designated as S_i, the sum bit, and C_{i+1}, the carry to the next higher full adder. Figure 4-21

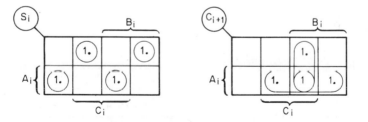

Figure 4-21. K-maps of the Sum and Carry Outputs from a Full Adder

shows the K-maps for the sum and carry outputs. The S_i function cannot be simplified in either SOP or POS form. The sum-of-minterms form for S_i is

$$S_i = \bar{A}_i\bar{B}_iC_i + \bar{A}_iB_i\bar{C}_i + A_i\bar{B}_i\bar{C}_i + A_iB_iC_i \qquad (4\text{-}16)$$

Equation (4-16) can be reexpressed as

$$S_i = A_i \oplus B_i \oplus C_i \qquad (4\text{-}17)$$

The C_{i+1} function, in simplified SOP form, is

$$C_{i+1} = B_iC_i + A_iC_i + A_iB_i \qquad (4\text{-}18)$$

The simplified POS form for C_{i+1} is

$$C_{i+1} = (A_i + B_i)(A_i + C_i)(B_i + C_i) \qquad (4\text{-}19)$$

Both forms for C_{i+1} require nine cost units for implementation.

The sum output can be implemented with NAND/NOR gates, but Eq. (4-17) shows that EXCLUSIVE-OR gates can be used to advantage. With RTL EXCLUSIVE-OR gates, the sum can be implemented as shown in Fig. 4-22. Positive-TRUE input and output symbolism is assumed. Keeping the positive-TRUE input and output symbolism shown in Fig. 4-22 makes the OR/AND form for C_{i+1} preferable for use with RTL gating. The logic diagram for C_{i+1} is given in Fig. 4-23.

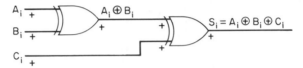

Figure 4-22. An EXCLUSIVE-OR Implementation for S_i

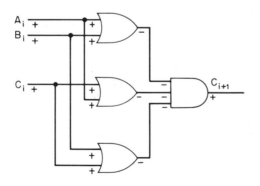

Figure 4-23. An RTL Implementation for C_{i+1}

The *best* solution to the implementation of the full-adder operation is the use of a MSI version of the full adder. Such devices are available in RTL, TTL, and most other IC families, and are usually packaged as pairs. The generalized symbol for a full adder is shown in Fig. 4-24. All MSI full adders can be assigned either fully positive-TRUE or fully negative-TRUE input and output voltage symbolism. This is a consequence of the logical characteristics of the full-addition operation, as shown in Prob. 3-25.

MSI full adders can be interconnected to add binary numbers of any length. The interconnection of three full adders and one half-adder in order to add two four-bit binary numbers is shown in Fig. 4-25. There is no carry into the low-order bit, so a *half*-adder can be used. The carry out of the high-order full adder can be used as the fifth sum bit. It has a binary weight of $2^4 = 16$.

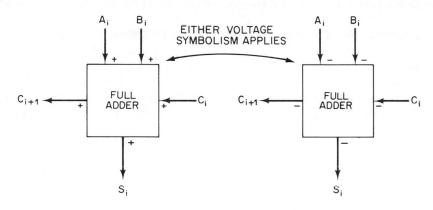

Figure 4-24. The MSI Full Adder

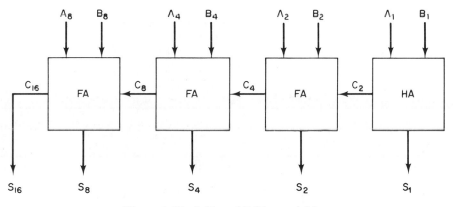

Figure 4-25. A Four-bit Binary Adder

4.4 Redundancies: Don't-care and Can't-happen Conditions

Both of the combinational design examples given in Sec. 4.3.6 are *fully specified,* meaning that a specific TRUE or FALSE value is defined for every combination of input variables. In many applications, however, there are input combinations that, for some reason or other, do not require a specific logical output value. These cases, commonly called *redundancies*, fall into two general categories: *don't-care* input conditions wherein the output value of the function is of no concern to the designer and *can't-happen* input conditions that, because of some constraint, can never appear at the input to the combinational device.

Combinational problems that have either don't-care or can't-happen inputs are said to be *incompletely specified*. The redundancies are useful in simplifying the combinational logic that is required to implement the Boolean expression, since the output values that are produced in the presence of a redundant input combination can be freely chosen as TRUE or FALSE, depending on which values are most useful in simplifying the logical system.

Redundancies are normally plotted as an X on a K-map, with the X being freely used as *either* a 0 or a 1 (or as both) during the grouping process. Other symbols that are used for redundancies include $\emptyset$ and D; only the X symbolism is used in this text.

EXAMPLE 4-31: Given $F(A,B,C) = \sum 2, 3, 7$ and that input conditions $\bar{A}\bar{B}\bar{C}$ and $AB\bar{C}$ can never occur, simplify the expression for F.

Solution: The K-map for F is shown in Fig. E4-31. Note the redundancies at m_0 and m_6. If the X at m_6 is used as a 1, the grouping for F becomes

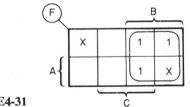

Figure E4-31

$F = B$. The redundancy at m_0 is best considered as a 0. Without the use of the redundancy at m_6 the expression for F would be

$$F = \bar{A}B + BC$$

The use of the can't-happen condition at m_6 saves 6 c.u.

The redundancies that are present in an incompletely specified function can be listed in compact form as

$$F(A,B,C,D) = \sum 0, 1, 5, 9, 10, 13, X_6, X_{11}, X_{14} \qquad (4\text{-}20)$$

showing that the function is TRUE for input conditions $m_0, m_1, m_5, m_9, m_{10}$, and m_{13} and has redundancies at m_6, m_{11}, and m_{14}. The K-map for the function of Eq. (4-20) is given in Fig. 4-26.

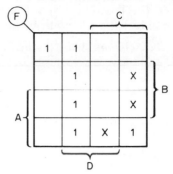

Figure 4-26. The K-map of Eq. (4-20)

The function given in Eq. (4-20) can also be expressed in tabular form:

A	B	C	D	F
0	0	0	0	1
0	0	0	1	1
0	0	1	0	0
0	0	1	1	0
0	1	0	0	0
0	1	0	1	1
0	1	1	0	X
0	1	1	1	0
1	0	0	0	0
1	0	0	1	1
1	0	1	0	1
1	0	1	1	X
1	1	0	0	0
1	1	0	1	1
1	1	1	0	X
1	1	1	1	0

where the Xs in the output column show the redundant conditions. This table, the map given in Fig. 4-26, and Eq. (4-20) all give the same information.

When an incompletely specified function is mapped, simplified, and expressed by logical equations that are in either simplified SOP or POS form, the input conditions that were originally redundant lose their unspecified output conditions. Any redundant input combinations whose Xs were grouped as if they were 1s will produce a TRUE output should that redundancy appear at the inputs to the logical system. The other Xs will produce a FALSE output. Thus, even though the redundant input conditions should never be of concern, it is a good design step to consider the output responses that they would produce if they did occur (remembering Murphy's law [7]).

The behavior of the function in the presence of redundancies can be determined by examining the K-map groupings that were used in simplification or by algebraically applying the redundant conditions to the simplified function and calculating the outputs that occur.

EXAMPLE 4-32: Given $F(A,B,C) = \sum 0, 5, X_1, X_4, X_7$. Simplify and give a complete truth table for the simplified function.

Solution: The original truth table for F is

A	B	C	F
0	0	0	1
0	0	1	X
0	1	0	0
0	1	1	0
1	0	0	X
1	0	1	1
1	1	0	0
1	1	1	X

showing redundancies at m_1, m_4, and m_7. The K-map of F is shown in Fig. E4-32, and can be grouped to give

$$F(A,B,C) = \bar{B}$$

Since the Xs at m_1 and m_4 are included in the $\bar{B}$ grouping, the simplified function will be TRUE in these cases. F will be FALSE at m_7, since the X at m_7 is not included in the group.

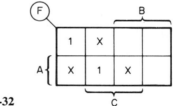

Figure E4-32

Algebraically, the function can be seen to be TRUE for $m_1 = \bar{A}\bar{B}C$ and $m_4 = A\bar{B}\bar{C}$, since both contain $\bar{B}$. The function is FALSE for $m_7 = ABC$, since B is TRUE in this case. The complete truth table for $F = \bar{B}$ is

A	B	C	F
0	0	0	1
0	0	1	1
0	1	0	0
0	1	1	0
1	0	0	1
1	0	1	1
1	1	0	0
1	1	1	0

which is correct for all the originally specified input conditions.

4.5 Design Examples

The presentation of the combinational design process is now complete. The process begins with a problem statement, proceeds through K-mapping, grouping, and implicant selection, and leads to the simplified SOP and POS forms. Implementation of either of these expressions with modern gating circuits is straightforward, with the voltage symbolism approach being used to implement A-O-N expressions with NAND/NOR gates. To complete the process, the response of the combinational system to its redundant input conditions should be determined (just in case). To illustrate this complete process, several examples of combinational design at the functional level are presented in this section.

4.5.1. Incremental Rotation Detector.   The digital device represented in Fig. 4-27 is a simple form of a shaft-position encoder. Its outputs are two Boolean variables that describe the quadrant to which the shaft has rotated.

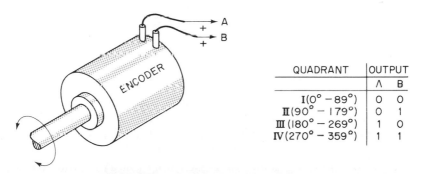

QUADRANT	OUTPUT	
	A	B
I $(0° - 89°)$	0	0
II $(90° - 179°)$	0	1
III $(180° - 269°)$	1	0
IV $(270° - 359°)$	1	1

Figure 4-27. A Simple Shaft-position Encoder

The outputs are generated with positive-TRUE symbolism. A digital system is to sample the values of A and B at intervals and to determine what incremental movements the shaft has made. The output signals from the system are designated as P (positive rotation), N (negative rotation), and Z (zero rotation). It is assumed that the values of A and B will be sampled often enough so that the shaft will not rotate more than 90° between successive readings.

At a *systems* level, the digital system must contain a timing device that will initiate each sampling operation, a storage device that will store the previous values of A and B, and a logical section that will compare the present values of the encoder's outputs with the previous values in order to generate the proper output signals.

The design of the timing and storage devices is covered in Chapters 5, 6, and 7. It is assumed at this point that the previous values of *A* and *B*, designated as *C* and *D*, are available to the logical portion of the system. The operation of the combinational comparator is truth-tabled below:

Present Values		Previous Values		Outputs		
A	B	C	D	P	M	Z
0	0	0	0	0	0	1
0	0	0	1	0	1	0
0	0	1	0	X	X	X
0	0	1	1	1	0	0
0	1	0	0	1	0	0
0	1	0	1	0	0	1
0	1	1	0	0	1	0
0	1	1	1	X	X	X
1	0	0	0	X	X	X
1	0	0	1	1	0	0
1	0	1	0	0	0	1
1	0	1	1	0	1	0
1	1	0	0	0	1	0
1	1	0	1	X	X	X
1	1	1	0	1	0	0
1	1	1	1	0	0	1

The values of *P*, *M*, and *Z* follow from comparison of the present values of *A* and *B* with their previous values, *C* and *D*. If $ABCD = 0000$, then the quadrant has not changed, since *AB* agrees with *CD*. Thus, the *Z* output (no change) must be TRUE. When $ABCD = 0001$, the previous quadrant was II and the present quadrant is I, indicating a negative rotation. This requires that $M = 1$.

The case where $ABCD = 0010$ implies a movement from quadrant III to I. But, since the sampling rate was assumed to be fast enough so that this much rotation could not occur between successive samples, this combination is a can't-happen. Hence, the values of *P*, *M*, and *Z* need not be specified. The remainder of the tabulation follows the same rationale.

The K-map for the *P* output function is plotted in Fig. 4-28a. One choice for the SOP groupings is shown. These groups give the following SOP expression for *P*:

$$P = \bar{A}B\bar{C}\bar{D} + ABC\bar{D} + \bar{A}CD + A\bar{C}D \qquad (4\text{-}21)$$

for a cost of 18 c.u.

The best POS groupings for *P* are found via the K-map of Fig. 4-28b. The groups shown in Fig. 4-28b give the simplified SOP expression:

$$\bar{P} = \bar{B}\bar{D} + \bar{A}\bar{C}D + \bar{A}BC + AB\bar{C} + ACD \qquad (4\text{-}22)$$

In POS form Eq. (4-22) becomes

$$P = (B + D)(A + C + \bar{D})(A + \bar{B} + \bar{C})(\bar{A} + \bar{B} + C)(\bar{A} + \bar{C} + \bar{D}) \qquad (4\text{-}23)$$

with a cost of 19 c.u. The SOP form is cheaper.

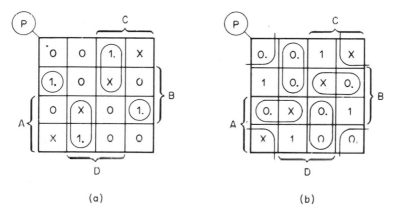

(a)   (b)

Figure 4-28. The SOP and POS Groupings on the K-map of the *P* Output of the Rotation Detector

The K-map and simplified forms for the *M* function are given in Fig. 4-29. Similarly, the minimization of the *Z* function is shown in Fig. 4-30.

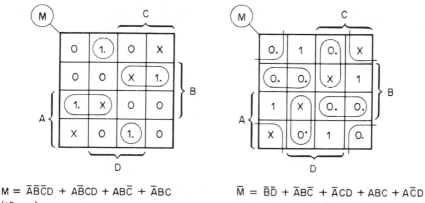

$M = \bar{A}\bar{B}\bar{C}D + \bar{A}BCD + AB\bar{C} + \bar{A}BC$
(18 c.u.)

$\bar{M} = \bar{B}\bar{D} + \bar{A}B\bar{C} + \bar{A}CD + ABC + A\bar{C}D$
$M = (B + D)(A + \bar{B} + C)(A + \bar{C} + \bar{D}) \cdot$
       $(\bar{A} + \bar{B} + \bar{C})(\bar{A} + C + \bar{D})$ (19 c.u.)

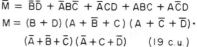

Figure 4-29. K-map Minimization of the *M* Output Function

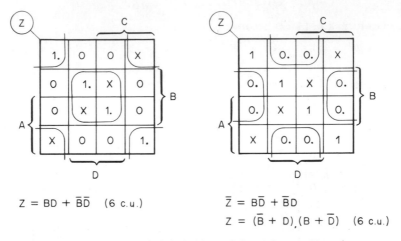

$$Z = BD + \bar{B}\bar{D} \quad \text{(6 c.u.)}$$

$$\bar{Z} = B\bar{D} + \bar{B}D$$

$$Z = (\bar{B} + D)(B + \bar{D}) \quad \text{(6 c.u.)}$$

Figure 4-30. K-map Minimization of the Z Output Function

The AND/OR expression for Z is the complement of the EXCLUSIVE-OR function. This is the **COINCIDENCE** function given in Eq. (3-9); i.e.,

$$Z = B \odot D \tag{4-24}$$

An RTL implementation for Eq. (4-24) is shown in Fig. 4-31. The complementary relationship between the EXCLUSIVE-OR operation and the COINCIDENCE operation is indicated. The SOP forms for P and M are easily implemented with any type of NAND/NOR gates.

Figure 4-31. An RTL Implementation of $Z = B \odot D$

The final set of combinational equations chosen herein is

$$P = \bar{A}B\bar{C}\bar{D} + ABC\bar{D} + \bar{A}CD + A\bar{C}D$$

$$M = \bar{A}\bar{B}\bar{C}D + A\bar{B}CD + AB\bar{C} + \bar{A}BC \tag{4-25}$$

$$Z = \bar{B}\bar{D} + BD = B \odot D$$

Consideration of the effect that the redundancies at m_2, m_7, m_8, and m_{13} have on these expressions gives the following table:

A	B	C	D	P	M	Z
0	0	1	0	0	0	1
0	1	1	1	1	1	1
1	0	0	0	0	0	1
1	1	0	1	1	1	1

Thus, all the can't-happen conditions will produce a Z (no change) output, and both m_7 (a IV to II rotation) and m_{13} (a II to IV rotation) will appear as a positive, negative, and zero change, all at the same time.

The total cost of implementing the comparison logic is 42 c.u. plus four inverters to generate $\bar{A}$, $\bar{B}$, $\bar{C}$, and $\bar{D}$. If an EXCLUSIVE-OR gate is used to implement Z, the cost is reduced below 42 c.u.

4.5.2. A Binary Comparator.

Two multibit binary numbers are to be compared, beginning with their least-significant bits (LSB) and proceeding to their most-significant bits (MSB). The general hardware arrangement for one possible comparison scheme is shown in Fig. 4-32. The two low-order

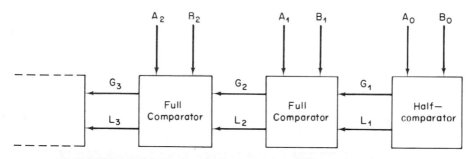

Figure 4-32. A Combinational Binary Comparator

bits are compared by a *half-comparator* that determines if $A_0 > B_0$ or $B_0 > A_0$. The output signals from the half-comparator are G_1 and L_1. G_1 indicates that A_0 exceeds B_0. L_1 indicates that A_0 is smaller than B_0. The logic for these two outputs is

$$G_1 = A_0\bar{B}_0, \qquad L_1 = \bar{A}_0 B_0 \qquad (4\text{-}26)$$

If neither G_1 nor L_1 is TRUE, then A_0 must equal B_0.

In all higher-ordered columns a *full-comparison* operation takes place. The G and L signals generated by the lower columns are combined with the A and B bits in the same column to generate the G and L signals that go to the next higher column.

The logic is as follows. In the ith column, if A_i is 1 and B_i is 0, then A

$> B$ thus far and $G_{i+1} = 1$, no matter what the values for L_i and G_i are. Similarly, if $A_i = 0$ and $B_i = 1$, then $A < B$ thus far and $L_{i+1} = 1$. If, however, $A_i = B_i$, then the values of L_i and G_i are passed on as L_{i+1} and G_{i+1} with no change. In truth table form this is

G_i	L_i	A_i	B_i	G_{i+1}	L_{i+1}
0	0	0	0	0	0
0	0	0	1	0	1
0	0	1	0	1	0
0	0	1	1	0	0
0	1	0	0	0	1
0	1	0	1	0	1
0	1	1	0	1	0
0	1	1	1	0	1
1	0	0	0	1	0
1	0	0	1	0	1
1	0	1	0	1	0
1	0	1	1	1	0
1	1	0	0	X	X
1	1	0	1	X	X
1	1	1	0	X	X
1	1	1	1	X	X

Note that G_i and L_i can never both be TRUE, making m_{12}, m_{13}, m_{14}, and m_{15} appear as can't-happen conditions.

The K-map and simplified forms for G_{i+1} are given in Fig. 4-33. Both forms for implementing G_{i+1} have a cost of 9 c.u.

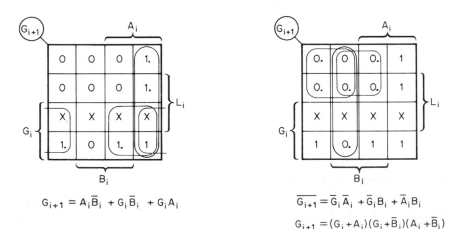

$$G_{i+1} = A_i\bar{B}_i + G_i\bar{B}_i + G_iA_i$$

$$\overline{G_{i+1}} = \bar{G}_i\bar{A}_i + \bar{G}_iB_i + \bar{A}_iB_i$$

$$G_{i+1} = (G_i + A_i)(G_i + \bar{B}_i)(A_i + \bar{B}_i)$$

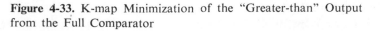

Figure 4-33. K-map Minimization of the "Greater-than" Output from the Full Comparator

Similarly, the K-map minimization of L_{i+1} is shown in Fig. 4-34. Again, both expressions cost 9 c.u. In TTL form, assuming double-rail inputs, the SOP expressions may be implemented as shown in Fig. 4-35. To indicate equality,

$$E_{i+1} - \overline{G_{i+1}} \cdot \overline{L_{i+1}} - \overline{G_{i+1} + L_{i+1}} \tag{4-27}$$

may be used.

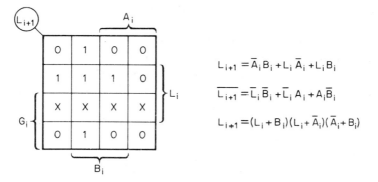

$$L_{i+1} = \overline{A}_i B_i + L_i \overline{A}_i + L_i B_i$$

$$\overline{L_{i+1}} = \overline{L}_i \overline{B}_i + \overline{L}_i A_i + A_i \overline{B}_i$$

$$L_{i+1} = (L_i + B_i)(L_i + \overline{A}_i)(\overline{A}_i + B_i)$$

Figure 4-34. K-map Minimization of the "Less-than" Output from the Full Comparator

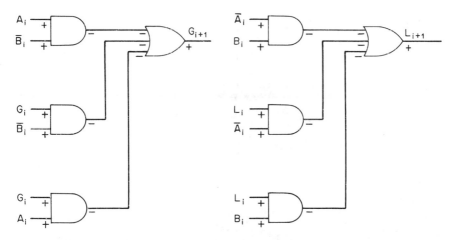

Figure 4-35. TTL Implementation of Full Comparator (AND/OR Form)

Should a comparison stage erroneously receive inputs that show *both* L_i and G_i to be TRUE, the SOP expressions for G_{i+1} and L_{i+1} dictate the behavior given in the following truth table:

G_i	L_i	A_i	B_i	G_{i+1}	L_{i+1}
1	1	0	0	1	1
1	1	0	1	0	1
1	1	1	0	1	0
1	1	1	1	1	1

The error is propagated if $A_i = B_i$ and is corrected if $A_i \neq B_i$.

4.5.3. Code-to-Code Conversion. A very common application of combinational logic is the conversion of data that are represented in one digital code into a representation in another code. Consider a decimal digit that is represented in the 8421 binary-coded-decimal (BCD) code.* This code uses the four-variable minterms from m_0 through m_9 to represent the decimal values 0–9. A combinational system can be designed that will convert data from the 8421 BCD code into the excess-three code, a code that uses m_3–m_{12} to represent the ten decimal values. A tabulation of this code-to-code conversion process is given below:

Decimal Value	Input A B C D				Output W X Y Z			
0	0	0	0	0	0	0	1	1
1	0	0	0	1	0	1	0	0
2	0	0	1	0	0	1	0	1
3	0	0	1	1	0	1	1	0
4	0	1	0	0	0	1	1	1
5	0	1	0	1	1	0	0	0
6	0	1	1	0	1	0	0	1
7	0	1	1	1	1	0	1	0
8	1	0	0	0	1	0	1	1
9	1	0	0	1	1	1	0	0

The four input bits are named A, B, C, and D. The output signals are W, X, Y, and Z. The six input combinations that do not appear in the table (m_{10}–m_{15}) are to be used as redundancies.

The K-maps for the W, X, Y, and Z output functions are shown in Fig. 4-36. The simplified SOP and POS forms for each output function are also

*For more discussion of the properties of this and other codes used as examples in Chapters 4 and 6, see Chapter 10.

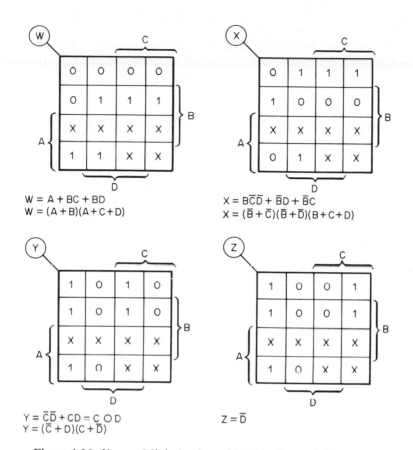

$$W = A + BC + BD$$
$$W = (A + B)(A + C + D)$$

$$X = B\overline{C}\overline{D} + \overline{B}D + \overline{B}C$$
$$X = (\overline{B} + \overline{C})(\overline{B} + D)(B + C + D)$$

$$Y = \overline{C}\overline{D} + CD = C \odot D$$
$$Y = (\overline{C} + D)(C + \overline{D})$$

$$Z = \overline{D}$$

Figure 4-36. K-map Minimization of 8421-to-Excess-3 Converter

given in Fig. 4-36. The combinational expressions for the four output functions are easily implemented by using NAND/NOR gates. An EXCLUSIVE-OR gate can be used in implementing the Y output.

The analysis of the output signals that are produced by the six redundant inputs is left to Prob. 4-19.

4.5.4. The Seven-Segment Display. Many devices have been developed for displaying the output signals that are produced by digital systems. One of the more recent developments is the seven-segment display unit which can indicate all the decimal digits and several alphabetic characters. A typical arrangement for the segments is shown in Fig. 4-37. Seven-segment display devices are available in a wide variety of sizes and types. Illumination is produced by incandescent bulbs, fluorescence, electroluminescence, light-emitting diodes, and other techniques.

To use a seven-segment display, a code conversion system must be designed. This converter, commonly called a *decoder*, changes the logical input signals that describe the decimal value into the seven signals that turn each segment ON or OFF so as to display the proper character. This example considers the design of a decoder that accepts a digit encoded in 8421 BCD code and converts that input code into the seven output signals, **a** to **g**, that display each digit.

There is some latitude in selecting the segmented representations for the various digits. For example, the digit 1 can be represented by segments **b** and **c** or by segments **e** and **f**. The **b-c** representation is used herein. The representation for 6 is also somewhat open. The **c**, **d**, **e**, **f**, and **g** segments *must* be ON, but the **a** segment can be either ON or OFF, since either of the symbols given in Fig. 4-38 can represent a 6. Similar don't-care segments occur with the **d** segment for a 9 and the **f** segment for a 7, as shown in Fig. 4-39.

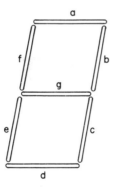

Figure 4-37. A Typical Seven-segment Display

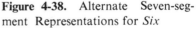

Figure 4-38. Alternate Seven-segment Representations for *Six*

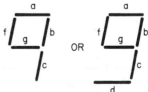

Figure 4-39. Don't-care Segments Present in *Seven* and *Nine*

The complete specification of the segments used to represent each digit is tabulated as follows:

Inputs				Outputs							Symbol
A	B	C	D	a	b	c	d	e	f	g	
0	0	0	0	1	1	1	1	1	1	0	0
0	0	0	1	0	1	1	0	0	0	0	1
0	0	1	0	1	1	0	1	1	0	1	2
0	0	1	1	1	1	1	1	0	0	1	3
0	1	0	0	0	1	1	0	0	1	1	4
0	1	0	1	1	0	1	1	0	1	1	5
0	1	1	0	X	0	1	1	1	1	1	6
0	1	1	1	1	1	1	0	0	X	0	7
1	0	0	0	1	1	1	1	1	1	1	8
1	0	0	1	1	1	1	X	0	1	1	9

There are six redundant input conditions.

The K-map and simplified expressions for controlling the **a** segment are given in Fig. 4-40. The redundancy at m_6 is included in the grouping for the C implicant, making the representation for the numeral 6 have the **a** segment.

The K-map and simplified expressions for the **b** segment are shown in

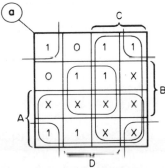

$$a = A + C + BD + \bar{B}\bar{D}$$
$$= A + C + (B \odot D)$$

Figure 4-40. The K-map Minimization of the **a** Output Function

Fig. 4-41. The RTL implementation for the simplified SOP expression for **b** is given in Fig. 4-42. Negative-TRUE input and output symbolism are assumed, and an RTL EXCLUSIVE-OR gate is used to simplify the gating.

Continuing the combinational design process for the other five segments gives the simplified expressions of Eq. (4-28). Only simplified SOP forms are given.

$$\begin{aligned}
c &= B + \bar{C} + D \\
d &= \bar{B}\bar{D} + B\bar{C}D + \bar{B}C + C\bar{D} \\
e &= \bar{B}\bar{D} + C\bar{D} \\
f &= A + B + \bar{C}\bar{D} \\
g &= A + C\bar{D} + (B \oplus C)
\end{aligned} \qquad (4\text{-}28)$$

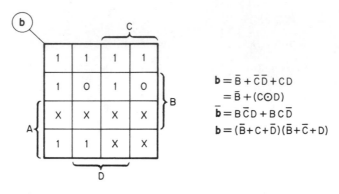

$$b = \bar{B} + \bar{C}\bar{D} + CD$$
$$= \bar{B} + (C \odot D)$$
$$\bar{b} = B\bar{C}D + BC\bar{D}$$
$$b = (\bar{B}+C+\bar{D})(\bar{B}+\bar{C}+D)$$

Figure 4-41. The K-map Minimization for the **b** Output Function

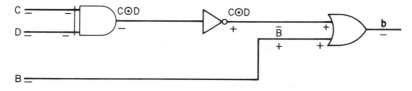

Figure 4-42. An RTL Implementation of $b = B + (C \odot D)$

The consideration of simplified POS forms for the decoder is left to Problem 4-21.

The logical expression for the **d** output is FALSE when the input combination is $m_9 = A\bar{B}CD$. As a result, the representation for the numeral 9 will not have the bottom segment. Similarly, the expression for the **f** segment is TRUE for $m_7 = \bar{A}BCD$, making the displayed image for a 7 have the hook on its left-hand side.

Thus far, the design of the decoder has assumed that the six minterms, $m_{10}–m_{15}$, are redundancies of the can't-happen variety. Of concern, however, is the effect that one of these inputs would have on the displayed image. It would certainly be desirable that an erroneous input to the decoder not produce a valid decimal character. Otherwise, the observer would not detect the error.

To determine the effect that the redundancies will have on the display, the TRUE/FALSE values of the logical expression for *each* segment must be determined. Consider $m_{10} = A\bar{B}C\bar{D}$ first. The **a** segment will be TRUE, since $a = A + C + BD + \bar{B}\bar{D}$, and both A and C are TRUE. Similarly, **b**, **c**, **d**, **e**, **f**, and **g** are also TRUE, as seen by applying m_{10} to the expressions given in Eq. (4-28). Thus, the representation for m_{10} has all of the segments TRUE, thereby forming an 8. Similar consideration of the other five redundant inputs gives the following results:

Input	a	b	c	d	e	f	g	Symbol
$m_{11} = A\bar{B}CD$	1	1	1	1	0	1	1	9
$m_{12} = AB\bar{C}\bar{D}$	1	1	1	0	0	1	1	9
$m_{13} = AB\bar{C}D$	1	0	1	1	0	1	1	5
$m_{14} = ABC\bar{D}$	1	0	1	1	1	1	1	6
$m_{15} = ABCD$	1	1	1	0	0	1	1	9

All of these appear as valid numerical symbols.

There are many approaches to eliminating these falsely correct characters. One of the simplest is to recognize that no valid decimal character has both segments **c** and **d** missing. Thus, by making these two segments (**c** and **d**) FALSE for input combinations m_{10}–m_{15}, it is assured that no valid characters will be produced by an invalid input combination. This requires that minterms m_{10}–m_{15} be entered as 0, rather than don't-cares, on the K-maps for the **c** and **d** output functions. The modified K-maps and simplified SOP expressions for **c** and **d** are shown in Fig. 4-43. These modified expressions prevent the appearance of a falsely valid character. They increase the cost of the decoder by 9 c.u. over the cost of the expressions for **c** and **d** that are given in Eq. (4-28).

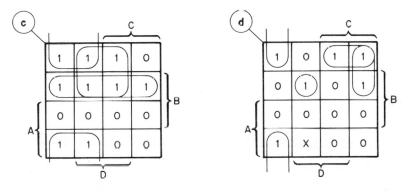

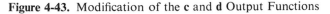

$$\mathbf{c} = \bar{B}\bar{C} + \bar{A}B + \bar{A}D$$

$$\mathbf{d} = \bar{A}B\bar{C}D + \bar{A}\bar{B}C + \bar{A}C\bar{D} + B\bar{C}D$$

Figure 4-43. Modification of the **c** and **d** Output Functions

In some applications it is useful to add other "frills" to the decoder. A common addition is the provision of a LAMP TEST input that, when made TRUE, will cause all seven of the segments to be illuminated. This is useful in maintenance of the display.

A second added capability is the provision for *zero blanking* at the high-order end of long decimal words. The zero-blanking operation makes a four-digit representation for 57 appear as 57 rather than 0057. The logic behind the zero-blanking operation requires that each decoder recognize whenever it is receiving the input code combination that represents 0. This condition alone is not enough to make the blanking decision, however, since only *leading* 0s should be blanked out. Thus, a logical signal that indicates that all higher-ordered decades are 0 must be included in the zero-blanking circuitry. This extra input signal is commonly called *ZBI*, for *zero-blanking input*.

The decoder should blank out all its segments whenever the code for 0 is received and *ZBI* is TRUE. With the 8421 BCD code, the logic for blanking the segments is

$$BLANK = ZBI \cdot \bar{A}\bar{B}\bar{C}\bar{D} \qquad (4\text{-}29)$$

The decoder should also generate a signal that acts as the *ZBI* to the next lower decade. This signal is commonly called *ZBO*, for *zero-blanking output*. When TRUE, *ZBO* indicates that all higher-ordered decades *and* the present decade are 0. The logic for *ZBO* is the same as the logic for *BLANK* that is given in Eq. (4-29).

To add zero blanking to the decoder, the logical expressions for the **a**, **b**, **c**, **d**, **e**, and **f** segments must be modified as shown in Fig. 4-44. The gating

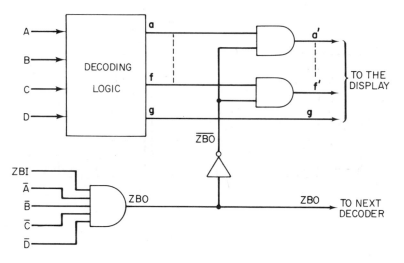

Figure 4-44. Zero Blanking Added to a Seven-segment Decoder

arrangement of Fig. 4-44 assures that if *ZBI* is TRUE *and* the input combination is m_0, then segments **a**–**f** are turned off. The **g** segment will already be turned off, since **g** is not present in the representation for 0. The logic for turning off the segments is used to generate the *ZBO* signal. To disable the zero-blanking operation, the *ZBI*s to all decoders are made FALSE. The complete connection for zero blanking over four decades is shown in

Fig. 4-45. The low-order decade is not blanked, so that the all-zeros number is displayed as a single low-order 0.

For illustration, the complete RTL logic diagram for the **a** segment (with zero blanking and *LAMP TEST*) is given in Fig. 4-46.

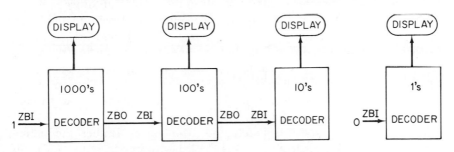

Figure 4-45. Zero Blanking over Four Decades

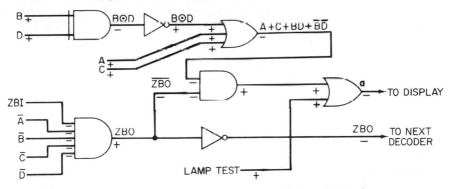

Figure 4-46. The Complete Logic for Decoding the **a** Output Function

The usefulness of the 8421 BCD/seven-segment decoding operation has led to the manufacturing of MSI decoders (single IC packages that perform the complete decoding operation, including *LAMP TEST* and zero blanking). Some decoders include an output transistor stage that is capable of switching the 60–100-mA current associated with incandescent displays [4]. These devices are called *decoder-drivers*.

4.6 Tabular Minimization

The K-map can be expanded to represent functions having more than five arguments by continuing the fold-out process introduced in Sec. 4.3. Pictorial representations of larger maps are given in Refs. [3] and [5]. The minimization process becomes increasingly difficult as the size of the map increases, however, since the various groupings become more and more disjointed.

When working with larger functions, the tabular reduction method developed by Quine [12] and modified by McCluskey [10] is an alternative to the K-map method. The Quine-McCluskey minimization method involves simple, repetitive operations that compare each minterm that is present in a sum-of-minterms expression for a Boolean function to all other minterms with which it may form a combinable grouping. The reduced implicants that result from this process are compared among themselves to identify further reductions (groups of four and the like) until no further reductions are possible. This process identifies all the prime implicants that may be used to represent the function in simplified SOP form. A selection process is then required to determine which set of prime implicants does the best (cheapest) job of covering the function.

The tabular method is definitely not a work-saver. Rather, the method involves a great deal of tedium and is loaded with opportunities for error. In fact, the tabular method is discussed herein *only* to familiarize the reader with the basics of the grouping technique and to introduce prime implicant selection techniques. The use of computer-aided reduction of functions having more than five arguments is recommended, and a FORTRAN program that does this is included in this chapter. The manual tabular method should be used only as a last resort.

4.6.1. The Decimal Grouping Process.

Like the K-map process, tabular reduction begins with the sum-of-minterms expression for the function that is to be reduced. The minterms are first separated by *index*, where the index of a minterm is the number of its variables that are TRUE. The groups of equal-index minterms are then arranged in ascending order, in column 1 of the tabulation. Each minterm is represented by its decimal subscript.

EXAMPLE 4-33: List the four-variable minterms in groups of equal index.

Answer:

Minterm Number	Index
0	0
1, 2, 4, 8	1
3, 5, 6, 9, 10, 12	2
7, 11, 13, 14	3
15	4

EXAMPLE 4-34: Given $F(5) = \sum 0, 3, 6, 9, 11, 14, 15, 17, 21, 25, 29, 30, 31$. Separate F into its equal-index groups.

Solution:

Index	Groups
0	0
1	—
2	3, 6, 9, 17
3	11, 14, 21, 25
4	15, 29, 30
5	31

Once the minterms have been separated into the equal-index groups and listed in ascending order in column 1, the manual tabular minimization process can be carried out according to the simplified algorithmic process that follows.*

Step 1.

The decimal number representing each minterm in a given equal-index group is compared with the numbers of all minterms in the group having the next-higher index. Only those minterms having higher numbers as well as higher indices need be considered for comparison. For example, m_9 (index 2) need not be compared to m_7 (index 3), since 7 is less than 9. Two minterms will combine (as a group of two) if their decimal representations differ by 2^N. The group forms a reduced implicant, as per Eq. (4-0). Each minterm that is combined into a group of two is checked off of the original list. A given minterm may combine into several groups, but any minterms that remain unchecked throughout the tabulation process are essential implicants by themselves, since they are not included in any of the groupings.

Step 2.

The reduced implicants that are produced by pairing minterms during step 1 are entered into a second column (column 2) in the form

$$I(2^N) \tag{4-30}$$

where I is the *lower* minterm number of the pair and 2^N is the power of 2 by which the minterm numbers differed. Only the lower minterm number needs to be recorded. The entries in the second column should be separated into the groups of reduced implicants that result from the various cross-group combinations of step 1. A horizontal bar is useful in separating the equal-index groups in column 1 and the groups of reduced implicants that appear in column 2 and other higher-ordered columns.

*For further discussion of manual tabular reduction, see Refs. [2] and [3].

EXAMPLE 4-35:   Given $F(A,B,C,D) = \sum 1, 4, 10, 11, 12, 14, 15$,   show   the results of steps 1 and 2.

Solution: The equal-index groups are

Index	Minterm
0	—
1	1, 4
2	10, 12
3	11, 14
4	15

The results of steps 1 and 2 are

Column 1	Column 2
1	4(8)
4✓	10(1)
10✓	10(4)
12✓	12(2)
11✓	11(4)
14✓	14(1)
15✓	

The 4(8) implicant results from combining minterms 4 and 12. It is the only implicant that results from comparing terms from the first group of column 1 with terms in the second group. Note that minterm 1 is unchecked, indicating that it is an essential implicant by itself.

The tabular minimization process continues as follows.

Step 3.

The implicants in each group of column 2 are compared with all members of the next group that have the same power of 2 in parentheses. If the minterm numbers differ by a power of 2, then the two implicants can be combined. This corresponds to two groups of two minterms being combined to form a group of four. The newly formed implicant is entered into a third column as

$$I(2^N, 2^M) \tag{4-31}$$

where I is the lowest minterm number and 2^N and 2^M are the powers of 2 that are identified in forming the grouping.

Step 3 is closely related to steps 1 and 2. The implicants that are combined to form the entries in column 3 should be checked off in column 2. The entries in column 3 should then be compared to see if further groupings result, leading to a fourth column (groups of eight minterms). The process continues until no further groupings are possible. The entries in the last column and the unchecked implicants that remain in all previous columns form the set of prime implicants that can be used to represent the function.

EXAMPLE 4-36: Complete the tabulation of Ex. 4-35.

   Solution:

Column 1	Column 2	Column 3
1	4(8)	10(1, 4)
4√	10(1)√	
10√	10(4)√	
12√	12(2)	
11√	11(4)√	
14√	14(1)√	
15√		

Note that four implicants in column 2 combine to produce the 10(1,4) term in column 3. On a K-map this corresponds to

$$AC = A\bar{B}C + ABC + ACD + AC\bar{D}$$

showing that a group of four minterms begins as four groups of two, as shown in Fig. E4-36. In larger problems it can be seen that *six* pairings in

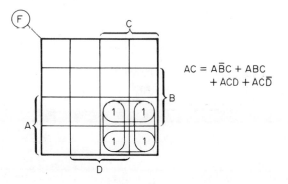

$$AC = A\bar{B}C + ABC$$
$$+ ACD + AC\bar{D}$$

Figure E4-36

column 3 will produce the same entry in column 4 and so on. This "overlap" is useful in detecting combinations that were missed in previous columns.

   The final set of prime implicants is 1, 4(8), 12(2), and 10(1,4).

   The prime implicants that appear in column 2 and any higher-ordered columns are expressed in shorthand form [as in Eqs. (4-30) and (4-31)]. It is very easy, however, to convert these contracted expressions into normal

Boolean form. First, the minterm whose number appears on the left of the contracted form is written in TRUE/FALSE form. Then the terms corresponding to the powers of 2 given in the parentheses are crossed out. The remaining variables describe the prime implicant.

EXAMPLE 4-37: Convert the prime implicants found in Ex. 4-36 into Boolean form.

Solution:

$$1 \longrightarrow \bar{A}\bar{B}\bar{C}D$$
$$4(8) \longrightarrow \cancel{A}B\bar{C}\bar{D} \longrightarrow B\bar{C}\bar{D}$$
$$12(2) \longrightarrow AB\cancel{C}\bar{D} \longrightarrow AB\bar{D}$$
$$10(1,4) \longrightarrow A\cancel{B}C\cancel{D} \longrightarrow AC$$

For comparison, the K-map minimization of the original function (given in Ex. 4-35) is shown in Fig. E4-37:

$$F = \bar{A}\bar{B}\bar{C}D + B\bar{C}\bar{D} + AC$$

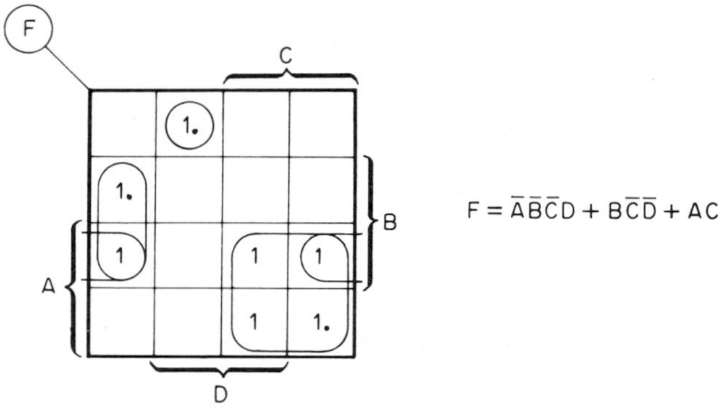

$$F = \bar{A}\bar{B}\bar{C}D + B\bar{C}\bar{D} + AC$$

Figure E4-37

The K-map shows that the $AB\bar{D}$ implicant is not actually needed. The tabular minimization technique finds *all possible* prime implicants, however, without regard to which implicants are actually necessary to describe the function. The identification of nonessential implicants is covered in Sec. 4.6.2.

The following examples serve to further illustrate the tabular method for finding the prime implicants.

EXAMPLE 4-38: Find the prime implicants for $F(5) = \sum 0, 1, 2, 3, 10, 16,$ 17, 18, 19, 28, 29.

Solution:

Column 1	Column 2	Column 3	Column 4
0 ✓	0(1) ✓	0(1, 16) ✓	0(1, 2, 16)
1 ✓	0(2) ✓	0(1, 2) ✓	
2 ✓	0(16) ✓	0(2, 16) ✓	
16 ✓	1(2) ✓	1(2, 16) ✓	
3 ✓	1(16) ✓	2(1, 16) ✓	
10 ✓	2(1) ✓	16(1, 2) ✓	
17 ✓	2(8)		
18 ✓	2(16) ✓		
19 ✓	16(1) ✓		
28 ✓	16(2) ✓		
29 ✓	3(16) ✓		
	17(2) ✓		
	18(1) ✓		
	28(1)		

The prime implicants are

$$0(1,2,16) = \bar{A}\bar{B}\bar{C}\bar{D}\bar{E} = \bar{B}\bar{C}$$
$$2(8) = \bar{A}\bar{B}\bar{C}D\bar{E} = \bar{A}\bar{C}D\bar{E}$$
$$28(1) = ABC\bar{D}\bar{E} = ABC\bar{D}$$

Note that all six terms in column 3 combine to form the single term given in column 4.

EXAMPLE 4-39: A combinational device is to receive a six-bit binary number and indicate whenever the number is divisible by 8. Design the combinational logic for this device.

Solution: $G(A,B,C,D,E,F) = \sum 0, 8, 16, 24, 32, 40, 48, 56$:

Column 1	Column 2	Column 3	Column 4
0 ✓	0(8) ✓	0(8, 16) ✓	0(8, 16, 32)
8 ✓	0(16) ✓	0(8, 32) ✓	
16 ✓	0(32) ✓	0(16, 32) ✓	
32 ✓	8(16) ✓	8(16, 32) ✓	
24 ✓	8(32) ✓	16(8, 32) ✓	
40 ✓	16(8) ✓	32(8, 16) ✓	
48 ✓	16(32) ✓		
56 ✓	32(8) ✓		
	32(16) ✓		
	24(32) ✓		
	40(16) ✓		
	48(8) ✓		

The reduced form for G contains only one term:

$$0(8,16,32) = \bar{A}\bar{B}\bar{C}\bar{D}\bar{E}\bar{F}$$

Thus, $G = \bar{D}\bar{E}\bar{F}$, which is the expected answer, since any binary number having three low-order 0s *must* be divisible by 8.

4.6.2. Prime-Implicant Selection.

The tabular minimization technique has one major drawback: it finds *all possible* prime implicants, with no regard as to which prime implicants are the best choice for covering the function. The K-map technique allows the designer to use his own "smarts" in selecting the best covering. The methodology of the tabular process defeats this heuristic selection process and hence necessitates a tabular selection technique.

To pick out the best set of prime implicants, a *prime-implicant table* must be constructed. This table lists the minterms that represent the function (from column 1) across its top and lists the prime implicants that are found by the tabulation process down its left side. An X is placed at the intersection of each row and column wherein the prime implicant (row) contains the minterm (column) in its grouping. Since the prime implicants are given in shorthand form, it is first necessary to reconstruct the numbers of the minterms that are combined to make the implicant.

If the prime implicant is a minterm (an unchecked term from column 1), then no reconstruction is necessary. Consider, however, a term such as 4(1) from column 2. This term represents the combination of *two* minterms into one reduced implicant. One of the minterms is m_4, the term whose number appears outside of the parentheses. The other minterm is numbered $4 + 1$ $= 5$, since the two minterm numbers had to differ by 1 in order to be combined. Thus, for a four-variable problem,

$$4(1) = \bar{A}B\bar{C}\bar{D} = \bar{A}B\bar{C} = (m_4 + m_5) \tag{4-32}$$

The partial prime-implicant table for the 4(1) term is given in Table 4-1. Since minterms m_4 and m_5 are covered by the prime implicant $4(1) = \bar{A}B\bar{C}$, Xs are entered at the intersections of their columns with the 4(1) row.

Table 4-1:   A Simple Prime-Implicant Table

	4	5
$4(1) = \bar{A}B\bar{C}$	X	X

A term such as 0(4,8) from column 3 is the combination of *four* minterms. They are m_0, m_4, m_8, and m_{12}. The minterm numbers are obtained by adding the various combinations of the numbers in the parentheses to m_0, the lowest-numbered minterm. The prime-implicant table for the 0(4,8) implicant is given in Table 4-2.

Table 4-2: Another Simple Prime-Implicant Table

	0	4	8	12
$0(4, 8) = \bar{C}\bar{D}$	X	X	X	X

EXAMPLE 4-40: The prime implicants that describe $F(A,B,C,D) = \sum 1, 4, 10, 11, 12, 14, 15$ (Ex. 4-35) are found in Ex. 4-37. They are

$$m_1 = \bar{A}\bar{B}\bar{C}D = \mathbf{a}$$
$$4(8) = B\bar{C}\bar{D} = \mathbf{b}$$
$$12(2) = AB\bar{D} = \mathbf{c}$$
$$10(1,4) = AC = \mathbf{d}$$

Construct the prime-implicant table for this function.

Solution: The prime implicants are labeled for easy reference. The prime-implicant table begins as

		1	4	10	11	12	14	15 ⟵ Minterms
Prime	a	X						
Implicants ⟶	b							
	c							
	d							

The X at the intersection of the first row and the first column shows that prime implicant **a** (m_1) covers itself. Prime implicant **b** covers *two* minterms. The reconstruction of the minterm numbers begins with the shorthand descriptor of prime implicant **b**, 4(8). The two minterms that are covered are m_4 and m_{12}, since they differ by 8. Four minterms are covered by the 10(1,4) implicant. Their numbers are m_{10}, m_{11}, m_{14}, and m_{15}. The complete prime-implicant table is

	1	4	10	11	12	14	15
a	X						
b		X			X		
c					X	X	
d			X	X		X	X

The minterms that make up a function represent the input combinations for which the function *must* be TRUE. The prime implicants are the logical components that can be combined to represent the function in a form that is more minimal than the sum-of-minterms form. If any minterm is covered by only one prime implicant, then that implicant is an *essential implicant*, since none of the other prime implicants will make the function TRUE when the combination that corresponds to the singly covered minterm appears as the input to the function. Essential implicants are indicated whenever a column in the prime-implicant table contains only one X. The selection process begins with the identification of all essential implicants. The Xs which define them can be circled. These implicants *must* be included in the expression for the function. All minterms that are covered by the essential implicants can then be removed from the prime-implicant table.

EXAMPLE 4-41: Select the essential implicants from the prime-implicant table given in Ex. 4-40.

Solution: The prime-implicant table is repeated below. The bottom line is a summation of the number of Xs in each column.

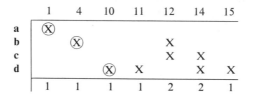

Prime implicants **a**, **b**, and **d** are essential. Note that **a** covers m_1, **b** covers m_4 and m_{12}, and **d** covers the rest of the minterms, making prime implicant **c** unnecessary. Thus, the reduced form for the original function is

$$F = \mathbf{a} + \mathbf{b} + \mathbf{d}$$
$$= \bar{A}\bar{B}\bar{C}D + B\bar{C}\bar{D} + AC$$

Whenever the essential implicants cover all the minterms, the problem is completed. In many cases, however, the removal of the essential terms and the minterms that they cover leaves a reduced prime-implicant table that has two or more Xs in every column. This table corresponds to the *cyclic map* described in Sec. 4.3.4.

The presence of a cyclic prime-implicant table indicates that there will be more than one set of prime implicants that will cover the remaining minterms. The simplest possible cyclic prime-implicant table is given in Table 4-3. Here, the minterm m_i can be covered by either prime implicant. The decision as to which implicant to choose must be based upon some criterion. Cost is the measure that is normally used.

Table 4-3:   The Simplest Cyclic Prime-Implicant Table

$$m_i$$

a	X
b	X

To show the cost of each implicant, an extra column is added to the prime-implicant table. The cost that is shown for each prime implicant is modified to include *both* the first- and second-level gating costs for implementing the prime implicant. The first-level gating cost will be equal to the number of variables present in the implicant. The second-level cost will add one more cost unit, representing the gate input of the OR gate that combines the implicants at the second level. The exception to this is an implicant that has only one remaining variable. Its total cost is 1 c.u., since no first-level gating is required.

EXAMPLE 4-42:   Give the costs for each of the following implicants: A, $X\bar{Y}$, $B\bar{C}\bar{E}$, and $A\bar{B}D\bar{E}$.

Answer: 1, 3, 4, and 5 c.u.

Once the cost of each implicant is computed, the various sets of implicants that can be used to cover the minterms which are present in the cyclic prime-implicant table can be determined and their costs compared. The lowest-cost covering can then be selected.

EXAMPLE 4-43:   Given the following prime-implicant table, select the best representation for the function.

	I	J	K	L	M	Cost
a	X			X		2
b				X	X	3
c	X				X	4
d		Ⓧ	X			5

Solution: Implicant **d** is essential; it covers m_J and m_K. Removing that row and both columns yields the following prime-implicant table. Note that it is cyclic.

	I	L	M	Cost
a	X	X		2
b		X	X	3
c	X		X	4

Examination of the reduced prime-implicant table shows that there are three ways to cover the three minterms that remain. These coverings are

$$\mathbf{a} + \mathbf{b} \qquad (5 \text{ c.u. total})$$
$$\mathbf{a} + \mathbf{c} \qquad (6 \text{ c.u. total})$$
$$\mathbf{b} + \mathbf{c} \qquad (7 \text{ c.u. total})$$

The best selection is $\mathbf{a} + \mathbf{b}$, making the best covering for the overall function become

$$F = \mathbf{a} + \mathbf{b} + \mathbf{d}$$

for 10 c.u. total.

Often the designer does not care if all possible sets of implicants that may be used to cover the function are considered. Rather, he only wants to be sure that the set of implicants that is finally chosen is as cheap as any other set. If this is the case, simplification of the cyclic table may be possible.

One type of simplification is the use of *row dominance*. If one row in a prime-implicant table contains Xs in the same positions as another row, plus at least one more X, then the first row is said to *dominate* the second row. This indicates that the implicant that defines the first row covers all the minterms present in the second row and at least one other term. Since the dominating row covers everything that the other row covers and more, the dominated row, if it has a cost that is equal to or greater than the cost of the dominating row, can be removed from the prime-implicant table.

EXAMPLE 4-44:   Simplify the following prime-implicant table.

	I	J	K	Cost
a	X	X	X	3
b		X		3
c	X		X	3

Solution: Row **a** dominates rows **b** and **c**. Since it costs the same as the others, the chart can be reduced to

	I	J	K	Cost
a	X	X	X	3

If two or more rows contain Xs in identical locations, the rows are said to be *equal*. All but the lowest-cost row of equal rows can be removed from the table.

A second useful simplification concept is *column dominance*. The same definition applies as in row dominance. To simplify the map, however, the *dominating* column is removed, with cost not being of concern.

EXAMPLE 4-45: Simplify the following cyclic prime-implicant table.

	I	J	K	L	M	N	Cost
a	X		X			X	3
b		X		X			3
c	X		X	X			3
d		X		X	X	X	3
e			X		X		3

Solution: Column K dominates column I, so K can be removed. Likewise, L dominates J, so L can be removed. This leaves

	I	J	M	N	Cost
a	⊗			X	3
d		⊗	X	X	3

This is no longer cyclic, since both implicants are essential. Comparison with the original prime-implicant table shows that **a** + **d** covers all the original minterms.

In some cases the application of row and column dominance does not eliminate the cyclic nature of a prime-implicant table. The designer must then consider all the possible sets of the prime implicants that remain in the table in order to determine which set is the cheapest way to cover the function. There are two methods for developing the different sets of implicants: *branching* and Petrick's method [11]. Both methods are illustrated in Ex. 4-46.

EXAMPLE 4-46: Given the following cyclic prime-implicant table, find the cheapest covering for the function.

	I	J	K	L	Cost
a	X	X			2
b		X	X		3
c			X	X	3
d	X			X	4

Solution: Row and column dominance cannot be used to simplify this table. Thus, all the various sets of prime implicants must be considered. The branching method begins by recognizing that m_I can be covered by *either*

a or **d**. First, prime implicant **a** is selected as if it were essential. This implicant covers both m_I and m_J. This selection reduces the prime-implicant table to

	K	L	Cost
b	X		3
c	X	X	3
d		X	4

Since row **c** now dominates both **b** and **d** (at equal or less cost), the table reduces to

	K	L	Cost
c	X	X	3

Thus, the first covering that the branching method identifies is

$$F = \mathbf{a} + \mathbf{c}$$

for a cost of 5 c.u.

If prime implicant **d** is now selected as a cover for m_I, the prime-implicant table reduces to

	J	K	Cost
a	X		2
b	X	X	3
c		X	3

Since **b** now dominates **c** with equal cost, **c** may be removed. Note that row *a* cannot be removed, even though it is dominated by **b**, since row **a** has a lower cost. The prime-implicant table is now reduced to

	J	K	Cost
a	X		2
b	X	Ⓧ	3

Implicant **b** is now essential, making the function become

$$F = \mathbf{b} + \mathbf{d}$$

for a cost of 7 c.u. The first set of implicants is cheaper.

In some cases the branching process will have to be applied several times in succession in order to obtain the various coverings for a complex prime-implicant table.

Alternative Solution: To use Petrick's method, the implicants that make up each column in the cyclic table are combined in OR/AND form. The first column in the original prime-implicant table is $\mathbf{a} + \mathbf{d}$. The second column is $\mathbf{a} + \mathbf{b}$. The complete expression is

$$F = (\mathbf{a} + \mathbf{d})(\mathbf{a} + \mathbf{b})(\mathbf{b} + \mathbf{c})(\mathbf{c} + \mathbf{d})$$

This expression, when expanded and simplified, will give all the possible coverings for F. The Boolean algebra for reducing the expression is

$$\begin{aligned} F &= (\mathbf{a} + \mathbf{d})(\mathbf{a} + \mathbf{b})(\mathbf{b} + \mathbf{c})(\mathbf{c} + \mathbf{d}) \\ &= (\mathbf{a} + \mathbf{bd})(\mathbf{c} + \mathbf{bd}) \qquad \text{from Eq. (2-18)} \\ &= \mathbf{ac} + \mathbf{bd} \qquad\qquad\; \text{from Eq. (2-18) again} \end{aligned}$$

Thus, F can be covered by $\mathbf{a}$ and $\mathbf{c}$ (cost of 5 c.u.) or $\mathbf{b}$ and $\mathbf{d}$ (cost of 7 c.u.), as shown by the branching method. These techniques, plus a lot of blood, sweat, tears, and toil, can be used to minimize any Boolean function.

4.6.3. POS Forms and Tabular Minimization. The tabular minimization method results in simplified SOP forms. It is simple to obtain reduced POS forms, however, by following the approach given in Sec. 4.3.5 for use with K-maps. The complement of the function is expressed in sum-of-minterms form, minimized, and complemented via De Morgan's theorem, thereby producing the reduced POS expression.

4.6.4. Redundancies and Tabular Reduction. It is a simple matter to include don't-care and can't-happen cases in the tabular reduction process, although their inclusion increases the amount of work that must be performed during the initial tabulation process. The redundancies are included in column 1 of the tabulation, just as if they were normal minterms. Thus, the tabular minimization process begins with both the usual minterms *and* the redundancies.

The redundancies are *not* included in the prime-implicant selection table, however. As a result, they have no effect upon the selection of implicants. Their only use is in *forming* the prime implicants, wherein they allow larger groupings and correspondingly lower costs of implementation.

EXAMPLE 4-47: A combinational device is to detect whenever a digit encoded in 8421 BCD code is evenly divisible by 3. Solve the problem in SOP and POS form. Use both K-mapping and tabular minimization.

Solution: The output function is TRUE for m_0, m_3, m_6, and m_9 and has the usual 8421 BCD redundancies:

$$F(A,B,C,D) = \sum 0, 3, 6, 9, X_{10}, X_{11}, X_{12}, X_{13}, X_{14}, X_{15}$$

Using K-map minimization gives (see Fig. E4-47a)

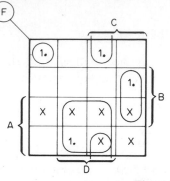

$$F = \overline{A}\,\overline{B}\,\overline{C}\,\overline{D} + \overline{B}CD + BC\overline{D} + AD$$

Cost = 16 c.u.

Figure E4-47a

and (see Fig. E4-47b)

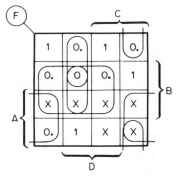

$$\overline{F} = \overline{A}\,\overline{C}D + \overline{B}C\overline{D} + B\overline{C} + BD + A\overline{D}$$
$$F = (A + C + \overline{D})(B + \overline{C} + D)\cdot$$
$$(\overline{B} + C)(\overline{B} + \overline{D})(\overline{A} + D)$$

Cost = 17 c.u.

Figure E4-47b

The simplified SOP form is 1 c.u. cheaper.

Alternative Solution: Tabular minimization in SOP form begins by listing both the TRUE minterms and the redundancies in equal-index groups in column 1. The usual combination process then follows:

Column 1	Column 2	Column 3
0	3(8)	9(2, 4)
—	6(8)	10(1, 4)
3 √	9(2) √	12(1, 2)
6 √	9(4) √	
9 √	10(1) √	
10 √	10(4) √	
12 √	12(1) √	
11 √	12(2) √	
13 √	11(4) √	
14 √	13(2) √	
15 √	14(1) √	

The implicants, their costs, and the minterms that each covers are

Implicant	Decimal Form	Boolean Form	Cost	Minterms Covered
a	9(2, 4)	AD	3	9, 11, 13, 15
b	10(1, 4)	AC	3	10, 11, 14, 15
c	12(1, 2)	AB	3	12, 13, 14, 15
d	3(8)	$\bar{B}CD$	4	3, 11
e	6(8)	$BC\bar{D}$	4	6, 14
f	0	$\bar{A}\bar{B}\bar{C}\bar{D}$	5	0

The prime-implicant table is

	0	3	6	9	Cost
a				X	3
b					3
c					3
d		X			4
e			X		4
f	X				5

Only the TRUE minterms are used in forming the prime-implicant table. The redundant minterms are omitted. Implicants **b** and **c** do not cover any minterms listed in the table. Rather, they are groups that cover only redundancies. The table shows that implicants **a, d, e,** and **f** are essential, giving

$$F = AD + \bar{B}CD + BC\bar{D} + \bar{A}\bar{B}\bar{C}\bar{D}$$

The POS solution begins by minimizing the sum-of-minterms expression for $\bar{F}$:

$$\bar{F} = \sum 1, 2, 4, 5, 7, 8, X_{10}, X_{11}, X_{12}, X_{13}, X_{14}, X_{15}$$

Column 1	Column 2	Column 3
1✓	1(4)	4(1,8)
2✓	2(8)	8(2,4)
4✓	4(1)✓	5(2,8)
8✓	4(8)✓	10(1,4)
5✓	8(2)✓	12(1,2)
10✓	8(4)✓	
12✓	5(2)✓	
7✓	5(8)✓	
11✓	10(1)✓	
13✓	10(4)✓	
14✓	12(1)✓	
15✓	12(2)✓	
	7(8)✓	
	11(4)✓	
	13(2)✓	
	14(1)✓	

The prime implicants that result are described in the following table:

Implicant	Decimal Form	Boolean Form	Cost	Minterms Covered
a	4(1,8)	$B\bar{C}$	3	4, 5, 12, 13
b	8(2,4)	$A\bar{D}$	3	8, 10, 12, 14
c	5(2,8)	BD	3	5, 7, 13, 15
d	10(1,4)	AC	3	10, 11, 14, 15
e	12(1,2)	AB	3	12, 13, 14, 15
f	1(4)	$\bar{A}\bar{C}D$	4	1, 5
g	2(8)	$\bar{B}C\bar{D}$	4	2, 10

The prime-implicant table is

	1	2	4	5	7	8	Cost
a			X	X			3
b						X	3
c				X	X		3
d							3
e							3
f	X			X			4
g		X					4

Implicants *a*, *b*, *c*, *f*, and *g* are *all* essential. They cover all the minterms. Thus

$$\bar{F} = B\bar{C} + A\bar{D} + BD + \bar{A}\bar{C}D + \bar{B}C\bar{D}$$

which is the same as the K-map result.

4.6.5. Computer-Aided Minimization.   As the number of variables involved in a combinational problem increases, the work involved in the minimization process increases as well. In fact, each new variable doubles the magnitude of the problem. Thus, for problems involving six or more variables, the minimization process can be very tedious and time-consuming.

The use of the computer as a minimization tool is well advised. Appendix A presents a FORTRAN computer program* that can be used for minimization. The program examines the TRUE, FALSE, and redundant minterms that describe a single-output combinational function. It uses a programmed algorithm developed by Carroll† to find the prime implicants that describe the function and to select and indicate all essential implicants.

When cyclic conditions arise, an approximate selection technique developed by Bowman and McVey† is used to select a suitable set of prime impli-

---

*The program requires a logical AND function, called IAND, that is not available in standard FORTRAN. It can be provided by an assembly language subroutine, however.
†These bibliographic references are given at the beginning of the listing of the program.

cants from among the cyclic set. The approximate method does not *guarantee* minimal cost, but it offers a reasonable compromise between computational expense and reduction in gating cost.

The program, named MINTE, is given in subroutine form. The input data are presented to the subroutine as an array that has 2^N elements, where N is the number of variables. The version of MINTE that is given in the Appendix is dimensioned for up to ten variables; it can be expanded, however, if sufficient computer storage is available.

Prior to calling MINTE into execution, the data array must be filled with integer values that describe the TRUE, FALSE, or redundant values of each of the 2^N minterms that describe the function that is to be minimized. The "code" for this is

$$0 = \text{FALSE minterm}$$
$$1 = \text{TRUE minterm}$$
$$-1 = \text{redundant minterm}$$

One approach to this encoding process is to truth-tabulate the behavior of the function and replace all Xs with -1. The array members are numbered from 1 to 2^N, rather than from 0 to 2^{N-1} as the minterm indices are numbered. Thus, the value of m_0 is stored at the first array location, say $C(1)$, and so on.

The subroutine processes the data array and prints out the results of the minimization. This output begins with a listing of the input data. The complete set of prime implicants is listed next, along with their cost (first-level cost only) and a notation indicating which implicants are essential.

If a cyclic condition occurs, the approximate selection method is used to choose a set of prime implicants to cover the TRUE minterms that are not covered by the essential implicants. These are noted as CHOSEN in the prime implicant listing. The minimized function is represented by the ORing of those implicants marked ESSENTIAL or CHOSEN.

Also, if a cyclic condition does arise, a table showing the cyclic conditions is listed. This CONSTRAINT TABLE lists the TRUE minterms that remain after the essential implicants have been removed from the prime-implicant tabulation, along with the prime implicants that cover each of these minterms. The cyclic prime-implicant table can be constructed from this information. Then, manual selection techniques (branching and the like) can be used to check the results of the approximate selection technique if desired.

The program minimizes in SOP form. POS minimization can be obtained by minimizing the complement of the function. This amounts to replacing the 0s in the data array by 1s, the 1s by 0s, and repeating the minimization. De Morgan's theorem is then used to convert to POS form, as shown in Sec. 4.3.5.

EXAMPLE 4-48: The listing of MINTE given in Appendix A includes the results obtained from minimization of

$$F(A,B,C,D,E) = \sum 0, 3, 4, 6, 7, 15, 21, 23, 26, 28,$$
$$X_2, X_8, X_{12}, X_{14}, X_{17}, X_{24}, X_{31}$$

The minimized expression for F that is found by the program is

$$F = \bar{A}\bar{D}\bar{E} + \bar{A}\bar{B}D + CDE + B\bar{D}\bar{E} + A\bar{B}\bar{D}E + ABC\bar{E}$$

As a comparison, consider a K-mapped minimization of this problem. The K-map, with essential and chosen prime implicants marked, is shown in Fig. E4-48.

The CONSTRAINT TABLE (shown in Appendix A) can be used to implement Petrick's method of prime-implicant selection. The essential implicants cover all the TRUE minterms but m_0, m_4, m_{15}, m_{21}, and m_{23}. The table shows that m_0 can be covered by prime implicants number 1 or number 2 and so on. Thus, Petrick's formulation for obtaining all possible coverings of the remaining minterms is

$$\text{Coverings} = (1 + 2)(1 + 2 + 4)(5 + 6)(8 + 9)(6 + 9)$$

where the numbers are the prime-implicant numbers given in the tabulation of the implicants. Expansion of this expression gives the following three-implicant coverings:

$$\text{Coverings} = 1 \cdot 5 \cdot 9 + 1 \cdot 6 \cdot 8 + 2 \cdot 5 \cdot 9 + 1 \cdot 6 \cdot 9 + 2 \cdot 6 \cdot 8 + 2 \cdot 6 \cdot 9$$

All these coverings cost 10 c.u. The MINTE program selected $1 \cdot 6 \cdot 8$.

For reference, the cyclic prime-implicant table is

	0	4	15	21	23
1	X	X			
2	X	X			
4		X			
5			X		
6			X		X
8				X	
9				X	X

Note that the m_4 column can be removed since it is dominated by the m_0 column. The MINTE program does not actually consider dominance in making its selection.

EXAMPLE 4-49: A combinational comparator receives two decimal digits that are encoded in excess-three code. It is to indicate whenever the input numbers are equal. This is an eight-variable problem. The input variables can be named A, B, C, and D (first number) and E, F, G, and H (second

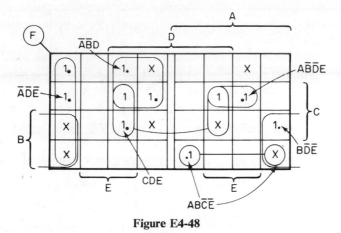

Figure E4-48

number). There are 10 TRUE minterms, 90 FALSE minterms, and 156 redundancies.

Solution: The results of SOP minimization are

$$COMP = \bar{B}\bar{C}\bar{D}\bar{F}\bar{G}\bar{H} + \bar{A}C\bar{D}\bar{E}G\bar{H} + \bar{A}\bar{B}E\bar{F}$$
$$+ \bar{B}C\bar{D}\bar{F}G\bar{H} + \bar{A}C\bar{D}E\bar{G}H$$
$$+ BCDFGH + ACDEGH + ABEF$$

for a cost of 66 c.u. The POS results are

$$COMP = (\bar{D} + H)(\bar{C} + G)(\bar{B} + F)(\bar{A} + E)(D + \bar{E} + \bar{H}) \cdot$$
$$(\bar{B} + D + \bar{H})(C + \bar{F} + \bar{G})(C + \bar{E} + \bar{G})(B + \bar{F} + \bar{G}) \cdot$$
$$(B + \bar{E} + \bar{F})(A + \bar{E})$$

for a cost of 39 c.u. Computer solution of both these problems required less than 20 sec (IBM 360/65, WATFIV Compiler).

4.7   Multiple-Function Minimization

Many combinational design problems have two or more output signals that are to be generated from the same set of input variables. All the examples given in Sec. 4.5 are of this type. The simplest approach to the multiple-output design problem is to consider each output function separately. Each one of the output functions is minimized and implemented without regard to the other output functions.

It is often possible to reduce the cost of the overall combinational system by sharing logical capability between output functions, however. This can be accomplished informally by comparing the various output functions to identify common implicants or formally by using multiple-function minimization techniques.

4.7.1. Informal Techniques.

Consider the combinational device described in Fig. 4-47. This simple combinational system has two output signals, F_1 and F_2. The K-map minimizations for F_1 and F_2, each considered separately, are shown in Fig. 4-48. The K-mapping process given in Fig. 4-48 shows that the simplified SOP forms for F_1 and F_2 are

$$F_1 = \bar{A}\bar{B} + \bar{B}C, \qquad F_2 = A\bar{B}C + AB\bar{C} \qquad (4\text{-}33)$$

giving a total cost of 14 c.u. for separated implementation of F_1 and F_2.

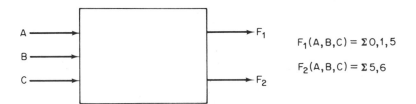

$$F_1(A,B,C) = \Sigma\, 0,1,5$$
$$F_2(A,B,C) = \Sigma\, 5,6$$

Figure 4-47. A Simple Multiple-output Combinational System

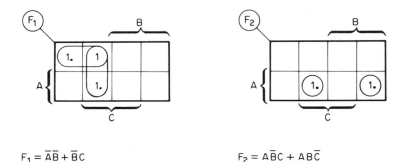

$$F_1 = \bar{A}\bar{B} + \bar{B}C \qquad\qquad F_2 = A\bar{B}C + AB\bar{C}$$

Figure 4-48. Separated K-map Minimization of F_1 and F_2

It can be recognized, however, that the $\bar{B}C$ implicant in the expression for F_1 can be replaced by $A\bar{B}C$ with no change in logical behavior. Considering F_1 alone, this would be a poor choice, since it increases the cost of implementing F_1 by 1 c.u. The $A\bar{B}C$ implicant is essential for implementing F_2, however, and can be used in implementing F_1 with no additional first-level cost. The complete logic diagram for the joint implementation of F_1 and F_2 is shown in Fig. 4-49. The joint implementation scheme has a cost of 12 c.u., 2 less than the cost of the separated implementation method.

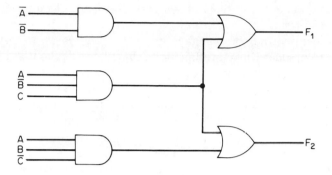

Figure 4-49. Joint Implementation of F_1 and F_2

Once an essential implicant such as $A\bar{B}C$ in the expression for F_2 has been generated by a gate, that implicant can be used in generating other output functions. Such terms are called *cost-free terms* and can be marked by an underline:

$$F_1 = \bar{A}\bar{B} + A\bar{B}C \qquad \text{(7 c.u.)}$$
$$F_2 = AB\bar{C} \mid \underline{A\bar{B}C} \qquad \text{(5 c.u.)} \tag{4-34}$$

The underline shows that the secondary use of the $A\bar{B}C$ implicant is free of first-level cost.

EXAMPLE 4-50:   Given

$$F_1 = AB + A\bar{B}C\bar{D} + \bar{A}D$$
$$F_2 = \bar{A}\bar{B}C + \bar{A}D + CD$$
$$F_3 = A\bar{B}C\bar{D} + AB + CD$$

Mark cost-free terms and give the cost of implementing each function.

Solution:

$$F_1 = AB + A\bar{B}C\bar{D} + \bar{A}D \qquad \text{(11 c.u.)}$$
$$F_2 = \bar{A}\bar{B}C + \underline{\bar{A}D} + CD \qquad \text{(8 c.u.)}$$
$$F_3 = \underline{A\bar{B}C\bar{D}} + \underline{AB} + \underline{CD} \qquad \text{(3 c.u.)}$$

The repeated use of cost-free terms is allowed until the loading limit of the gate that generates the implicant is reached. Thereafter the implicant must be generated again or a buffer used for power amplification.

The use of the Wired-OR (or Wired-AND) at the second level of gating prevents the sharing of cost-free terms. The wired-logic functions are implemented by joining the outputs of the first-level gates. As a result, none of the first-level terms are available individually.

The identification of shared terms is not always easy. Section 4.7.2 discusses techniques whereby all possible shared cost reductions are considered, but such techniques require a considerable amount of work, especially if there are more than four output functions. The following suggestions are offered as a compromise between separated minimization and the complete investigation techniques. They apply only to K-map minimization.

1. Once the various output functions have been K-mapped, look for the smallest prime implicant that appears on any of the maps.

2. Circle that implicant on all maps, even though it may not be a *prime* implicant on all the maps. This indicates that the implicant can be used as a cost-free term in covering 1s on other maps.

3. Continue the grouping process, working from the smallest essential implicants to the largest ones.

4. Where choices of implicants are available on one map, consider the usefulness that each choice may have in covering minterms on other maps.

5. Select the best set of implicants for covering each function.

EXAMPLE 4-51: Design combinational logic for implementing

$$F_1(A,B,C,D) = \sum 0, 4, 5, 6, 7, 8, 10, 12, 14$$
$$F_2(A,B,C,D) = \sum 5, 8, 12$$
$$F_3(A,B,C,D) = \sum 0, 4, 10, 14$$

Solution: The K-maps for F_1, F_2, and F_3 are shown in Fig. E4-51a–c. The smallest essential prime implicant is m_5 on the map for F_2. This implicant is marked on both the F_2 map *and* the map for F_1. The next smallest implicants are $A\bar{C}\bar{D}$ on F_2, $\bar{A}\bar{C}\bar{D}$ on F_3, and $AC\bar{D}$ on F_3. These are marked on *all* maps as shown in Figs. E4-51d–f. F_1 is completely covered except for m_6 and m_7. These two minterms are best covered by the $\bar{A}B$ implicant, eliminating the need for the m_5 term in the logic for F_1. The complete expressions for all three functions are

$$F_1 = \bar{A}\bar{C}\bar{D} + A\bar{C}\bar{D} + AC\bar{D} + \bar{A}B \quad \text{(15 c.u.)}$$
$$F_2 = \bar{A}B\bar{C}D + \underline{A\bar{C}\bar{D}} \quad \text{(6 c.u.)}$$
$$F_3 = \underline{\bar{A}\bar{C}\bar{D}} + \underline{AC\bar{D}} \quad \text{(2 c.u.)}$$

for a total cost of 23 c.u. Note that the use of the $\bar{C}\bar{D}$ implicant in F_1 (rather than $\bar{A}\bar{C}\bar{D}$ and $A\bar{C}\bar{D}$) will *increase* the overall cost by 1 c.u., since $\bar{A}\bar{C}\bar{D}$ and $A\bar{C}\bar{D}$ can be used as cost-free terms when implementing F_3.

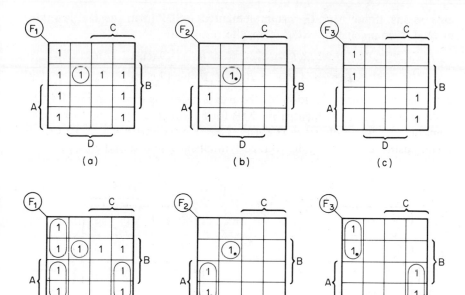

Figure E4-51a, b, c, d, e, f

Separated minimization would give

$$F_1 = \bar{C}\bar{D} + \bar{A}B + A\bar{D} \qquad \text{(9 c.u.)}$$
$$F_2 = \bar{A}B\bar{C}D + A\bar{C}\bar{D} \qquad \text{(9 c.u.)}$$
$$F_3 = \bar{A}\bar{C}\bar{D} + AC\bar{D} \qquad \text{(8 c.u.)}$$

for a total cost of 26 c.u.

The shared use of cost-free terms applies to POS forms as well as SOP forms. In fact, in some cases it is possible to obtain a cost reduction by implementing some output functions in SOP form, others in POS form, and by sharing terms between them.

EXAMPLE 4-52:   Consider the joint minimization of the G_{i+1} and L_{i+1} outputs from the binary comparison circuit designed in Sec. 4.5.2.

Solution: The K-maps for G_{i+1} and L_{i+1} are shown in Fig. E4-52a and b. Separated minimization of G_{i+1} and L_{i+1} gives a total cost of 18 c.u. in both SOP and POS forms. There are no common implicants in either pair of expressions. If, however, G_{i+1} is implemented in SOP form and L_{i+1} is implemented in complemented SOP form, the results are

$$G_{i+1} = A_i\overline{B_i} + G_iA_i + G_i\overline{B_i} \qquad \text{(9 c.u.)}$$
$$\overline{L_{i+1}} = \underline{A_i\overline{B_i}} + \overline{L_i}A_i + \overline{L_i}\,\overline{B_i} \qquad \text{(7 c.u.)}$$

The joint TTL implementation for these expressions is shown in Fig. E4-52c. Note that $\overline{L_{i+1}}$ is implemented in SOP form, which is equivalent to implementing L_{i+1} in simplified POS form. The conversion from $\overline{L_{i+1}}$ to L_{i+1} is accomplished by changing the voltage symbolism assigned to the $\overline{L_{i+1}}$ output.

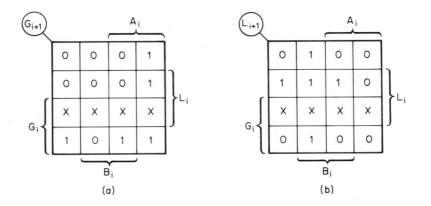

(a)   (b)

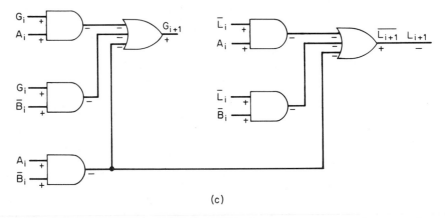

(c)

Figure E4-52a, b, c

If G_{i+1} and L_{i+1} must have the same symbolism, an inverter is required. Even so, the shared term permits a cost reduction from 18 to 17 c.u.

4.7.2. Formal Multiple-Function Minimization.   The informal techniques for simplifying multiple-output combinational systems are useful

with small problems (four or five input variables and four or five output functions). They can lead to cost reductions, but they do not in any way guarantee minimal cost, even in very simple problems. To illustrate this, consider the following two-output problem (a classic example):

$$F_1(A,B,C) = \sum 1, 3, 7; \qquad F_2(A,B,C) = \sum 2, 6, 7 \qquad (4\text{-}35)$$

The K-maps for these two functions are given in Fig. 4-50. All the prime implicants are groups of two and none of them overlap, so that the suggestions of Sec. 4.7.1 are of no use. Separated reduction gives

$$F_1 = \bar{A}C + BC \quad (6 \text{ c.u.}); \qquad F_2 = AB + B\bar{C} \quad (6 \text{ c.u.}) \qquad (4\text{-}36)$$

for a total cost of 12 c.u.

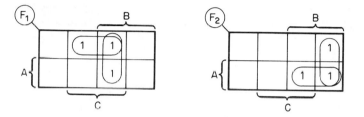

Figure 4-50. The K-maps for F_1 and F_2

The two functions share m_7, however, and a reduction in system cost can be obtained by implementing the pair of outputs as:

$$F_1 = \bar{A}C + ABC \quad (7 \text{ c.u.}); \qquad F_2 = B\bar{C} + \underline{ABC} \quad (4 \text{ c.u.}) \qquad (4\text{-}37)$$

for a total of 11 c.u. The key to the simplification given in Eq. (4-37) is the use of m_7 as a cost-free term, even though m_7 is not a prime implicant for either F_1 or F_2.

To recognize the presence of implicants that, even though they are not prime implicants for any particular output function, are of use in reducing overall cost, the *intersections* (ANDing together) of all the output functions must be considered along with the functions themselves. These intersections can be studied in either K-map or tabular form. All possible intersections should be considered, making the multiple-function minimization process very involved, especially when more than four outputs are present.

The formal multiple-function minimization problem resolves itself into two parts: the *identification* of useful implicants and the *selection* of the sets of implicants that are used to cover each output function. The first task can be accomplished by using either K-maps or tabular techniques. The second task is best accomplished by using a prime-implicant table.

The K-map Method.

If K-maps are used, the maps for each separate output function and all possible intersections between output functions must be drawn.

EXAMPLE 4-53: Plot the K-maps for the following functions *and* their various intersections.

$$F_1(A,B,C) = \sum 0, 1, 3, 7$$
$$F_2(A,B,C) = \sum 3, 6, 7$$
$$F_3(A,B,C) = \sum 0, 4, 7$$

Solution: See Fig. E4-53a–c.

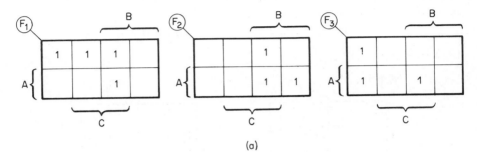

(a)

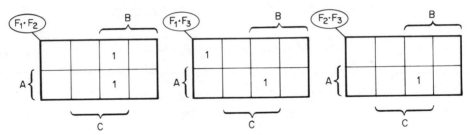

(b)

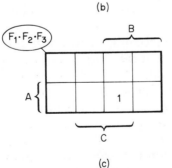

(c)

Figure E4-53a, b, c

When redundancies are present, they will normally appear at the same locations on all the K-maps, since they are associated with particular combinations of the input variables. The Xs that mark the redundancies should be shown on all the maps to which they apply, including the intersection maps.

Once the maps have been constructed, the usual K-map grouping process is used to identify the various prime implicants that can be used to form the output functions. The final intersection map (the one involving all the output functions) should be considered first. The 1s on that map should be grouped as well as possible.

The intersection maps at the next level are then considered, with their various prime implicants being found. Prime implicants that are identified on the first map need not be shown on any other maps. Only new implicants formed by new groupings need to be marked. This process continues until all the prime implicants on all the maps have been found.

The implicants should then be listed in a table that gives their Boolean form, the minterms that they cover, the map from which they come, and their cost (including second-level inputs).

EXAMPLE 4-54:   Find the multiple-function prime implicants for the three functions given in Ex. 4-53.

Solution: The K-maps are repeated in Fig. E4-54a–c. The prime implicants are marked on each map. Note that m_7 is not used as an implicant on

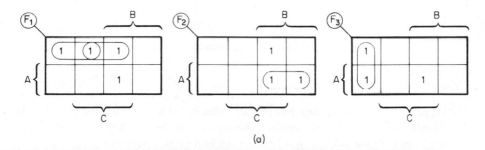

(a)

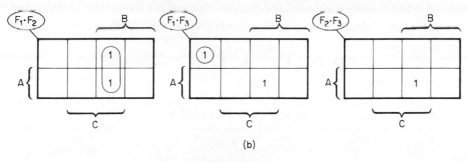

(b)

Figure E4-54a, b

the $F_2 \cdot F_3$, the $F_1 \cdot F_3$, or the F_3 maps, since it already appears as a prime implicant on the $F_1 \cdot F_2 \cdot F_3$ map. Similarly, since the BC implicant appears on the $F_1 \cdot F_2$ map, it is not shown as a prime implicant on either the F_1 or the F_2 map.

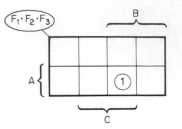

Figure E4-54c

The tabulation of the prime implicants is

Implicant	Minterms Covered	Map	Cost*
ABC	7	$F_1 \cdot F_2 \cdot F_3$	4
BC	3, 7	$F_1 \cdot F_2$	3
$\bar{A}\bar{B}\bar{C}$	0	$F_1 \cdot F_3$	4
$\bar{A}\bar{B}$	0, 1	F_1	3
$\bar{A}C$	1, 3	F_1	3
AB	6, 7	F_2	3
$\bar{B}\bar{C}$	0, 4	F_3	3

*Includes both first- and second-level costs.

Once all the implicants have been found, a composite prime-implicant table is constructed in order to identify the cheapest covering for the various output functions. An example of the composite prime-implicant tabulation is given in Table 4-4, using the functions and implicants from Ex. 4-52.

The minterms that make up each function are listed across the top of the table. Minterms that appear in more than one function (m_7 for example)

Table 4-4: A Multiple-Function Prime-Implicant Table

Implicant	Functions Covered	F_1 0	1	3	7	Cost	F_2 3	6	7	Cost	F_3 0	4	7	Cost
1. $\bar{A}\bar{B}$	1	X	X			3				—				
2. $\bar{B}\bar{C}$	3					—				—	X	X		3
3. $\bar{A}C$	1		X	X		3				—				—
4. BC	1, 2			X	X	3	X		X	3				—
5. AB	2					—		X	X	3				—
6. $\bar{A}\bar{B}\bar{C}$	1, 3	X				4				—	X			4
7. ABC	1, 2, 3				X	4			X	4			X	4

are listed with each function. A separate cost column is included with each function.

The implicants are listed down the left-hand side of the chart, along with a column giving the functions that are involved in the generation of each implicant. The Xs in the table mark the minterms that are covered by each implicant. Only those functions that are involved in the generation of a given implicant are considered in locating the Xs. For example, implicant $\bar{A}C$ covers minterms m_1 and m_3. No X is placed under m_3 in the F_2 column, however, since the $\bar{A}C$ implicant applies to F_1 alone. The cost of each implicant is entered into the cost columns of all functions to which that implicant applies.

A multiple-function prime-implicant table can be simplified by using row and column dominance as long as the following restrictions are met:

1. Column dominance can be used only for columns within the same function.

2. Entire rows (across all functions) must be considered in identifying row dominance.

There is no row dominance in the Table 4-4. Column dominance applies to columns 3 and 7 in F_2 and to columns 0 and 4 in F_3. The removal of the dominating columns (m_7 in F_2 and m_0 in F_3) gives Table 4-5. With the removal

Table 4-5:   Prime-Implicant Table 4-4 with Dominating Columns Removed

Implicant	Functions Covered	F_1					F_2			F_3		
		0	1	3	7	Cost	3	6	Cost	4	7	Cost
1. $\bar{A}\bar{B}$	1	X	X			3			—			—
2. $\bar{B}\bar{C}$	3					—				X		3
3. $\bar{A}C$	1		X	X		3			—			—
4. BC	1, 2			X	X	3	X		3			—
5. AB	2					—		X	3			—
~~6. $\bar{A}\bar{B}\bar{C}$~~	~~1, 3~~	~~X~~				~~4~~						~~4~~
7. ABC	1, 2, 3				X	4			—		X	4

of m_0 from F_3, row 1 dominates row 6. This allows row 6 to be removed; it is crossed out in Table 4-5.

Once row and column dominance have been applied to exhaustion, the essential prime implicants should be identified. There are five essential implicants in Table 4-5: m_0 (F_1), m_3 and m_6 (F_2), and m_4 and m_7 (F_3). When an essential implicant is identified, it and all the columns that it covers are removed from the table. If it involves more than one function, only the columns that are covered in the function to which it is essential are removed.

To represent the effect that an essential implicant has on the other functions, the cost of that implicant is reduced to 1 c.u. in the cost columns of all other functions to which that implicant applies. This indicates the cost-free nature of the essential implicant when it is used in covering minterms in other functions. The single cost unit that is left is the cost of the second-level gate input. As before, cost-free sharing of an implicant is allowed only up to the fan-out limitation of the gate that forms the implicant.

The selection of the essential prime implicants that are present in Table 4-5 completely covers both F_2 and F_3. Their expressions are

$$F_2 = BC + AB \quad (6\text{ c.u.}); \qquad F_3 = \bar{B}\bar{C} + ABC \quad (7\text{ c.u.}) \qquad (4\text{-}38)$$

When the essential implicants are removed from Table 4-5, the prime-implicant table can be reduced to Table 4-6. The single-unit costs of the BC and

Table 4-6.   Table 4-4 Reduced to Its Final Form

	F_1		
Implicant	3	7	*Cost*
3. $\bar{A}C$	X		3
4. BC	X	X	1
7. ABC		X	1

ABC implicants indicate that they are essential to either F_2 or F_3 and, as a result, can be used as cost-free implicants in generating F_1. Rows 3 and 7 are dominated by row 4. Their removal makes row 4 essential; it covers both m_3 and m_7. The final expressions for the three functions are

$$\begin{aligned}
F_1 &= \bar{A}\bar{B} + BC && (6\text{ c.u.}) \\
F_2 &= AB + \underline{BC} && (4\text{ c.u.}) \\
F_3 &= \bar{B}\bar{C} + \overline{ABC} && (7\text{ c.u.})
\end{aligned} \qquad (4\text{-}39)$$

The Tabular Method.

The tabular method for identifying the multiple-function implicants is closely related to the normal tabular reduction process. The minterms that make up the various output functions are sorted by index (as usual) to form column 1 of the tabulation. Each minterm is given a subscript that identifies the functions that contain that minterm. Redundancies are entered into column 1 as usual. The redundant minterms will normally have *all possible* subscripts.

EXAMPLE 4-55:   Give the column 1 entries for the three functions listed in Ex. 4-53.

Solution: The functions, in sum-of-minterms form, are repeated below:

$$F_1(A,B,C) = \sum 0, 1, 3, 7$$
$$F_2(A,B,C) = \sum 3, 6, 7$$
$$F_3(A,B,C) = \sum 0, 4, 7$$

The starting column for the tabular reduction is

Column 1

$0_{1,3}$
$\overline{1_1}$
4_3
$\overline{3\ _{1,2}}$
6_2
$\overline{7_{1,2,3}}$

The tabular minimization process proceeds as usual, with the following exceptions:

1. Two terms can be combined only if their subscripts contain at least one common term.

2. The next-column entry that results from a combination of terms is given a subscript that equals the intersection of the subscripts of the terms that were combined.

3. A term is checked off only if it combined to form a next-column entry whose composite subscript is the same as the subscript of the term.

EXAMPLE 4-56: Show the tabular minimization of

$$F_1 = \sum 0, 1$$
$$F_2 = m_2$$

Solution:

Column 1   *Column 2*

$0_1\checkmark$ $0_1(1)$
$1_1\checkmark$
2_2

Note that 0_1 and 2_2 cannot be combined since their subscripts differ.

EXAMPLE 4-57: Show the tabular minimization of

$$F_1 = \sum 0, 1$$
$$F_2 = m_1$$

Solution:

Column 1	Column 2
$0_1 \checkmark$	$0_1(1)$
$\overline{1_{1,2}}$	

The $1_{1,2}$ term is not checked off since the column 2 entry does not have both 1 and 2 as subscripts.

EXAMPLE 4-58: Show the tabular minimization for

$$F_1 = \sum 0, 1, 2$$
$$F_2 = \sum 0, 1, 2, 4$$
$$F_3 = m_1$$

Solution:

Column 1	Column 2
$0_{1,2} \checkmark$	$0_{1,2}(1)$
$\overline{1_{1,2,3}}$	$0_{1,2}(2)$
$2_{1,2} \checkmark$	$0_2(4)$
$4_2 \checkmark$	

The unchecked terms and the terms in the last column are the implicants that can be used for multiple-output implementation. The subscripts tell which functions are involved in the generation of the particular implicant. A tabulation of the properties of each implicant should be made, as was shown in the K-map procedure.

EXAMPLE 4-59: Find the multiple-output implicants for Ex. 4-53. Use tabular techniques.

Solution: The tabulation is given below:

Column 1	Column 2
$0_{1,3}$	$0_1(1)$
$\overline{1_1 \checkmark}$	$\overline{0_3(4)}$
$4_3 \checkmark$	$\overline{1_1(2)}$
$3_{1,2} \checkmark$	$\overline{3_{1,2}(4)}$
$6_2 \checkmark$	$6_2(1)$
$\overline{7_{1,2,3}}$	

The implicants and their properties are listed below:

Implicant	Boolean Form	Minterms Covered	Functions Covered	Cost
$0_1(1)$	$\bar{A}\bar{B}$	0, 1	1	3
$0_3(4)$	$\bar{B}\bar{C}$	0, 4	3	3
$1_1(2)$	$\bar{A}C$	1, 3	1	3
$3_{1,2}(4)$	BC	3, 7	1, 2	3
$6_2(1)$	AB	6, 7	2	3
$0_{1,3}$	$\bar{A}\bar{B}\bar{C}$	0	1, 3	4
$7_{1,2,3}$	ABC	7	1, 2, 3	4

These agree with the implicants that were found by using the K-map process.

Once the multiple-function prime implicants have been identified by either the tabular method or the K-map method, the prime implicant selection process is carried out.

The formal multiple-function minimization technique is tedious and time-consuming. The work increases dramatically as the number of output functions increases. The method does provide for minimal, two-level implementation of a multiple-output system, however, and can be used whenever the expected cost saving justifies the effort.

4.7.3. An Example: Code-to-Code Conversion. As an example of the multiple-function minimization technique, consider the design of a logical device that is to convert from 8421 BCD code to 6, 3, 1, −1 BCD code. A block diagram for this device and its input/output table are given in Fig. 4-51. An examination of the input/output table shows that the sum-of-minterms expressions for the output functions are

$$F_1(A,B,C,D) = \sum 5, 6, 7, 8, 9, X_{10}, X_{11}, X_{12}, X_{13}, X_{14}, X_{15}$$
$$F_2(A,B,C,D) = \sum 2, 3, 4, 8, 9, X_{10}, X_{11}, X_{12}, X_{13}, X_{14}, X_{15}$$
$$F_3(A,B,C,D) = \sum 1, 4, 6, 7, 9, X_{10}, X_{11}, X_{12}, X_{13}, X_{14}, X_{15} \quad \text{(4-40)}$$
$$F_4(A,B,C,D) = \sum 2, 5, 6, 8, 9, X_{10}, X_{11}, X_{12}, X_{13}, X_{14}, X_{15}$$

The K-map method for identifying the implicants is illustrated first. The maps for the output functions and their various intersections are given in Fig. 4-52. The various implicants are marked. The implicants that are present on the maps given in Fig. 4-52 are listed in Table 4-7. The complete tabulation for prime-implicant selection is given in Table 4-8.

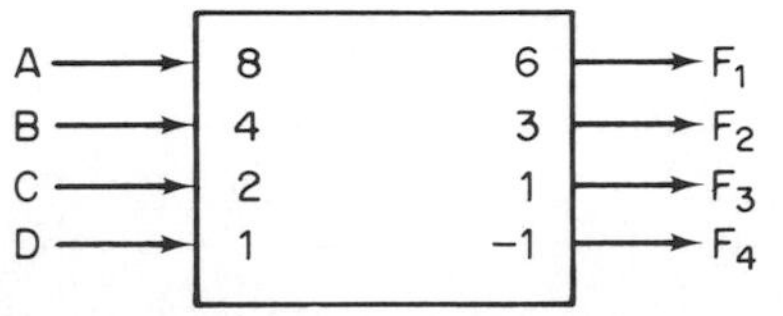

Input				Output			
A	B	C	D	F_1	F_2	F_3	F_4
8	4	2	1	6	3	1	-1
0	0	0	0	0	0	0	0
0	0	0	1	0	0	1	0
0	0	1	0	0	1	0	1
0	0	1	1	0	1	0	0
0	1	0	0	0	1	1	0
0	1	0	1	1	0	0	1
0	1	1	0	1	0	1	1
0	1	1	1	1	0	1	0
1	0	0	0	1	1	0	1
1	0	0	1	1	1	1	1

A → | 8   6 | → F_1
B → | 4   3 | → F_2
C → | 2   1 | → F_3
D → | 1   -1 | → F_4

Figure 4-51. An 8421 BCD to 6, 3, 1, -1 BCD Converter

Observation of Table 4-8 shows that column dominance removes column 9 from F_1, since it is dominated by column 8. Column 9 can also be removed from F_2 and F_3 for the same reason. Implicant number 3 is essential to F_1, F_2, and F_4, with no change in cost since it costs only 1 c.u. to begin with.

Table 4-7: **The Multiple-Function Prime Implicants for the 8421 to 6, 3, 1, -1 Code Converter**

Implicant	Function Covered	Minterms Covered	Cost
1. AD	1, 2, 3, 4	9	3
2. $BC\bar{D}$	1, 3, 4	6	4
3. A	1, 2, 4	8, 9	1
4. $\bar{B}C\bar{D}$	2, 4	2	4
5. $B\bar{C}\bar{D}$	2, 3	4	4
6. $B\bar{C}D$	1, 4	5	4
7. BC	1, 3	6, 7	3
8. BD	1	5, 7	3
9. $\bar{B}C$	2	2, 3	3
10. $B\bar{D}$	3	4, 6	3
11. $\bar{B}\bar{C}D$	3	1, 9	4
12. CD	4	2, 6	3

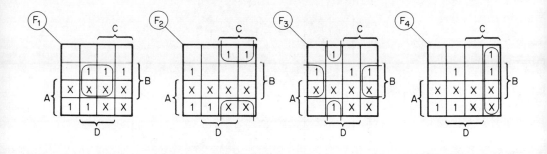

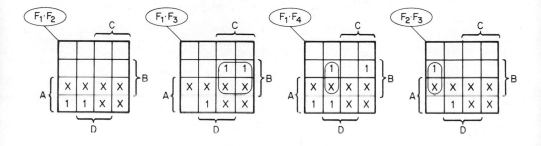

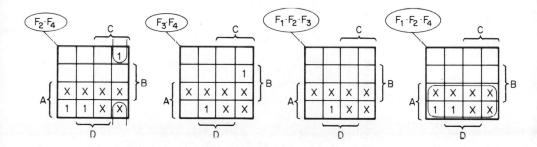

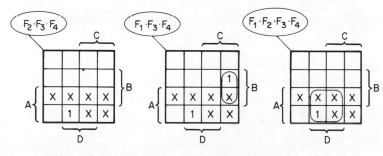

Figure 4-52. The Multiple-function K-maps for the Code-to-code Converter

Table 4-8: The Multiple-Function Prime-Implicant Table for the 8421 to 6, 3, 1, −1 Code Converter

		F1						F2						F3						F4					
Implicant	Functions	5	6	7	8	9	Cost	2	3	4	8	9	Cost	1	4	6	7	9	Cost	2	5	6	8	9	Cost
1. AD	1, 2, 3, 4					X	3					X	3					X	3					X	3
2. $B\bar{C}\bar{D}$	1, 3, 4		X				4						—			X			4			X			4
3. A	1, 2, 4				X	X	1				X	X	1						—				X	X	1
4. $\bar{B}C\bar{D}$	2, 4						—	X					4						—	X					4
5. $\bar{B}\bar{C}\bar{D}$	2, 3						—			X			4		X				4						—
6. $\bar{B}\bar{C}D$	1, 4	X					4						—						—		X				4
7. BC	1, 3		X	X			3						—			X	X		3						—
8. BD	1	X		X			3						—						—						—
9. $\bar{B}C$	2						—	X	X				3						—						—
10. $B\bar{D}$	3						—						—		X	X			3						—
11. $\bar{B}\bar{C}D$	3						—						—	X			X	X	4						—
12. $C\bar{D}$	4						—						—						—	X		X			3

Other essential implicants are

1. Number 9 in F_2.
2. Number 5 in F_2 (reducing the cost of implicant number 5 to 1 c.u. in F_3).
3. Number 11 in F_3.
4. Number 7 in F_3 (reducing the cost of implicant number 7 to 1 c.u. in F_1).
5. Number 6 in F_4 (reducing the cost of implicant number 6 to 1 c.u. in F_1).

When the minterms covered by these essential implicants are removed, the prime-implicant table is reduced to Table 4-9.

Table 4-9: The Reduced Form of Table 4-8

Implicant	F_1				F_3		F_4		
	5	6	7	Cost	4	Cost	2	6	Cost
1. AD				—		—			—
2. $BC\bar{D}$		X		4		—		X	4
3. A				—		—			—
4. $\bar{B}C\bar{D}$				—		—	X		4
5. $B\bar{C}\bar{D}$				—	X	1			—
6. $B\bar{C}D$	X			1		—			—
7. BC		X	X	1		—			—
8. BD	X		X	3		—			—
9. $\bar{B}C$				—		—			—
10. $B\bar{D}$				—	X	3			—
11. $\bar{B}\bar{C}D$				—		—			—
12. $C\bar{D}$				—		—	X	X	3

Row 4 is dominated by row 12. Removing row 4 makes implicant number 12 essential to F_4. Since implicant number 12 covers both m_2 and m_6, the coverage of F_4 is completed. Implicant number 5 is the cheapest cover for m_4 in F_3, and implicants 6 and 7 cover the remainder of F_1 at a cost of only 2 c.u.

The final expressions for the output functions are

$$
\begin{aligned}
F_1 &= 3 + 6 + 7 = A + B\bar{C}D + BC & \text{(8 c.u.)} \\
F_2 &= 3 + 5 + 9 = A + B\bar{C}\bar{D} + \bar{B}C & \text{(8 c.u.)} \\
F_3 &= 7 + 11 + 5 = BC + \bar{B}\bar{C}D + B\bar{C}\bar{D} & \text{(6 c.u.)} \\
F_4 &= 3 + 6 + 12 = A + B\bar{C}D + C\bar{D} & \text{(5 c.u.)}
\end{aligned}
\tag{4-41}
$$

The total cost is 27 c.u.

The multiple-function prime implicants can also be found by tabular minimization. The complete tabulation for the example under consideration is given in Table 4-10. The redundancies are included with the minterms in

Table 4-10: **Tabular Minimization of the Code-to-Code Converter**

Column 1	Column 2	Column 3	Column 4
$1_3\checkmark$	$1_3(8)$	$2_2(1,8)$	$8_{1,2,4}(1,2,4)$
$2_{2,4}\checkmark$	$2_2(1)\checkmark$	$2_4(4,8)$	
$4_{2,3}\checkmark$	$2_{2,4}(8)$	$4_3(2,8)$	
$\underline{8_{1,2,4}\checkmark}$	$4_{2,3}(8)$	$8_{1,2,4}(1,2)\checkmark$	
$3_2\checkmark$	$8_{1,2,4}(1)\checkmark$	$8_{1,2,4}(1,4)\checkmark$	
$5_{1,4}\checkmark$	$8_{1,2,4}(2)\checkmark$	$\underline{8_{1,2,4}(2,4)\checkmark}$	
$6_{1,3,4}\checkmark$	$\underline{8_{1,2,4}(4)\checkmark}$	$5_1(2,8)$	
$9_{1,2,3,4}\checkmark$	$3_2(8)\checkmark$	$6_{1,3}(1,8)$	
$10_{1,2,3,4}\checkmark$	$5_1(2)\checkmark$	$9_{1,2,3,4}(2,4)$	
$\underline{12_{1,2,3,4}\checkmark}$	$5_{1,4}(8)$	$10_{1,2,3,4}(1,4)$	
$\underline{7_{1,3}\checkmark}$	$6_{1,3}(1)\checkmark$	$12_{1,2,3,4}(1,2)$	
$11_{1,2,3,4}\checkmark$	$6_{1,3,4}(8)$		
$13_{1,2,3,4}\checkmark$	$9_{1,2,3,4}(2)\checkmark$		
$\underline{14_{1,2,3,4}\checkmark}$	$9_{1,2,3,4}(4)\checkmark$		
$15_{1\ 2,3,4}\checkmark$	$10_{1,2,3,4}(1)\checkmark$		
	$10_{1,2,3,4}(4)\checkmark$		
	$12_{1,2,3,4}(1)\checkmark$		
	$\underline{12_{1,2,3,4}(2)\checkmark}$		
	$\underline{7_{1,3}(8)\checkmark}$		
	$11_{1,2,3,4}(4)\checkmark$		
	$13_{1,2,3,4}(2)\checkmark$		
	$14_{1,2,3,4}(1)\checkmark$		

column 1. Since each redundancy applies to all four output functions, the redundancies are given all four subscripts.

The implicants that are found in Table 4-10 are listed in Table 4-11. If the redundant minterm numbers are removed from the *Minterms Covered* column and if the AC and AB implicants are omitted (they cover *only* redun-

Table 4-11: **The Multiple-function Prime Implicants as found by the Tabular Method**

Implicant	Boolean Form	Functions Covered	Minterms Covered	Cost
$8_{1,2,4}(1,2,4)$	A	1, 2, 4	8, 9, 10, 11, 12, 13, 14, 15	1
$2_2(1,8)$	$\bar{B}C$	2	2, 3, 10, 11	3
$2_4(4,8)$	$C\bar{D}$	4	2, 6, 10, 14	3
$4_3(2,8)$	$B\bar{D}$	3	4, 6, 12, 14	3
$5_1(2,8)$	BD	1	5, 7, 13, 15	3
$6_{1,3}(1,8)$	BC	1, 3	6, 7, 14, 15	3
$9_{1,2,3,4}(2,4)$	AD	1, 2, 3, 4	9, 11, 13, 15	3
$10_{1,2,3,4}(1,4)$	AC	1, 2, 3, 4	10, 11, 14, 15	3
$12_{1,2,3,4}(1,2)$	AB	1, 2, 3, 4	12, 13, 14, 15	3
$1_3(8)$	$\bar{B}\bar{C}D$	3	1, 9	4
$2_{2,4}(8)$	$\bar{B}C\bar{D}$	2, 4	2, 10	4
$4_{2,3}(8)$	$B\bar{C}\bar{D}$	2, 3	4, 12	4
$5_{1,4}(8)$	$B\bar{C}D$	1, 4	5, 13	4
$6_{1,3,4}(8)$	$BC\bar{D}$	1, 3, 4	6, 14	4

dancies), Table 4-11 is identical to the implicant list that was developed by using K-map methods (Table 4-7).

It should be noted that the proper application of the informal minimization suggestions given in Sec. 4.7.1 gives the same results as the formal minimization techniques for this example. The K-maps for the four output functions are repeated in Fig. 4-53. The smallest prime implicants are all groups of two. They are shown in the maps given in Fig. 4-53. The $B\bar{C}D$

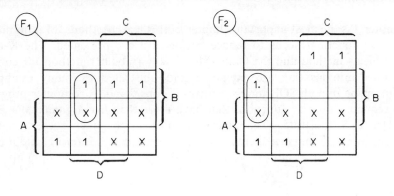

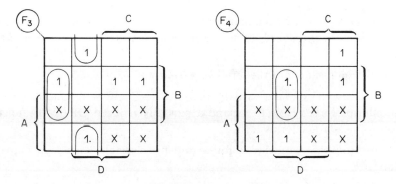

Figure 4-53. K-maps for the 8421 to 6, 3, 1, −1 Code Converter

implicant from F_4 is of use in covering m_5 on F_1. Similarly, the $\bar{B}\bar{C}\bar{D}$ implicant from F_2 can be used, cost-free, to cover m_4 on F_3. The BC implicant can be used on both F_1 and F_3. No other overlaps are possible. This gives the following results:

$$F_1 = A + BC + B\bar{C}D$$
$$F_2 = A + B\bar{C}\bar{D} + \bar{B}C$$
$$F_3 = \underline{BC} + \underline{B\bar{C}\bar{D}} + \bar{B}\bar{C}D$$
$$F_4 = A + C\bar{D} + \underline{B\bar{C}D}$$

$$(4\text{-}42)$$

which agrees with the earlier results.

The informal approach certainly does not always give the same results as the formal minimization process, but it is recommended as a good compromise between low cost and little design effort. For problems that have more inputs or more outputs, the application of any of these techniques is quite difficult. Computerized assistance is advised [1].

4.8 Summary

Chapter 4 is the most important chapter in this text. Methods for minimizing Boolean functions so as to reduce gating cost are presented. The K-map (graphical) method and the Quine-McCluskey (tabular) method for combinational minimization are explained and illustrated by many examples. Minimizing in both SOP and POS forms is considered. A computer program for reducing Boolean functions that have more than five arguments is given and its use is illustrated. The minimization of systems having several output functions by sharing logic between outputs is discussed. Informal and formal (both graphical and tabular) techniques for multiple-function minimization are discussed and illustrated by example.

4.9 Did You Learn?

1. How minterms and maxterms can be paired to form a single reduced term? How these reduced terms can be combined to obtain further simplification?

2. The meaning of the terms *implicant* and *implicate*?

3. How to expand a partially reduced AND/OR function into its sum-of-minterms form? How to expand a partially reduced OR/AND function into product-of-maxterms form?

4. How the two-, three-, four-, and five-variable Karnaugh maps are arranged?

5. How to plot on a K-map a function that is given in sum-of-minterms form? Product-of-maxterms form? How to plot minterms according to their numbered location?

6. How to plot partially reduced AND/OR and OR/AND expressions directly on a K-map?

7. How to extract the sum-of-minterms expression from a plotted K-map? The product-of-maxterms expression?

8. How to recognize groups of combinable minterms on a K-map? The Boolean algebra that underlies the combination process?

9. How to identify *essential implicants*?

10. How to select the best set of implicants?

11. How to use a K-map to express a function in both simplified SOP and simplified POS forms?

12. What a *redundancy* is? How don't-care and can't-happen conditions arise? How to use them to simplify combinational expressions? How to determine the response to redundant inputs, once the combinational expressions have been simplified?

13. How to use tabular reduction techniques?

14. How to construct a prime-implicant table? How to remove dominated rows and dominating columns? How to select prime implicants to cover a cyclic table?

15. How multiple-function minimization techniques can reduce the cost of combinational systems that have multiple outputs?

16. How to use the informal and the formal (K-map and tabular) multiple-function minimization techniques?

4.10 References

[1] Dietmeyer, D. L., *Logic Design of Digital Systems*. Boston, Mass.: Allyn and Bacon, Inc., 1971.

[2] Givone, Donald D., *Introduction to Switching Circuit Theory*. New York: McGraw-Hill Book Company, 1970.

[3] Hill, Fredrick J., and Gerald R. Peterson, *Introduction to Switching Theory and Logical Design*. New York: John Wiley & Sons, Inc. 1968.

[4] Hillenbrand, Bill, "Special Report: Decoder/Drivers for Digital Displays," *Electronic Products*, 13, No. 1 (January 18, 1970), 114–118.

[5] Hurley, R. B., *Transistor Logic Circuits*. New York: John Wiley & Sons, Inc., 1961.

[6] Karnaugh, M., "The Map Method for Synthesis of Combinational Logic Circuits," *Trans. of AIEE, Communications and Electronics*, 72, Part I (November 1953), 593–599.

[7] Klipstein, D. L., "The Contributions of Edsel Murphy to the Understanding of the Behavior of Inanimate Objects," *EEE* (August 1967).

[8] Lackey, Robert B., Letter to the Editor, *Computer Design*, 6, No. 6 (June 1967), 16.

[9] Levine, Robert I., "Logic Minimization Beyond the Karnaugh Map," *Computer Design*, 6, No. 3 (March 1967), 40–43.

[10] McCluskey, E. J., Jr., "Minimization of Boolean Functions," *Bell System Technical Journal*, 35, No. 6 (November 1956), 1417–1444.

[11] Petrick, S. R., "A Direct Determination of the Irredundant Forms of a Boolean Function from the Set of Prime Implicants," *Technical Report No. 56–110*. Bedford, Mass.: AF Cambridge Research Center, April 1956.

[12] Quine, W. V., "A Way to Simplify Truth Functions," *American Math. Monthly*, 62, No. 9 (November 1955), 627–631.

[13] Vietch, E. W., "A Chart Method for Simplifying Truth Functions," *Proc. of ACM* (Pittsburgh) (May 1952), 127–133.

4.11 Problems

4-1. Simplify each of the following expressions by using algebraic techniques. Compute the cost savings (in cost units) that the reduction permits.

a. $F(A,B,C) = \sum 2, 3, 6, 7$

b. $F(W,X,Y,Z) = \sum 2, 6, 4$

c. $F(A,B,C,D,E) = \sum 0, 1, 2, 4, 8, 16$

d. $G(w,x,y,z) = \prod 0, 2, 8, 10$

e. $G(A,B,C) = \prod 0, 2, 4, 6$

f. $G(v,w,x,y,z) = \prod 0, 4, 8, 12, 16, 20, 24, 28$

4-2. Expand each of the following partially reduced AND/OR expressions into sum-of-minterms form. Use both algebraic manipulation and truth tabulation. Then, if possible, reduce the sum-of-minterms expression to a form that is cheaper to implement than the original expression. Compute the cost saving (in cost units) that results.

a. $F(A,B,C) = \bar{A}\bar{B} + \bar{A}BC$

b. $F(W,X,Y,Z) = \bar{W}\bar{Y}\bar{Z} + \bar{W}X\bar{Y}Z + X\bar{Y}\bar{Z} + W\bar{X}\bar{Z}$

c. $F(A,B,C,D) = AD + \bar{A}\bar{C}D + ABC\bar{D}$

d. $G(w,x,y,z) = \bar{y}\bar{z} + \bar{w}\bar{x}\bar{z} + \bar{w}xy\bar{z} + wy\bar{z}$

4-3. Expand each of the following partially reduced OR/AND expressions into product-of-maxterms form. Use both algebraic manipulation and truth tabulation.

a. $F(A,B,C) = (A + B)C$

b. $F(A,B,C,D) = (\bar{A} + B)(\bar{C} + D)$

c. $G(w,x,y,z) = (w + \bar{x} + y)(x + \bar{y} + z)(w + z)$

d. $F(W,X,Y,Z) = W(X + Y)(\bar{X} + \bar{Y} + \bar{Z})$

4-4. Plot each of the following functions on a K-map.

  a. $F(A,B,C,D) = \sum 0, 2, 5, 6, 9, 11, 12, 15$

  b. $F(A,B,C,D,E) = \sum 1, 3, 6, 8, 11, 14, 17, 21, 23, 29, 30, 31$

  c. $G(W,X,Y) = \sum 1, 4, 7$

  d. $F(w,x,y,z) = \prod 0, 2, 5, 6, 9, 11, 12, 15$

  e. $F(A,B,C,D,E) = \prod 0, 1, 5, 7, 11, 16, 18, 22, 25, 29, 31$

  f. $G(W,X,Y,Z) = \prod 4, 6, 10, 11, 12, 14, 15$

4-5. Plot each of the partially reduced functions given in Prob. 4-2 on a K-map. Extract the numbers of the TRUE minterms from the map and compare with the sum-of-minterms expression that was found algebraically in Prob. 4-2.

4-6. Plot each of the partially reduced functions given in Prob. 4-3 on a K-map. Extract the numbers of the FALSE minterms from the map. Convert these to a product-of-maxterms form for each function. Compare the results to the product-of-maxterms expression that was found algebraically in Prob. 4-3.

4-7. Plot each of the following partially reduced functions on a K-map.

  a. $F(V,W,X,Y,Z) = \bar{V}WX + W\bar{X}\bar{Y} + WZ + \bar{X}\bar{Z}$

  b. $F(A,B,C,D,E) = (A + \bar{B} + C)(\bar{A} + D + \bar{E})(\bar{C} + \bar{D} + E)(B + \bar{C})$

  c. $G(r,s,t,u,v) = \bar{u}\bar{v} + rstu + s + \bar{t}\bar{u}v$

4-8. Use a K-map to simplify each of the expressions given in Prob. 4-1. Mark all essential implicants on the K-map. Consider both SOP and POS forms. Implement both forms with RTL gates, assuming positive-TRUE input and output. Show a logic diagram. Count cost (in cost units) and estimate worst-case delay.

4-9. Repeat the analysis procedure given in Prob. 4-8 using the expressions given in Prob. 4-2. Assume TTL gates and positive-TRUE inputs and output.

4-10. Repeat the analysis procedure given in Prob. 4-8 using the expressions given in Prob. 4-3. Assume TTL gates and negative-TRUE inputs and output.

4-11. Repeat the analysis procedure given in Prob. 4-8 using the expressions given in Prob. 4-4. Assume TTL gates and negative-TRUE inputs and output.

4-12. Find simplified SOP and POS forms for the expressions given in Prob. 4-7. Give cost (in cost units, exclusive of inverters) for each form.

4-13. The functions given below form cyclic K-maps. Plot each function and give alternative SOP and POS expressions for implementing each function.

a. $F(A,B,C,D) = \sum 2, 3, 4, 6, 9, 11, 12, 13$

b. $F(W,X,Y,Z) = \sum 0, 2, 13, 15$

c. $G(v,w,x,y,z) = \sum 0, 2, 5, 10, 13, 15, 16, 20, 21, 26, 27, 31$

4-14. Prove that Eqs. (4-16) and (4-17) are equivalent.

*4-15. Design a combinational device that will receive a five-bit binary number and will indicate whenever the binary number is divisible by 6, 7, or 9. Consider both SOP and POS expressions. Show a logic diagram for RTL implementation. Assume positive-TRUE inputs and output.

*4-16. Use K-map techniques to design a five-input minority gate (output follows *minority* of inputs). Implement with TTL; assume positive-TRUE inputs and output.

4-17. Plot each of the following incompletely specified functions on a K-map. Simplify in both SOP and POS forms. Then, by using the simplified expressions, determine the response that would be produced if a redundant input did occur.

a.

A	B	C	D	F
0	0	0	0	1
0	0	0	1	X
0	0	1	0	0
0	0	1	1	0
0	1	0	0	0
0	1	0	1	1
0	1	1	0	0
0	1	1	1	X
1	0	0	0	0
1	0	0	1	1
1	0	1	0	0
1	0	1	1	X
1	1	0	0	1
1	1	0	1	1
1	1	1	0	0
1	1	1	1	0

b. $F(A,B,C,D) = \sum 0, 2, 5, 8, 10, 13, 14, 15,$ X_1, X_{11}, X_{12}

c. $F(v,w,x,y,z) = \sum 1, 3, 6, 9, 11, 14, 18, 19,$ $23, 25, 27, X_0, X_5, X_{10}, X_{16},$ X_{24}, X_{30}, X_{31}

d. $F(W,X,Y,Z) = \prod 1, 3, 5, 9, 12, 14,$ with redundancies at $m_4, m_5, m_9,$ $m_{11},$ and m_{15}

*4-18. Redesign the binary comparator of Sec. 4.5.2 to make its comparison process begin with the MSB and move to the LSB of the binary numbers.

4-19. Investigate the response of the code-to-code converter of Sec. 4.5.3 to the redundant input combinations. Consider both the SOP and the POS expressions given in Fig. 4-36.

*4-20. Design a converter that receives code I and generates code II in accordance with the following table. Minimize each output function separately, considering both SOP and POS forms. Determine the response to the redundant input conditions. Show a TTL implementation.

Code I				Code II			
A	B	C	D	W	X	Y	Z
0	1	1	1	1	1	0	0
1	0	0	1	1	0	1	1
0	1	1	0	1	0	1	0
0	0	0	1	1	0	0	1
1	1	0	1	1	0	0	0
1	1	1	1	0	1	1	1
0	1	0	1	0	1	1	0
0	0	0	0	0	1	0	1
0	1	0	0	0	1	0	0
0	0	1	0	0	0	1	1

4-21.   Develop the simplified POS expressions for controlling the segments of the display given in Sec. 4.5.4.

*4-22.   Design a seven-segment decoder that receives code II from Prob. 4-20 and generates the logic for controlling the **a–g** segments of the display shown in Fig. 4-37. Investigate the response to redundant inputs.

4-23.   The display unit shown in Fig. P4-23 is an alternative to the seven-segment display shown in Fig. 4-37. Design a decoder that will convert 8421 BCD into the control logic for the segments of this display. Note that some of the numbers take on rather strange shapes. Investigate the response to redundant inputs.

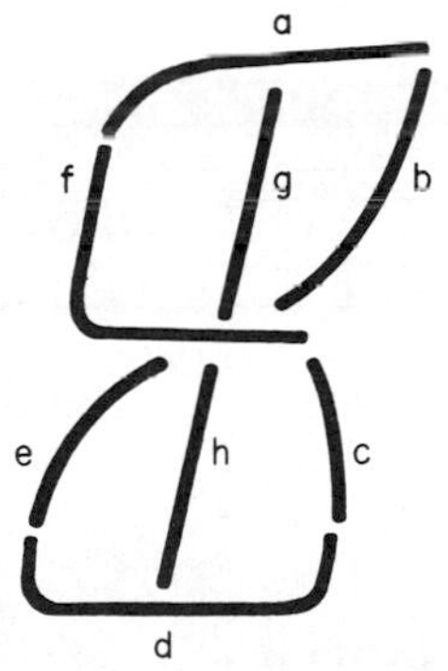

Decimal Value	Segments
0	c, d, e
1	g, h
2	a, b, d, c
3	a, b, c, d
4	f, g, h
5	a, f, c, d
6	c, d, e, g
7	a, b, h
8	a, b, c, d, e, f
9	a, b, f, h

Figure P4-23

*4-24.   Redesign the seven-segment decoder given in Sec. 4.5.4 so that the character E (for $Error$) is displayed whenever an erroneous 8421 BCD combination appears at the input to the decoder.

*4-25. The following BCD code is commonly called a *two-out-of-five code*, for obvious reasons. Its encoded arrangement makes the detection of errors somewhat straightforward. Design a combinational system that receives a code word and indicates whenever an erroneous combination is present. Consider both SOP and POS forms.

Decimal Value	Code Word				
	V	W	X	Y	Z
0	0	0	0	1	1
1	0	0	1	0	1
2	0	1	0	0	1
3	1	0	0	0	1
4	0	0	1	1	0
5	0	1	0	1	0
6	1	0	0	1	0
7	0	1	1	0	0
8	1	0	1	0	0
9	1	1	0	0	0

*4-26. A combinational system is to receive a decimal value encoded according to code I. It also receives a control signal, J, that affects the output of the system. When J is TRUE, the input is converted to its equivalent in code II (8421 BCD). When J is FALSE, the output is chosen from code III (excess-three code). Design the combinational system for implementing this process. Consider both SOP and POS forms. Show an RTL logic diagram. Investigate the response to redundancies for both $J = 1$ and $J = 0$.

Decimal Value	Code I				Code II				Code III			
	A	B	C	D	W	X	Y	Z	W	X	Y	Z
0	0	0	0	1	0	0	0	0	0	0	1	1
1	0	0	1	0	0	0	0	1	0	1	0	0
2	0	1	0	0	0	0	1	0	0	1	0	1
3	1	0	0	0	0	0	1	1	0	1	1	0
4	1	0	0	1	0	1	0	0	0	1	1	1
5	1	0	1	0	0	1	0	1	1	0	0	0
6	1	1	0	0	0	1	1	0	1	0	0	1
7	1	1	0	1	0	1	1	1	1	0	1	0
8	1	1	1	0	1	0	0	0	1	0	1	1
9	1	1	1	1	1	0	0	1	1	1	0	0

4-27. Simplify the functions given in Prob. 4-1 by using tabular techniques. Consider only SOP forms.

4-28. Simplify the partially reduced expressions of Prob. 4-2 by using tabular techniques. Start with the sum-of-minterms expressions found in Prob. 4-2 or 4-5. Consider only SOP forms.

4-29. Simplify the functions given in Prob. 4-4 by using tabular techniques. Consider only SOP forms.

4-30. Simplify each of the following functions by using tabular techniques. Consider only SOP forms. Rework considering every third minterm (or maxterm) as a redundancy.

   a. $G(A,B,C,D,E,F) = \sum 0, 1, 5, 6, 9, 15, 16, 18, 21, 25, 29, 30, 35, 37,$
$$39, 40, 45, 48, 49, 50, 55, 58, 60, 61, 63$$

   b. $F(U,V,W,X,Y,Z) = \prod 1, 2, 3, 5, 6, 9, 13, 14, 17, 19, 20, 21, 24, 29,$
$$35, 38, 39, 40, 48, 50, 57, 58, 60, 62, 63$$

4-31. Rework Prob. 4-16 by using tabular techniques. Consider both SOP and POS forms.

4-32. Rework the two-out-of-five error detector (Prob. 4-25) by using tabular techniques. Consider both SOP and POS forms.

4-33. Rework Probs. 4-27–4-30 by using tabular techniques to find simplified POS forms.

4-34. Rework Prob. 4-20 as a multiple-function minimization problem. Use both the informal technique (Sec. 4.7.1) and the formal technique (Sec. 4.7.2). Use both K-map *and* tabular techniques to find the multiple-function prime implicants.

4-35. Rework Prob. 4-24 as a multiple-function minimization problem, both formally (K-map and tabular methods) and informally.

4-36. Rework Prob. 4-26 as a multiple-function minimization problem, both formally and informally.

4-37. (Optional) If computer service is available, use the computer program given in Appendix A to rework Probs. 4-30 and 4-32. Also, design a combinational device that will receive two four-bit code words, one from code II and one from code III (Prob. 4-26) and that will make its output signal TRUE whenever the decimal values represented by the two code words are equal. Consider the simplification of this eight-variable problem in both SOP and POS forms.

4-38. Prove that for a Boolean function of n variables, the most cost units that can ever be *required* for implementing the function are

$$\text{Max. c.u.} = 2^{n-1} \cdot (n + 1) \qquad \text{(exclusive of inverters)}$$

or

$$\text{Max. c.u.} = [2^{n-1} \cdot (n + 1)] + n \qquad \text{(including inverters)}$$

Suggestion: Consider the K-map arrangement that has the most 1s without allowing any combinations.

4-39. Can an EXCLUSIVE-OR gate be used to reduce gating costs for the functions found in Prob. 4-38? For any K-map that has a checkerboard-like appearance?

4-40. The K-map shown in Fig. P4-40 is an alternative to the five-variable K-map that is shown in Fig. 4-7. Complete the filling in of the numbered minterm locations. Mark representative groupings of two, four, eight, and sixteen minterms on the new map. Use this map to rework Probs. 4-12, 4-17c, and 4-25.

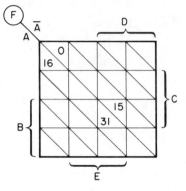

Figure P4-40

5 Introduction to Sequential Systems

5.0 Sequential Systems Versus Combinational Systems

Chapters 1–4 have introduced the concepts and design procedures that are associated with a class of digital systems that are said to be *combinational* in their operation. Chapters 5 and 6 consider a new class of systems, *sequential systems*. Combinational and sequential systems are different in many ways. Yet in some ways they are the same. Thus, as an introduction to sequential systems, it is appropriate to define combinational systems with more formality.

5.0.1. Combinational Systems Defined.

All combinational digital systems can be represented by the general diagram shown in Fig. 5-0. To

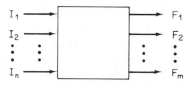

Figure 5-0. A General Combinational System

be classified as combinational, a system must meet the following restrictions:

1. The TRUE/FALSE values of the output signals must depend only on the current TRUE/FALSE values of the input signals.

2. There can be no feedback from the output signals to the input signals.

These two restrictions make the design and analysis of combinational systems a straightforward task. Each output signal can be expressed as a Boolean function of the input variables:

$$F_i = \phi_i(I_1, I_2, ..., I_n), \qquad i = 1, 2, ..., m \qquad (5\text{-}0)$$

where $I_1, I_2, ..., I_n$ are the Boolean input signals and the ϕ_is are the Boolean expressions that generate the F_i output signals. By applying the techniques that are developed in Chapters 1–4, the functions given in Eq. (5.0) can be minimized and implemented with any collection of gating circuits that will act as a complete set of Boolean connectives.

During the period immediately following changes in the input conditions, the outputs from a combinational system will assume their new values. These new values will appear within a time interval determined by the delays inherent in the gating circuitry that is used to generate the output signals. Ideally, the gating devices would be free of delay and the output signals would instantaneously follow any changes in the input signals. In actuality, however, each level of gating between input and output introduces its own time delay (as discussed in Sec. 3.10.3). This delay, commonly called the *settling time* of the combinational system, never needs to exceed $3t_g$, where t_g is the delay through a single gate or inverter. These three delays allow for two levels of gating (AND/OR, for example) and one level of inversion. The total delay can be reduced to $2t_g$ if double-rail inputs are available and may be increased beyond $3t_g$ if higher levels of gating are used to decrease gating costs.

A common technique for describing the behavior of nonideal gating circuitry is to lump the delays for all the gating levels and to separate that delay from the logical operation of the gates. This produces a system model such as that shown in Fig. 5-1, wherein a *delay-free* combinational system is followed by delay elements. The outputs from the idealized gating system, $F_1, F_2, ..., F_m$, follow the input conditions with no delay, while the actual outputs from the combinational system, $f_1, f_2, ..., f_m$, lag behind the changes in the input signals due to the gating delays. Thus, at a given time, the f outputs may not agree with the F signals, but these two sets of signals will

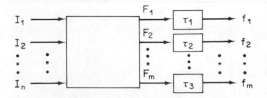

Figure 5-1. A Combinational System with Delays Included

always settle into agreement providing that the input signals are kept constant for a sufficiently long period.

The absence of feedback (restriction 2) assures that each change in the input signals will produce a single change in the output signals, with no instabilities or uncertainties being possible. This condition allows the behavior of a combinational system to be described with great simplicity. The 2^n input combinations can be listed and the value of each of the m output functions can be determined and tabulated for each possible input combination. True, the outputs may not agree with the input conditions immediately following an input change, but this condition cannot exist for longer than the worst-case gating delay.

5.0.2. Sequential Systems: What and Why.   In many logical systems the restrictions given above are not met. As a result, the outputs from the system cannot be determined from knowledge of the present input values alone. Rather, the past history of the input and output signals must be included in their determination. Such systems are fundamentally different from combinational systems, although they may be combinational in part. The term *sequential* is commonly used to describe this distinction.

Sequential systems are generally characterized as having two properties:

1. There is at least one feedback path from the output of the system to the input of the system.

2. The system has a memory capability for holding past information, so that previous input and output values can be used in determining the current (and future) output signals.

A generalized sequential system is shown in Fig. 5-2. The upper part of the system receives logical inputs and generates logical outputs by using

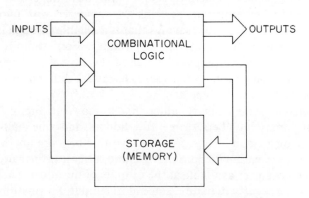

Figure 5-2. A Generalized Sequential System

purely combinational logic. In addition, some output information is stored in the memory portion of the system and used, later, as part of the input information. Thus, the current inputs and the information that is held in the system's memory are used to generate the current output signals and to determine what new information must be transferred into the memory. The memory capability shown in Fig. 5-2 can be provided in two ways: by using *storage devices* (magnetic cores, latching relays, electronic memory devices, etc.) or through *delay*, whereby the present values of input and output signals are delayed so as to reach the input of the sequential system at a later time.

The elimination of the combinational restrictions broadens the application of digital systems. The provisions for memory and feedback allow *time* to be included as a parameter in determining system behavior. Thus, information concerning past events can be used to calculate the present output signals, and information from both the past and the present can be used to specify future activities.

A common example of a sequential system is a dial combination lock. If such a lock has a combination of 11-15-30, then the dial must first be rotated several times to clear any previous inputs, then rotated to 11, 15, and 30 in order for the lock to open. Mechanical tumblers or cogwheels are used to "remember" the previous settings of the dial in order to identify the correct combination.

Another common sequential system is the TIME portion of a digital TIME-TEMPERATURE sign. The memory portion of the sequential system holds the current time. Once each second, usually in response to a timing signal that is generated by an oscillator, the stored information is changed so as to represent the newer time. This repetitive updating operation is common to many types of sequential digital systems.

5.0.3. The Space/Time Equivalence.

One advantage that becomes readily apparent when sequential systems are compared with purely combinational systems is the hardware savings that is available through the sequential (repetitive) use of the same logical circuitry. For example, in dialing a local telephone number a single dial is used seven times rather than seven dials being used at the same time. This saves hardware, but it does take longer, since the operation must be repeated.

Consider the four-bit binary adder that is shown in Fig. 4-25. To add two four-bit binary numbers, three full adders and one half-adder are required, as shown again in Fig. 5-3. The two numbers are applied as input signals to the combinational addition logic and, after a propagational delay, the five-bit sum becomes available at the outputs of the adder. Such an adder may be said to be *parallel* in nature, since all the input information is applied at the same time and, except for propagational delays, all the output signals become available at the same time.

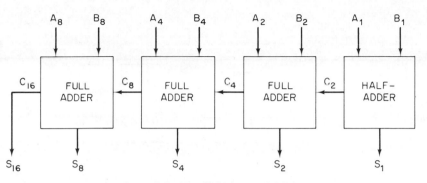

Figure 5-3. *Parallel* Binary Addition

The parallel adder may be reorganized into a sequential form, as shown in Fig. 5-4, in order to reduce the amount of combinational logic that is required to add the binary numbers. Only *one* full adder is needed for implementing the sequential addition process, since the single full adder can be used repeatedly. The sequential addition process, commonly called *serial* addition (as contrasted with the parallel addition circuitry shown in Fig. 5-3), begins by adding A_1 and B_1, the least-significant bits of the two binary numbers. The resultant sum and carry are stored. The stored carry is then added to the next bits, A_2 and B_2. This process is repeated until all the bits of the input numbers have been added and all the bits of the sum are available.

An example of this serial addition process is shown in Fig. 5-5, with $A = 1011_2$ and $B = 1010_2$. The sequence of diagrams illustrates the use of the storage portion of the sequential system to "remember" information that will be needed at a later time. The carry that is generated during each successive addition cycle is held in the storage area for use during the next

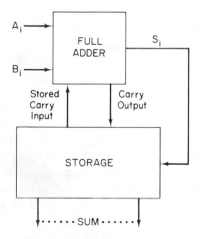

Figure 5-4. *Serial* Binary Addition

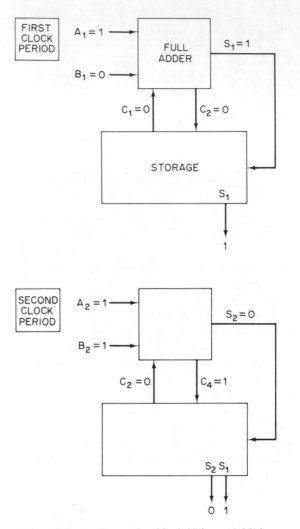

Figure 5-5. An Example of Serial Binary Addition

addition. The sum bits are also stored as they are generated, so that the entire sum will be available at the conclusion of the serial addition process. Note that the high-order sum bit is identically equal to the carry resultant from the addition of the most-significant bits, A_8 and B_8.

The relationship between a serial system and a parallel system for performing the same operation is often referred to as the *space/time equivalence*. If a combinational system can be separated into a sequence of identical circuits, with the output signals from one circuit acting as inputs to the next,

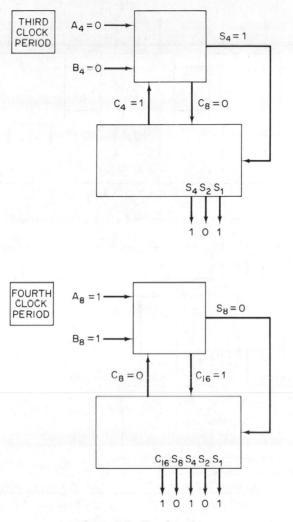

Figure 5-5. Continued

the system is said to be *iterative in space*. This type of circuit is shown in general form by Fig. 5-6. The circuit C_i is called an *iterative circuit*. Many copies of the circuit are used to build up the logical system shown in Fig. 5-6. Whenever such a space-iterative organization is available, there is a related serial, or *time-iterative*, organization, which, by using the storage capability present in a sequential system, performs the same logical operation through the repetitive use of only *one* of the iterative circuits. The time-iterative equivalent to the system of Fig. 5-6 is shown in Fig. 5-7.

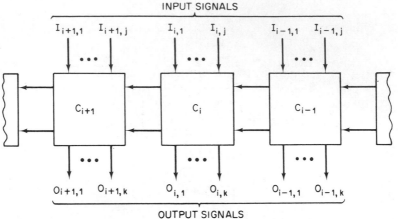

Figure 5-6. Space-iterative Circuit Organization

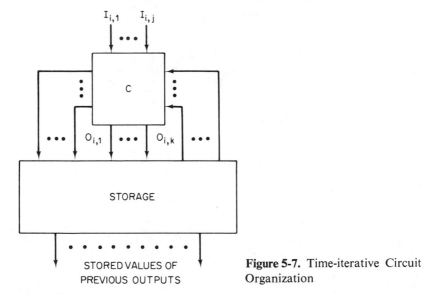

Figure 5-7. Time-iterative Circuit Organization

In later chapters of this text, several examples of this serial/parallel equivalence are given. For a more detailed treatment of this subject and of the general properties of iterative networks, see Ref. [5].

5.1 Sequential System Description

The advantages that are allowed by operating a digital system in the sequential mode are not without offsetting disadvantages. But before discussing these disadvantages and the techniques that are used to deal with them, it is necessary to introduce the formal notation with which sequential systems are described.

5.1.1. Inputs, Outputs, and System States.

A general diagram for a sequential system is given in Fig. 5-2. This diagram is repeated in more detail in Fig. 5-8. All the inputs to and outputs from the system are assumed

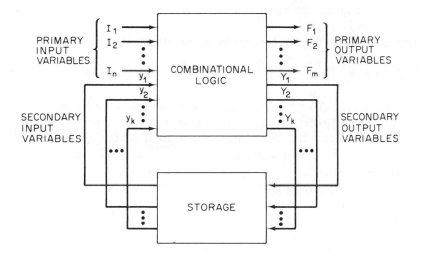

Figure 5-8. A Generalized Sequential System

to be Boolean in nature, although they may fall into two classes, *level* or *pulse* signals, as is discussed in Sec. 5.2.3. The combinational portion of the system receives two sets of input signals: the *primary input variables* (labeled as $I_1, I_2, ..., I_n$) and the *secondary input variables* $(y_1, y_2, ..., y_k)$. The secondary input variables are taken from the memory portion of the sequential system. The values of the k secondary input variables, taken in concert, are said to define the *present state* of the system. Hence, the secondary input variables are sometimes called the *present-state variables*. There are 2^k different combinations that these secondary input variables can take on, allowing the sequential system only 2^k different present states. For this reason sequential digital systems are often referred to as *finite-state* systems.

The values of the secondary input variables are a reflection of the past history of the sequential system's operation. The values of the primary input variables define the current input condition, one of 2^n different possible combinations. Taken together, the two sets of inputs are said to define the *total present state* of the system, with 2^{n+k} different *total* present states being possible.

EXAMPLE 5-0:   A dial combination lock has a three-digit combination. Each digit can take on 30 different values. Describe the behavior of the lock in sequential system terminology.

Answer: The lock must first be initialized by rotating the dial several times. This "clears" the mechanical memory of the lock. The cogwheels or tumblers of the lock must then be used to store the first two digits prior to the selection of the third. After the first number has been selected, the lock will be in one of 30 possible present states. After the second number has been selected and stored, the number of possible present states is $30 \times 30 = 900$. This information, when combined with the last input value (1 out of 30), leads to 27,000 different *total present states*, only one of which is the combination that will release the lock.

The outputs from the combinational portion of a sequential system are also separated into two categories. The *primary output variables* $(F_1, F_2, ..., F_m)$ are the signals that are sent from the sequential system into its associated environment. These signals may operate other systems, turn indicators on or off, or otherwise cause activity or transmit information from the system.

The other output signals are the *secondary output variables* $(Y_1, Y_2, ..., Y_k)$, which are returned to the memory portion of the sequential system. These signals describe the new values of the information that will be stored in the memory when the next cycle of the system's operation takes place. Thus, the secondary output signals are said to describe the *next state* of the sequential system. When the new values of the Y signals are transferred into the memory, thereby replacing the y signals, the sequential system moves from the present state into the next state.

The transfer from present state to next state takes place in one of two ways, depending on the type of memory that is used by the sequential system. If the memory is provided through delay, then the present-state-to-next-state change takes place when the end of the delay period is reached. If, on the other hand, the memory is provided by storage devices, the state change occurs whenever the control signals to the storage devices cause the new values of the state variables to replace the old.

During each present state, the values of the primary and secondary output variables are determined by combinational operation upon the primary and secondary input variables. Thus, given the present state of the system (as defined by $y_1, y_2, ..., y_k$) and the values of the primary input variables (one of 2^n different combinations), the combinational portion of the sequential system generates the appropriate primary output signals and, by generating the values of $Y_1, Y_2, ..., Y_k$, selects the next state to which the system will move. For each present state there can be as many as 2^n different next states and output combinations, corresponding to the 2^n different combinations that the n primary input variables can take on.

It should be noted that many types of storage devices cannot operate directly from the Y signals. Rather, to update the present state, they require a different set of control signals, commonly called *excitation variables*. These excitation variables are determined from the total present state, just as are the normal next-state variables. The use of excitation variables eliminates the need for explicitly determining the values of the Y variables, giving rise to the modified sequential system organization that is shown in Fig. 5-9. Note that the number of excitation variables may *exceed* the number of present-state variables.

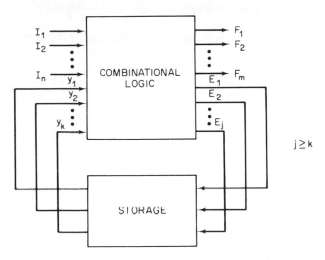

Figure 5-9. Generalized Sequential System that uses Excitation Variables

5.1.2. State Tables and State Diagrams.

The relationships among present-state variables, primary input variables, next-state (or excitation) variables, and primary output variables that describe the behavior of a sequential system can be specified in several ways. As an example, consider the simple sequential system that is shown in Fig. 5-10. This system has two primary input variables (having four different combinations of values), one primary output variable, and one state variable. It uses delay for memory. There are only two possible present states, $y = 0$ and $y = 1$. When combined

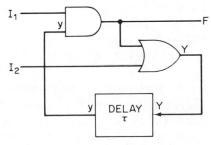

Figure 5-10. A Simple Sequential System

with the four input combinations, these give eight different *total* present states. The values of the next-state variable, Y, and the primary output variable, F, must be specified for each total present state. The tabular arrangement shown in Fig. 5-11 is a common method for presenting this information. This descriptive tool is called a *state table*.

Present State	Next State Y	Output F
y	$I_1 I_2 = 00\ \ 01\ \ 10\ \ 11$	$I_1 I_2 = 00\ \ 01\ \ 10\ \ 11$
0		
1		

Figure 5-11. A State Table Corresponding to the System of Fig. 5-10

The type of state table that is shown in Fig. 5-11 has as many rows as there are present states (two in the example) and one column for each of the 2^n combinations of the primary input variables. The intersection of each row and column defines a *total* present state, and the next-state and output values that correspond to that particular total present state are entered therein. The tabular form that is shown in Fig. 5-11 separates the next states and the primary output variables. Some authors (see Kohavi [5], for example) combine these data into one table.

Examination of the combinational logic given in Fig. 5-10 shows that

$$F = I_1 \cdot y \tag{5-1}$$

and

$$Y = (I_1 \cdot y) + I_2 \tag{5-2}$$

These expressions can be used to complete the state table. Consider present state $y = 0$ and inputs of $I_1 I_2 = 00$. Equations (5-1) and (5-2) show that, under these conditions, $Y = 0$ and $F = 0$. These data are entered into the state table as shown in Fig. 5-12a.

In this case, the values of Y (next state) and y (present state) are equal. Thus, no change in state will take place when the end of the delay period is reached. This condition, wherein all the next-state variables agree with their present-state counterparts, is referred to as *stability*. Stable states are indicated on the state table by circling their next-state location, as shown in Fig. 5-12a.

Now consider the total present state $y I_1 I_2 = 100$. Equations (5-1) and (5-2) give $Y = 0$ and $F = 0$. Since $Y \neq y$, the system is in an *unstable* state. At the end of the delay period y will change from TRUE to FALSE, taking on the value of Y. This change will move the system to the stable total present state, $y I_1 I_2 = 000$. This information, along with the completion of

Present State	Next State Y	Output F
y	$I_1I_2 = 00\ \ 01\ \ 10\ \ 11$	$I_1I_2 = 00\ \ 01\ \ 10\ \ 11$
0	⓪	0
1		

(a)

Present State	Next State Y	Output F
y	$I_1I_2 = 00\ \ 01\ \ 10\ \ 11$	$I_1I_2 = 00\ \ 01\ \ 10\ \ 11$
0	⓪ 1 ⓪ 1	0 0 0 0
1	0 ① ① ①	0 0 1 1

(b)

Figure 5-12. Completion of the State Table for the System of Fig. 5-10

the state table, is shown in Fig. 5-12b. All stable states are circled and all unstable present states are not.

Note that changes in the primary inputs move the system horizontally within the state table. If the total present state that is reached following an input change is unstable, however, one or more vertical state changes will take place following the horizontal change caused by the primary inputs. The system is constrained to vertical state changes as long as the primary input variables do not change again. The sequence of vertical state changes stops when a stable state is reached.

A second method for describing the behavior of a sequential system is the use of a *state diagram*. This method presents a pictorial representation of the present-state/next-state sequences that apply to the sequential device. State changes are marked with directed arrows, with the primary input and output conditions that apply to each state transfer given beside the arrows.

The state diagram for the system of Fig. 5-10 is given in Fig. 5-13. Each possible present state is represented by a circle, with the values of the present-state variables given within the circle. State-to-state transfers are indicated by arrows. Stable states are indicated by arrows that begin and terminate on the same state.

The primary input conditions that correspond to each transfer are indicated beside each arrow, along with the value of the output variable that corresponds to that present-state/primary input combination. The "key" to the notation of Fig. 5-13 is given to the right of the state diagram. The

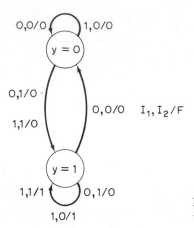

Figure 5-13. A State Diagram Describing the System of Fig. 5-10

primary input variables are given first, in I_1, I_2 order. A slash separates the input information from the output information. For a system with several primary input and output variables, this type of symbolism can be expanded as shown in Fig. 5-14. The sequential system shown in Fig. 5-14 moves from state X to state Y when

$$I_1 = 0, \qquad I_2 = 1, \qquad I_3 = 0 \tag{5-3}$$

and generates the following output combination:

$$F_1 = 1, \qquad F_2 = 0, \qquad F_3 = 1 \tag{5-4}$$

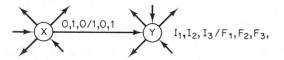

Figure 5-14. A State Transfer in a Complicated Sequential System

EXAMPLE 5-1: The logic diagram for a simple sequential system is given in Fig. E5-1a. Determine the state table and state diagram for the system.

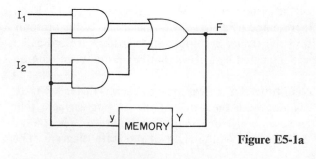

Figure E5-1a

Solution: See Fig. E5-1b.

Solution:

Present State	Next State Y*
y	$I_1I_2 = 00\ 01\ 10\ 11$
0	⓪ ⓪ ⓪ ⓪
1	0 ① ① ①

* The output, F, is equal to Y

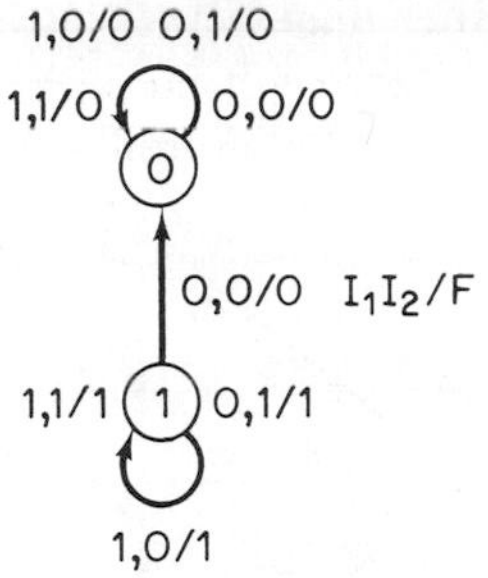

Figure E5-1b

A state diagram is a convenient way to present the overall behavior of a sequential system. It clearly describes the state-to-state sequences that make up the system's operational characteristics. State tables and state diagrams are essential tools in the analysis and design of sequential digital systems. These two techniques will be used extensively throughout the remainder of this text.

5.1.3. Generalized Sequential Systems.   Not all sequential systems are as completely specified as the example systems that are given in Sec. 5.1.2. In many cases a more general system specification is allowed, with the different present states represented by distinct names or symbols rather than by specific combinations of the present-state variables. This type of generalized system description can arise during the early stages of the sequential system design process.

EXAMPLE 5-2:   A sequential system is to act as a simple combination lock, allowing those who know the correct combination to enter a computer control room. The system receives four input signals, *R*, *A*, *B*, and *C*. The inputs come from push buttons that are mechanically interlocked so that only one button can be pushed at a time. The system is cleared (reset) by the *R* signal. Thereafter, the lock opens only on the *B,C,A* sequence. All other sequences transfer the system into an error state. The output signals are *OPEN* and *ERROR*. Develop the generalized state table and state diagram that describe this system.

Solution: Let the initial state of the system be named *a*, with the *R* signal always forcing the system into state *a*. Let the *ERROR* state be named *b*. Upon reaching state *b*, the system remains in state *b* until an *R* input is received. If the first input following the *R* signal is not *B*, the system moves to state *b*. Otherwise, it goes to a third state, say *c*, to await the rest of the combination. When in state c, any input other than C will move the system to state b. The C input will move the system to a fourth state, say d, from which an A input will move the system on into the *OPEN* state, say e. Any input other than A will move the system from state d to state b. Only the B,C,A sequence will cause the system to reach state e, the *OPEN* state. The *ERROR* and *OPEN* output signals can be taken directly from states b and e.

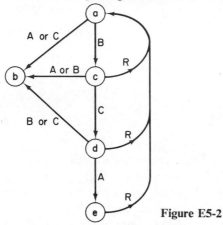

Figure E5-2

The state diagram for this system is shown in Fig. E5-2. Similarly, the state table for this system is

Present State	Next State					Outputs (OPEN, ERROR)				
	N^*	R	A	B	C	N	R	A	B	C
a	ⓐ	ⓐ	b	c	b	0,0	0,0	0,0	0,0	0,0
b	ⓑ	a	ⓑ	ⓑ	ⓑ	0,1	0,1	0,1	0,1	0,1
c	ⓒ	a	b	b	d	0,0	0,0	0,0	0,0	0,0
d	ⓓ	a	e	b	b	0,0	0,0	0,0	0,0	0,0
e	ⓔ	a	ⓔ	ⓔ	ⓔ	1,0	1,0	1,0	1,0	1,0

*N indicates the *no-input* condition. All present states are stable with no inputs present.

Prior to implementing a generalized system such as that of Ex. 5-2, it is necessary to assign to each possible present state a particular combination of present-state variables. If the system has N different possible present states, then

$$K \geq \log_2(N) \tag{5-5}$$

present-state variables are required, where Boolean state variables are assumed and K is an integer. This follows from the information representation concepts given in Chapter 1.

The relationships between the generalized present states and the specific combinations of present-state variables can be assigned as desired. Varying the assignments will usually affect the complexity of the combinational portion of the sequential system, since the combinational logic must interpret the present-state information in order to generate the next-state (or excitation) variables. A "good" state assignment is one that reduces the amount of combinational logic that is required to implement the sequential operation.

Many different approaches to this, the *state assignment problem*, have been developed. State assignment techniques are discussed as part of the sequential system design procedures that are presented in Chapter 6.

The number of different present states that are required during the operation of a sequential system determines the minimum number of present-state variables that the system must have, in accordance with Eq. (5-5). As a result, reducing the number of present states may allow a reduction in the number of present-state variables and, hence, a reduction in the memory requirements of the system.

EXAMPLE 5-3: A sequential system requires 27 different present states. What is the minimum number of present-state variables that it must have?

Answer: Five, since $\log_2(27) = 4.75$.

What effect will reducing the number of present states to 17 have on the number of present-state variables?

Answer: None, since $\log_2(17)$ is still greater than 4, necessitating five state variables. Reduction to *fewer than 17* present states will reduce the number of present-state variables, however.

Reduction of the number of present states, often called *state reduction*, is possible in many sequential systems. The basic concept involved in the state reduction process is the identification of *equivalent states*, i.e. (in very simple terms), two or more states that produce the same output signals and have the same next-state behavioral pattern. Once identified, equivalent states can be combined into a single state that acts in place of all the previously distinct states.

Like state assignment, state reduction is covered as part of the general sequential design process that is described in Chapter 6.

5.2 Problems (and Solutions) in Sequential Systems

Thus far in Chapter 5 the discussion has introduced sequential systems and, in a fairly brief manner, has attempted to show the advantages that accrue when sequential operation is allowed. These advantages have included the ability to use *time* as a parameter in determining system behavior, the presence of storage in the system so that information from the present can be carried into the future, and the saving in cost that is achieved through the repetitive use of the same logical circuitry.

These advantages are not obtained without offsetting disadvantages, however. Because a sequential system has feedback from its outputs to its inputs, certain types of instabilities and uncertainties may occur. When present, these conditions make the operation of circuit difficult or impossible to describe. They may even render the circuit useless, since its behavior may not be predictable or consistent. Several of these types of problems are listed and illustrated below.

    1. *The input or output conditions of the system may be indeterminant.*

EXAMPLE 5-4: Consider an ideal (delay-free) inverter connected as shown in Fig. E5-4. The feedback path indicates that this is a sequential system.

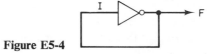

Figure E5-4

Assume that $I = 1$, thereby making $F = 0$. But, since F is returned to the input, $F = I$, and the values of F and I cannot differ. Thus, no logical interpretation of the behavior of this circuit can be made.

If an actual inverter is connected in this fashion it may settle to a voltage in between the high and low values, or it may oscillate.

    2. *The output condition of the system may be unstable, changing even though the external inputs do not change.*

EXAMPLE 5-5: Consider the logical circuit shown in Fig. E5-5. The delay can be that of the AND gate and the inverter or it can be a separate delay

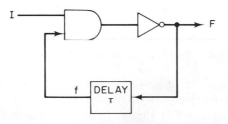

Figure E5-5

element. Assume that I has been kept FALSE long enough for the delay element to settle with $f = F = 1$ (since the output of the AND gate will be FALSE). If I is then made TRUE, F becomes FALSE, since $F = \overline{I \cdot f}$ and I and f are both TRUE. After a delay of τ sec, however, f assumes the new value of F, making the output of the AND gate become FALSE and F become TRUE. This cycle repeats itself with a period of 2τ.

The behavior of the circuit is described by the following state table:

Present State	Next State F	
f	$I = 0$	$I = 1$
0	1	1
1	①	0

Note that neither of the entries in the $I = 1$ column is stable.

The type of sequential system behavior described in Ex. 5-5 is called a *cycle*.

3. *The output condition of the system, even though stable, may not be predictable based upon the primary input conditions.*

EXAMPLE 5-6: A simple sequential system is shown in Fig. E5-6. Consider the behavior of the system when I_1 and I_2 are both TRUE. The partial state table for this condition is shown at the right of Fig. E5-6.

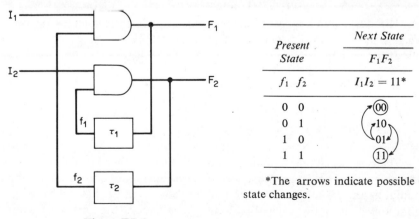

Present State	Next State F₁F₂
$f_1 \; f_2$	$I_1 I_2 = 11*$
0 0	⑳(00)
0 1	10
1 0	01
1 1	⑪

*The arrows indicate possible state changes.

Figure E5-6

Note that several conditions may occur. The system may be in a stable state at $f_1 f_2 = 00$ or $f_1 f_2 = 11$. If the present state is $f_1 f_2 = 01$ or $f_1 f_2 = 10$, however, then the system is unstable and its next-state behavior is unpredictable. The system may oscillate between the two unstable states or it may settle into one of the stable states. The actual result will depend on the values of the two delays, τ_1 and τ_2.

These (and other) types of erratic or unusual behavior limit the applicability of sequential systems. The important thing is, however, that these problems can be avoided by making certain restrictions in the way sequential systems are designed and used. These restrictions lead to reliable system performance without significantly limiting the utility of the sequential mode of operation.

5.2.1. Operational Restrictions in Sequential Systems.

The behavior of sequential systems has been analyzed by many researchers. Particularly notable contributions have been made by Huffman [3], who analyzed the general sequential system and developed much of the descriptive terminology that has been introduced thus far in Chapter 5, by Mealy [9], and by Moore [11].

The remainder of Sec. 5.2 presents a brief summary of some of the most useful aspects of the work of these and other researchers. The major emphasis of the discussion is to explain the reasons for the development of the *pulse mode* of sequential operation. Almost all the sequential design techniques that are presented in this text are based upon the pulse mode. Readers who are interested in the design and operation of non-pulse-mode sequential systems are referred to Refs. [1] and [2].

The aforementioned problems that may arise during sequential operation can normally be prevented or eliminated by satisfying one or more of the following restrictions:

1. Avoiding continuing instabilities (oscillations).

2. Allowing only *fundamental-mode* operation.

3. Allowing only *pulse-mode* operation.

The first restriction eliminates those conditions under which sequential systems exhibit oscillatory behavior. If a system is stable under some input conditions and exhibits oscillations under others, as in Ex. 5-5, then the input conditions that allow oscillation must be avoided. Systems that exhibit momentary instabilities but eventually settle into a stable state may cause problems, but they are not excluded under restriction 1. Oscillatory systems may be used in some special digital applications, most often to generate timing or sequencing signals, but such systems cannot be used for data processing or storage, since they change states at will and never give a constant output value.

The second restriction involves the way in which the primary inputs to a sequential system are allowed to change. A system is said to be operating in the *fundamental mode* if and only if its primary input variables are never allowed to change in logical value unless the system is in a stable state [8]. In other words, fundamental-mode operation assures that whenever a change in the primary inputs to a sequential system causes a change in its internal state, the primary inputs are kept constant thereafter until the system reaches a stable state. This settling process may involve a sequence of momentary instabilities, but fundamental-mode operation dictates that the primary inputs wait until stability is reached before they change again.

Fundamental-mode operation is a restriction on the way in which a sequential system is *used* rather than a restriction on the way it is *designed*. Fundamental-mode operation can be implemented by allowing only *one* of the primary input variables to change at a time and by separating successive input changes by a minimum time period that is sufficiently long to allow the system to settle into a stable state between the changes.

Unfortunately, in many systems the maintenance of fundamental-mode operation is difficult. Input signals may come from different sources and may change states at random times. Thus, the requirement that two or more signals must not change states too closely together may not be guaranteed. Special interface circuits may then be required to assure that the fundamental-mode restrictions are met. These circuits are often called *synchronizers* or *anticoincidence circuits*. [See Sec. 5.6.]

Multiple input changes that occur simultaneously or nearly simultaneously violate the fundamental mode and may lead to erratic operation. Without synchronization, the next state following this type of change may be indeterminant. As a result, most sequential systems are operated in either the fundamental mode or in the *pulse mode*, an alternative mode of sequential operation that eliminates almost all the problems that may arise during sequential operation.

As an illustration of the meaning of the fundamental-mode restriction (prior to describing the pulse mode), the next section introduces a very simple sequential system, the *flip-flop*, that is commonly used for data storage in larger sequential systems. The discussion covers the use of the flip-flop and the need for meeting the fundamental-mode restriction when using the flip-flop, and illustrates one type of problem that can arise even when the fundamental mode is maintained. The discussion of the last item points to the need for pulse-mode operation in many flip-flop applications.

5.2.2. The Basic Flip-flop. If all the data that must be stored by a sequential system are reexpressed in two-valued form, the storage requirements of the system can be met by using a set of two-valued storage devices, each of which holds the TRUE or FALSE value of one of the Boolean variables. These devices are sequential systems in themselves. Such storage devices can be constructed from an OR gate if its output terminal is connected back to one of its inputs, as shown in Fig. 5-15. The gate shown in Fig. 5-15 is

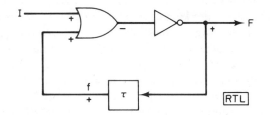

Figure 5-15. A Simple Boolean Storage Element

as a negative-TRUE NAND gate (RTL), necessitating the insertion of an inverter into the feedback path in order to change the voltage symbolism. The delays of the gate and the inverter are lumped into the single delay element that is shown in Fig. 5-15.

If the free input to the OR gate is made TRUE (high voltage for RTL), the output will become TRUE and the feedback path will hold the output of the circuit in the TRUE condition. Once it is set into this condition, the gate will remain in the TRUE state, commonly called *latched*, until the feedback path is broken. Obviously, this circuit can be used for the storage of Boolean variables having TRUE values.

To provide a useful method for breaking the feedback path, thereby facilitating the *unlatching* of the gate to store a FALSE value, the inverter shown in Fig. 5-15 can be replaced by an AND gate, as shown in Fig. 5-16.

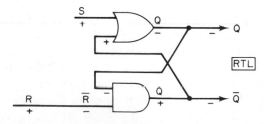

Figure 5-16. The NAND/NOR Latch

If the free input to the AND gate is made FALSE (also high voltage for RTL), the feedback path between the input and the output of the OR gate will be logically interrupted and the OR gate will return to its original state.

The simple logical device that is shown in Fig. 5-16 is commonly called a *NAND/NOR latch*, or simply a *latch*. Two output signals are available from the latch, each being the logical complement of the other. The outputs are commonly labeled as Q and $\bar{Q}$, with the Q output being TRUE when the device is in the latched or *set* state. The two controlling inputs are commonly called S (set) and R (reset), or *PRESET* and *CLEAR*. The voltage symbolism required to activate the control inputs is always the same as the input symbolism for the gate circuit when used as an OR gate, e.g., positive-TRUE for an RTL latch and negative-TRUE for a TTL latch.

The voltage symbolism that is assigned to the *outputs* of the latch is not fixed, however, since both the output signal, Q, and its complement, $\bar{Q}$, are available (double-rail output). By redefining the choice of which of the two outputs is Q and which is $\bar{Q}$, the complementary outputs of the latch can be obtained with *either* positive-TRUE or negative-TRUE symbolism.

A common symbol for the latch is shown in Fig. 5-17. The state table at the right of the figure shows the behavioral action of the latch when it is

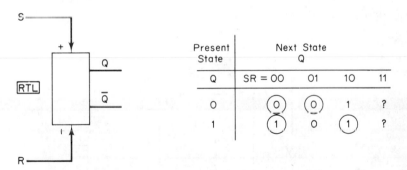

Figure 5-17. The NAND/NOR Latch Symbol and Behavior

given S and/or R signals. The table completely defines the action of the latch, which may be considered as a two-state sequential system. With neither control signal applied, the latch remains in one of its two stable states. Application of S will force the latch into the $Q = 1$ state. This will result in a change in state only if Q was FALSE prior to the setting signal. Removing the S input will leave the latch in the $Q = 1$ state. Similarly, an R signal forces $Q = 0$.

The state following the application of both S and R is indeterminant. During the time when both input signals are TRUE, both the Q and $\bar{Q}$ outputs will assume an output voltage equivalent to the TRUE output symbolism of an OR gate. This condition is logically inconsistent, since a Boolean variable and its complement can never have the same value. It is not a stable condition, however, and as soon as the simultaneous S and R signals are removed, the latch will assume either the SET or RESET state, with no predictability as to which stable condition will prevail. The simultaneous set and reset is useful in some specialized applications but, due to the uncertainty of the resultant next state, should generally be avoided. This condition, wherein the control inputs make a double change (from $SR = 11$ to $SR = 00$), violates the fundamental mode. Note that this is a restriction on the way the latch is *used*, not on the way it is designed.

For obvious reasons, the latch is commonly called a *flip-flop*. A more formal name is the *bistable multivibrator**, indicative of the two stable states that the latch can assume. The type of flip-flop shown in Fig. 5-17 can be specifically entitled a Direct-Coupled, Set-Reset flip-flop or DC S-R flip-flop. (DC here means direct-*coupled*, not direct current.) The simple latch circuit shown in Fig. 5-17 is not generally available from manufacturers because of the ease with which it can be constructed from two cross-coupled gates.

As an example of the use of this device, Fig. 5-18 shows how the DC S-R flip-flop can be used to store the value of a Boolean variable, A. Storage

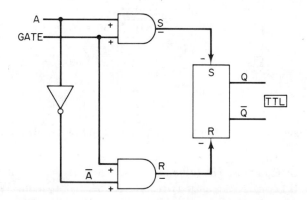

Figure 5-18. The D-C S-R Flip-flop Used to Store a Boolean Variable

*For an amusing variation on this device, see Ref. [13.]

devices of this type are often called *registers*, a name derived from digital computer applications. The TRUE or FALSE value of the variable is transferred to the flip-flop at the time the enabling signal, *GATE*, is made TRUE. If *A* is TRUE at that time, the *S* input will be enabled. When *GATE* is made FALSE, the flip-flop will remain in the state specified by *A*. Should the value of the variable *A* change during the time that *GATE* is TRUE, the flip-flop will follow the change and will be left in the state that *A* was in just prior to the time when *GATE* goes FALSE.

The arrangement shown in Fig. 5-18 is the type of sequential system that is shown, in general form, in Fig. 5-9. The primary input variables are *A* and *GATE*, the primary output variable is *Q*, and the excitation variables are *S* and *R*.

Despite its limited control capability, the DC S-R flip-flop can be used as a building block for useful sequential systems. As an example, consider the design of a digital combination lock.

EXAMPLE 5-7: A combination lock is to be built with digital components. A three-number combination is to be used, with input numbers ranging from 0 to 15 to be provided to the system by using four switches and binary numbering. After setting up the first number, the user must push a button marked *N1*. Next, the second number is set up and the *N2* switch is closed momentarily. The third number is then selected, and if all three are correct, the lock "opens" when *N3* is pressed.

Storage must be provided to hold the first two numbers. Eight flip-flops are required, labeled hereafter as A_1, A_2, A_4, and A_8 (first number) and B_1, B_2, B_4, and B_8 (second number). The logical connections for these flip-flops are shown in Fig. E5-7a. Switch *N1* acts as a gate signal to the A_1, A_2, A_4, and A_8 flip-flops; *N2* acts in the same way for B_1-B_8.

Prior to inserting the numbers, the system is given an *UNLOCK* signal, thereby clearing the *A* and *B* registers and the *RELEASE* flip-flop. Since all the A_i and B_i flip-flops are known to be reset, only those flip-flops that are to receive a TRUE variable need to be changed. Thus, the gating system shown in Fig. 5-18 can be simplified by eliminating both the inverter and the control gate that connect to the *R* input. The control gate for the *S* input is retained to permit a TRUE signal from one of the input switches to set its respective flip-flop. This gating system, the general reset followed by selective storage of TRUE values, uses only one control gate per bit of storage. It is somewhat serial in nature, however, since the *UNLOCK* signal must precede the gating signal in order to maintain the fundamental mode of operation. The gating system shown in Fig. 5-18 is a fully parallel system for inserting data into a flip-flop.

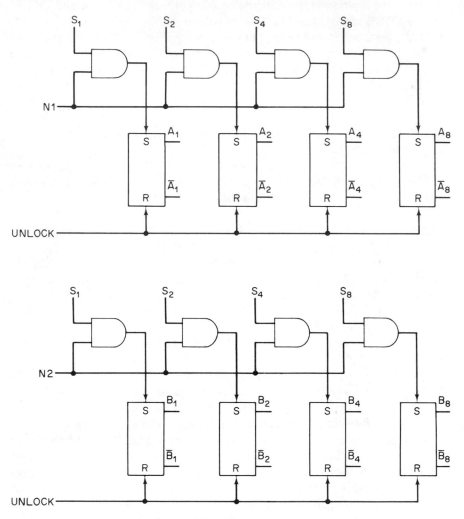

Figure E5-7a

The logic that detects the correct combination is shown in Fig. E5-7b. A combination of 8-11-3 is shown, although any other combination may be selected by changing the gating. When *N3* is closed, the *RELEASE* flip-flop will be set only if the proper combination has been given. The *UNLOCK* signal will reset the *RELEASE* flip-flop and the *A* and *B* storage registers to "close" the lock.

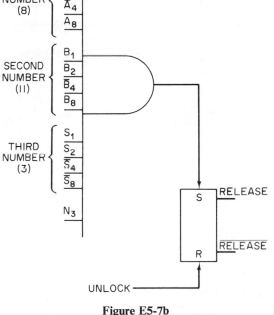

Figure E5-7b

Maintaining the fundamental mode of operation by not letting S and R both become TRUE at the same time will not solve all the problems inherent in the use of the flip-flop, however. In many counting and storage applications, for example, it is desired to have a flip-flop *toggle*, i.e., change from SET ($Q = 1$) to RESET ($Q = 0$) and back again on each successive control pulse. Figure 5-19 shows a method whereby a DC S-R flip-flop may be made to do this. The S and R inputs to the flip-flop have been gated with the output signals from the flip-flop so that the R input is enabled only when the flip-flop is in the SET state and the S input is enabled only in the RESET state.

This interconnection prevents the simultaneous application of the S and R control signals, thereby assuring fundamental-mode operation. The control input that can cause the flip-flop to change state is the only one that is enabled. Thus, when the *TOGGLE* signal becomes TRUE, the S or R input that is enabled will affect the desired change in the state of the flip-flop.

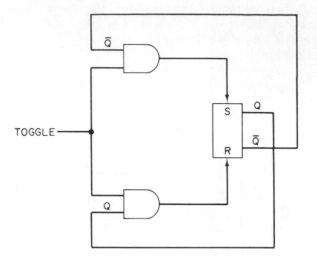

Figure 5-19. The D-C S-R Flip-flop in "Toggle" Connection

There is, of course, a delay between the time when the *TOGGLE* signal becomes TRUE and the Q and $\bar{Q}$ outputs from the flip-flop change. This delay is the sum of the delay through the control gate and the internal propagational delay of the flip-flop.

If the *TOGGLE* signal is made FALSE just after the change in state begins, then the flip-flop will make one state change. If, however, the *TOGGLE* signal remains TRUE long enough for the new output signals to appear at Q and $\bar{Q}$, the toggling action will repeat itself, since the change in output signals will change the enabled control input from S to R or R to S. Thus, as long as the *TOGGLE* signal is kept TRUE, the flip-flop will oscillate at a rate dictated by the net delay time in its feedback path. This oscillation will cease when *TOGGLE* is made FALSE, but the state in which the flip-flop will be left is difficult to predict.

This oscillatory behavior is certainly not the desired toggling behavior. It violates the first restriction given in Sec. 5.2.1, for one thing. Ideally, only one change of state should take place, no matter how long the *TOGGLE* signal is kept TRUE. This type of trouble is referred to as the *race* effect, whereby changes in the outputs of a sequential system create new input signals that, in turn, cause further changes in the outputs.

Several solutions to racing are available. One approach is to add additional delay between the outputs of the flip-flop and its gating. This delay is intended to keep the control signals constant until the *TOGGLE* signal goes FALSE. This delay will limit the speed with which the state of the flip-flop can be changed and, for any finite delay, an elongated *TOGGLE* signal will still bring on an oscillation.

A second solution is to use two flip-flops in a *double-rank* or *master-slave* configuration, as shown in Fig. 5-20. The *TOGGLE* signal is complemented so that when it is TRUE the inputs to the slave flip-flop are disabled and the inputs to the master are enabled. The gating shown in Fig. 5-20 will then transfer the complement of the state of the slave flip-flop into the master.

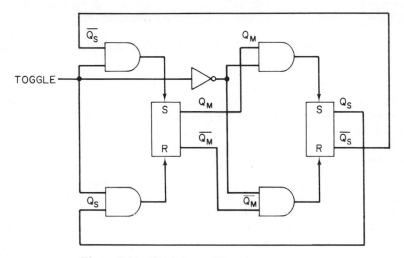

Figure 5-20. The Master/Slave Toggling Circuit

When *TOGGLE* is next made *FALSE*, the master flip-flop is disabled and its current value is transferred back to the slave, making the toggling action complete.

Double-rank counting systems and registers were common in early digital systems and have reappeared in the master-slave flip-flop configuration found in most TTL product lines. The integrated circuit master-slave flip-flops use internal biasing to enable and disable the master and slave as the clocking signal goes from FALSE to TRUE and back to FALSE.

5.2.3. The Pulse Mode.   The third and most widely used method for overcoming the racing hazard (and other related problems) is the use of the *pulse mode* of sequential operation. Unlike the fundamental mode (which is primarily a restriction upon the way in which a system is *used*), pulse-mode operation requires that the sequential system be *designed* in a specific way.

Pulse-mode sequential systems receive their primary input signals as before, but they require that one or more *pulse* inputs be included among them. A *pulse input* is one whose time duration is restricted: it must be long enough to adequately initiate the change from present state to next state and it must also be short enough so that it has come and gone before the secondary input variables have changed to their new values.

Pulse-mode operation also requires that all changes in the internal state of the sequential system take place only in response to a pulse signal. Pulse-mode systems normally use storage devices (rather than delay alone) as their memory, since the stored data that represents the present state must remain unchanged between pulses no matter how long the period between pulse inputs becomes. Delay devices cannot hold data indefinitely.

The pulse signals cause changes in the stored data to be made. These changes are dictated by the secondary output variables, or the excitation variables, if present. All state changes take place following the occurrence of a pulse input and are completed when the new values of the stored data appear at the secondary outputs. Only one state change will be made for each pulse input, since the width of the pulse is kept small enough so that the pulse is over before the new secondary output values appear.

Whenever no pulse signals are present at its inputs, commonly called the *null* input condition, a pulse-mode system *must* be in a stable present state, since the present-state variables are allowed to change only when a pulse is present. Thus, pulse-mode operation assures that cycles, races, and the other types of sequential instabilities that have been mentioned earlier do not occur. For this reason, most sequential digital systems are pulse-mode in nature.

The fundamental-mode restriction applies to pulse-mode systems as well, requiring that pulse inputs appear only when the system is in a stable state. Thus, two pulse inputs should not be allowed to occur so closely together that the sequential system has not completely settled from the first when the second appears. This requires that the null input state appear between successive pulse inputs.

The nonpulse input signals, commonly called *level* inputs, may change in logical value as desired, but the pulse-mode system will not change its present state so long as the null input condition (no pulse inputs) is maintained. The values of the level inputs must not change during the period while a pulse input is TRUE, however, since instabilities and erratic behavior may result from such changes. Pulse inputs are only TRUE for a short time, so that the level inputs are restricted from change only for brief periods. When required, special synchronizing circuitry can be used to assure that the level inputs to a sequential system do not change while a pulse input is TRUE. Synchronizing techniques are discussed in Sec. 5.6.

The basic arrangement for the storage portion of a pulse-mode sequential system is shown in Fig. 5-21, along with the general state table for this type of system. For simplicity, a system having only one pulse input is shown in the table. The pulse input enables all changes in the secondary input variables. Such changes are dictated by the values of the secondary output

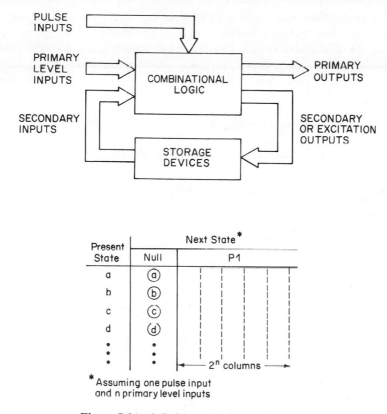

Figure 5-21. A Pulse-mode Sequential System

variables (or excitation variables), which, in turn, are developed by combinational action upon the primary and secondary input variables. Thus, for each present state (as represented by the present values of the secondary inputs), there will be 2^n possible next states, one for each combination of the n primary, nonpulse inputs. The pulse input is ANDed with the next-state information so that state changes take place only when the pulse is present.

The null input state is shown in the state table of Fig. 5-21. All present states are stable under the null input condition, as shown in the table. The present-state/next-state relationships that apply when the pulse input appears are shown in the remainder of the table. If more than one pulse input is present, then the state table must be expanded. Also, since all present states are stable when the null input condition is present, the null column is usually omitted from the table.

The primary outputs from a pulse-mode sequential system can be derived in either of two ways, as shown in general form in Fig. 5-22. If the primary outputs are derived from the present-state variables alone, then the system is said to be a *Moore machine* [11], Fig. 5-22a. The primary outputs from this type of sequential system are level signals, are independent of the primary input signals, and change values only when the state of the sequential system changes following a pulse input.

An alternative to the Moore machine is shown in Fig. 5-22b. The primary

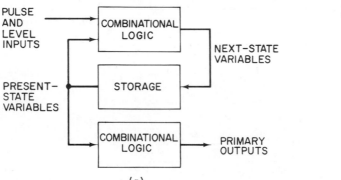

(a)

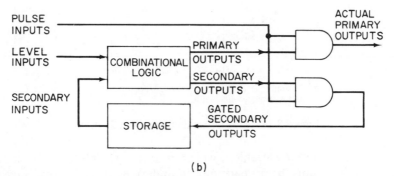

(b)

Figure 5-22. The Moore and Mealy Sequential Machines

outputs from this type of pulse-mode sequential system are determined from the *total* present state of the system. The pulse inputs are ANDed with the primary output signals so that the primary outputs are enabled only whenever a pulse input is TRUE. Thus, the output signals are pulses rather than levels. This type of sequential system is commonly called a *Mealy machine* [9].

Many sequential systems possess both Mealy and Moore characteristics. Conversions between the two forms can also be made (see Kohavi [5], p. 310, for example). Chapter 6 of this text gives design techniques that allow sequential systems of either (or mixed) type to be implemented.

In general use, the primary level inputs to a pulse-mode sequential system are used to *control* behavior, and the pulse inputs to such a system are used to *cause* the behavioral activity to take place. For example, a digital counter may receive an up/down control signal, say U, and a periodic timing signal. The U signal tells the counter whether to increment or decrement; it is a *level* signal. The timing signal is a pulse that tells the counter when to change states.

Many sequential systems have only *one* pulse input. Since this signal controls the times at which the system changes state, it is commonly called the *clock signal*, or simply the *clock*. Just as continuous variables are quantized for representation and processing in a digital system, *time* is quantized into a series of discrete intervals called *clock periods* in a clocked, pulse-mode system. Each occurrence of the clock pulse signifies the end of the current clock period, t_n or simply t, and the beginning of the next clock period, t_{n+1} or t_+. The clock signal may be periodic or aperiodic.

Changes in present state occur following each clock pulse. Such changes involve the storage of new information that will be required later and the removal from storage of any information that is no longer needed. During the period between clock pulses, the combinational portion of the sequential system logically combines the primary input signals and the stored information (the secondary or present-state inputs) and determines the primary output signals and the changes in stored information that should take place when the next clock pulse arrives. The changes in the stored information are controlled by the secondary output variables or the excitation variables.

The presence of only a single clocking pulse has definite advantages. Maintenance of the fundamental mode is simple, since coincidental or overlapping pulse inputs cannot occur. As a result, fundamental-mode operation is assured so long as two clocking pulses do not occur so close together that the sequential system cannot settle into stability during the null between them. This requirement fixes a maximum clocking rate (related to the minimum allowable time between pulses) above which the system may not operate properly. This rate is dictated by the settling time of the system, which, in turn, is determined by the propagational delays of the gates and storage devices that are used to construct the system. The estimation of maximum allowable clocking rates for single-pulse, pulse-mode, sequential systems is discussed briefly in Chapter 6.

EXAMPLE 5-8: A sequential system is to behave as a two-bit, up/down, binary counting device. Its general block diagram is shown in Fig. E5-8. The primary inputs to the system are the up/down control signal, U, and the pulse that acts as a counting signal. The secondary inputs to the system are y_1 (the 1 bit) and y_2 (the 2 bit), which represent the present state of the counter. There are no primary outputs. The secondary outputs are Y_1 and Y_2, which are the next-state values that will replace y_1 and y_2 when the next counting pulse appears.

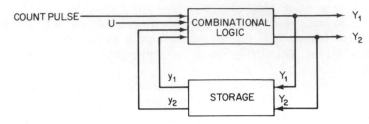

Figure E5-8

The behavior of this system can be described by the following state table:

Present State	Next States, Y_2 and Y_1		
$y_2\ y_1$	*Null*	*Pulse and $U = 0$*	*Pulse and $U = 1$*
0 0	0 0	1 1	0 1
0 1	0 1	0 0	1 0
1 0	1 0	0 1	1 1
1 1	1 1	1 0	0 0

When in the null input state, the system is always stable; i.e., the next state and the present state are the same. When a pulse is received and $U = 0$, the system counts downward, with the next state having a binary value that is one less than the binary value of the present state. When pulsed and $U = 1$, the system counts upward.

Since the present state is *always* stable as long as no pulses are present, the null input column is usually omitted from the state table of a pulse-mode system, as shown below:

Present State	Next States, Y_2 and Y_1	
$y_2\ y_1$	$U = 0$	$U = 1$
0 0	1 1	0 1
0 1	0 0	1 0
1 0	0 1	1 1
1 1	1 0	0 0

Also, as an alternative arrangement, the *total present state* can be used as the leftmost column of the state table. This type of table includes the values of the nonpulse primary inputs along with the usual present-state variables. The rearranged state table for the system of this example is shown below:

Total Present State			Next State	
U	y_2	y_1	Y_1	Y_2
0	0	0	1	1
0	0	1	0	0
0	1	0	0	1
0	1	1	1	0
1	0	0	0	1
1	0	1	1	0
1	1	0	1	1
1	1	1	0	0

The down-counting operation is shown in the top half of this table. The up-counting operation is shown in the bottom half.

5.2.4. The Transition Gate.

Often, the controlling input signals received by a sequential system are not in pulse form. In this case, they must be converted into pulse form, usually by a special pulse-narrowing circuit called a *pulse generator*. This device responds to a FALSE → TRUE change at its input, producing a narrow pulse when such a change in its input signal is received. (Some pulse generators respond to a TRUE → FALSE change.) After producing its output pulse, a pulse generator maintains a FALSE output until its input signal returns to FALSE and then changes back to TRUE again.

The period of time for which the controlling input remains TRUE is no longer of significance, except for the provision that it must be TRUE for long enough to initiate the pulse-generating operation. The pulse generation circuit responds to *changes* in its input signal rather than to the logical value of its input. This type of input is said to be *transition-sensitive*, or *edge-coupled*, indicating its sensitivity to changes in logical value.

Many pulse generation circuits include a level input that enables or disables the ability of a change of value on the transition-sensitive input to produce an output pulse. These two input types are often called *DC* inputs (for *direct-coupled*) and *AC* inputs (in contrast to the DC terminology). The use of the AC-input terminology is really a misnomer, since no alternating current considerations apply to a transition-sensitive input.

The common transition-sensitive gating scheme is shown in Fig. 5-23. The complete gate includes an enabling input that can block the action of

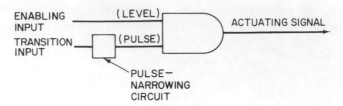

Figure 5-23. A Transition Gating Arrangement

the pulse from the pulse-narrowing circuit. For the transition gate to cause the flip-flop to change state, the enabling input must be at the proper voltage level *and* a transition in the proper direction (FALSE → TRUE or TRUE → FALSE) must appear at the transition input.

The transition gate may also be represented by the symbol shown in Fig. 5-24. Normal voltage symbolism can be used to define the TRUE voltage level required by the enabling input, but it cannot be used for the transition input, since that input responds to *changes* in level. For this input an arrow can be used to indicate that the gate is sensitive to positive-going or negative-going voltage transitions. The arrow points in the direction of the active transition. The arrow also indicates the "pulse" nature of the input.

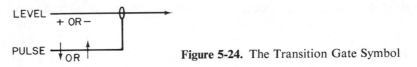

Figure 5-24. The Transition Gate Symbol

Changes in the enabling level at or near the occurrence of a clocking transition may violate the fundamental-mode restrictions and may or may not cause a change in state. Two important timing considerations arise:

1. *Setup time:* the period of time for which the enabling input must be TRUE prior to the occurrence of transition in order for the transition gate to affect the sequential system.

2. *Holdover time:* the time for which an enabling signal that changes from TRUE to FALSE will continue to cause the transition gate to be enabled.

These two timing factors are related to the time delays in the pulse-narrowing circuit of the transition gate. They affect the maximum allowable repetition rate of the clocking input.

The pulse-narrowing action of the pulse generator can be achieved in a variety of ways. In discrete component digital circuits, a resistance-capacitance (R-C) differentiating circuit is often used to generate a narrow pulse in response to a change in input value. The R-C transition gate is common to DTL integrated-circuit logic and some varieties of TTL as well. Its characteristics are discussed in Ref. [10].

In many integrated circuit configurations, however, the development of significant amounts of capacitance is costly, either in terms of processing steps or area upon the substrate. For these cases, special digital circuitry is used to provide the pulse-narrowing effect. The simple digital pulse-generating circuit described in Ex. 5-9 is used, in various forms, in RTL, TTL, and many other types of sequential integrated circuits.

EXAMPLE 5-9:   The digital circuit shown in Fig. E5-9 will produce a narrow pulse in response to a FALSE → TRUE change in its transition input, providing that the enabling level input is also TRUE. Describe the action of this circuit.

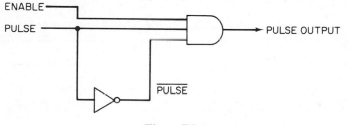

Figure E5-9

Answer: Assume that the *ENABLE* input is TRUE and that *PULSE* is FALSE. Thus, $\overline{PULSE}$ will be TRUE. The output from the gate will be kept FALSE, since *PULSE* is FALSE. When *PULSE* becomes TRUE, however, the delay of the inverter causes $\overline{PULSE}$ to remain TRUE for a short period. Thus, since all its inputs are momentarily TRUE, the output from the AND gate becomes TRUE. At the end of the delay period of the inverter the $\overline{PULSE}$ output changes to FALSE, making the output signal become FALSE as well. The pulse-generating activity can be reinitiated by returning the *PULSE* input to FALSE for a long enough period so that the $\overline{PULSE}$ output settles to a TRUE value.

EXAMPLE 5-10:   Reexpress the behavioral activity of the pulse-generating circuit of Ex. 5-9 in formal sequential systems terminology.

Answer: The pulse generation circuit can be redefined in accordance with the system model in Fig. E5-10. The delay of the inverter is separated and shown as a delay element. The output, *F*, is defined as

$$F = E \cdot P \cdot p$$

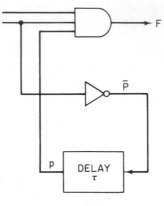

Figure E5-10

where p is the delayed value of $\bar{P}$. In state table form, the behavior of this circuit is

Present State	Next State				Output, F			
p	$EP = 00$	01	10	11	00	01	10	11
0	1	⓪	1	⓪	0	0	0	0
1	①	0	①	0	0	0	0	1

An output is produced only when the system is in total present state $EPp = 111$. This is an unstable state; it changes to $EPp = 110$ after a delay of τ sec. The 111 state can be reached from stable state $EPp = 101$ by changing the P input from FALSE to TRUE. This sequence is indicated by the arrows in the state table given above.

———————

With the inclusion of pulse-narrowing circuits between a sequential system and all its pulse-type inputs, pulse-mode operation is assured. Instabilities, races, cycles, and other sequential hazards cannot occur. For example, if the S-R toggling arrangement shown in Fig. 5-19 is modified to include a pulse generator between the latch and the *TOGGLE* signal, then correct toggling action is assured. This circuitry is shown in Fig. 5-25.

The pulse generator detects the FALSE $\rightarrow$ TRUE transition of the *TOGGLE* signal and sends a narrow pulse to the latch. This pulse, when ANDed with Q or $\bar{Q}$, produces the S or R signal which will cause the latch to change states. The narrow duration of the pulse assures that the control

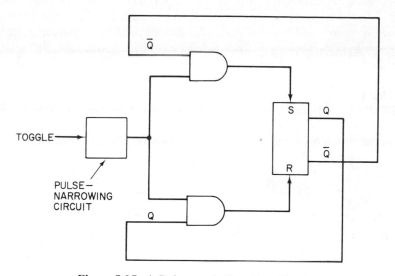

Figure 5-25. A Pulse-mode Toggling Circuit

inputs to the latch are disabled before any secondary state changes can be caused by the new conditions at the output to the latch.

The advantages of the pulse mode of operation are so great that most sequential digital systems are operated in the pulse mode. In fact, most types of sequential storage devices have internal pulse-narrowing circuitry, such as that shown in Fig. 5-25, built into their pulse inputs. (There are exceptions, e.g., the basic latch, the master-slave flip-flop, and some types of direct-coupled registers.) This circuitry assures that pulse-mode operation is obtained, even if the controlling input signals are not narrow pulses. Boolean storage devices of this type are called *clocked flip-flops, edge-triggered flip-flops,* or *AC flip-flops.* Several common types of clocked flip-flops are discussed in the next section.

5.3 *Clocked Flip-flop Types*

All the types of clocked flip-flops that are commonly used within the memory portion of sequential digital devices are built around the basic S-R latch. In addition, logical control circuitry is included between the external control inputs to the flip-flop and the S-R latch itself. These control inputs fall into three classes: *direct inputs,* which preset or reset the flip-flop through direct connection to the latch, *level inputs,* and *pulse inputs.* The last are also called *transition inputs.* The level and pulse inputs are combined by the logical and pulse-narrowing circuitry of the flip-flop. These inputs are most often used to initiate sequential behavior. The pulse inputs have internal pulse generators built into the circuitry of the storage device to assure pulse-mode operation.

Several common types of clocked flip-flops have been developed and are in general usage. These types of flip-flops have been given generally accepted names. Sections 5.3.1–5.3.5 introduce the four most common of these.

5.3.1. The AC S-R Flip-flop.
The simplest of the clocked or transition-sensitive flip-flop types is a direct extension of the DC S-R flip-flop. Two forms for this, the AC S-R flip-flop, are shown in Fig. 5-26. The type shown at the left of Fig. 5-26 has separated transition inputs, which may

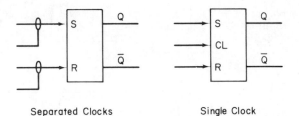

Separated Clocks   Single Clock

Figure 5-26. The AC S-R Flip-flop

be of use in sequential systems that receive more than one pulse input. The type shown on the right, common to integrated-circuit S-R flip-flops, has only one transition input (labeled *CL* for *CLOCK*) that is common to *both* the setting and resetting transition gates. The first type is the *split-clock* S-R flip-flop; the latter is the *single-clock* type.

The sequential behavior of the single-clock S-R flip-flop is shown in the following state table:

Present State	Next State Q_+			
Q	$SR = 00$	01	10	11
0	⓪	⓪	1	?
1	①	0	①	?

The present state of the flip-flop is shown as Q, with the state of the flip-flop following a clocking transition shown as Q_+. The state of the flip-flop is not changed by the clock pulse if both the S and R inputs are FALSE when the pulse occurs. If S is TRUE, however, the flip-flop is forced into the SET state $(Q - 1)$. It is forced into the RESET state if R is TRUE.

Note that if both the S and R inputs are TRUE when the flip-flop receives a clocking transition, then the next state of the flip-flop is indeterminant. This results from the fact that when both S and R are TRUE, the internal control circuitry passes the clocking pulse to both inputs of the latch. Thus, the setting and resetting inputs to the latch will receive nearly simultaneous pulses. This condition violates the fundamental-mode restriction, leading to uncertain behavior, as described in Sec. 5.2.2. The SR $= 11$ condition must be avoided.

5.3.2. The AC J-K Flip-flop. To make use of the undefined state that arises when both of the inputs to the S-R flip-flop are enabled, the outputs of the flip-flop can be fed back to the control inputs, as shown in Fig. 5-27. The extra gating serves to disable one of the set and reset transition

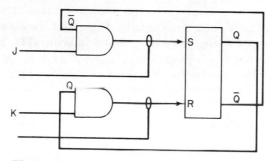

Figure 5-27. The Separated-clock AC J-K Flip-flop

gates. Whenever the flip-flop is SET, only the resetting gate is enabled, and vice versa. This does not limit the sequential behavior of the flip-flop, since the disabled input could not cause a change in state even if it were enabled. Thus, if both transition gates are enabled and clocked, only one of them will affect the flip-flop, and a single change in state (toggle) will take place. The complete sequential action of the J-K flip-flop is shown in the following behavioral table. Note the change in state when J and K are both enabled:

Present State	Next State Q_+			
Q	$JK = 00$	01	10	11
0	⓪	⓪	1	1
1	①	0	①	0

The inputs to this type of flip-flop are called J and K, as shown in Fig. 5-27, rather than S and R. An interesting comment on the origin of this nomenclature is given in Ref. [6]. The ability of the J-K flip-flop to give the toggling response when both J and K are enabled will often lead to a simplification of its control logic relative to that required by the S-R flip-flop. The pulse-narrowing and holdover effects of the transition gates are essential to the J-K operation, because of the racing problems that were previously described.

In addition to setup and holdover times, a third parameter is of concern when the J-K flip-flop is used. This is the *skew time*: the maximum spread between a *J* transition and a *K* transition that will still allow toggling action. Obviously, if the two signals are too far apart they will appear as separate SET and RESET commands. When a J-K flip-flop is constructed with integrated-circuit technology, the storage and holdover times within the transition gates are necessarily small, due to the low capacitance present in the miniaturized circuitry. As a result, the allowable skew between *J* and *K* inputs is so small as to force simultaneity. Thus, with the exception of a few devices (e.g., the *pulse-triggered binary* found in some DTL product lines), integrated-circuit J-K flip-flops are of the single-clock type, as shown in Fig. 5-28. The single-clock configuration is not a constraint on sequential systems

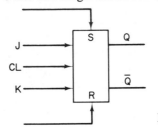

Figure 5-28. The Single-clock AC J-K Flip-flop

that receive a single clock pulse, since all flip-flops receive the same clocking signal, but it may limit the use of the IC J-K flip-flop in applications wherein multiple pulse sources are present.

5.3.3. The T Flip-flop. A third type of clocked flip-flop is shown in Fig. 5-29. This is the *toggle* or T flip-flop. Whenever its control input is enabled, each clocking pulse causes the flip-flop to change states, as shown in the following state table:

Present State	Next State Q_+	
Q	$T = 0$	$T = 1$
0	⓪	①↷
1	①	⓪

The T flip-flop can be constructed by joining the *J* and *K* inputs to a single-clock J-K flip-flop and using the pair as the *T* input.

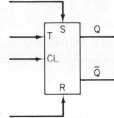

Figure 5-29. The T Flip-flop

5.3.4. The Use of the J-K Flip-flop. The S-R, J-K, and T flip-flop types are closely related. The J-K will behave exactly as an S-R flip-flop if its *J* input is used as *S* and its *K* input is used as *R*. The J-K will act as a T flip-flop if *J* and *K* are joined. Thus, the J-K flip-flop type alone can be used as a direct replacement to either the S-R or T types, and it still offers behavioral advantages over either. As a result, the J-K flip-flop type has become the most popular type of flip-flop configuration.

The single-clock J-K flip-flop is available in almost every type of IC or discrete-component type of digital logic. Its general usefulness and its widespread availability have made it the best choice as a memory device for sequential systems.

In keeping with this, most of the sequential design examples that are given in this text use the single-clock J-K flip-flop. General design techniques that can be used with any type of flip-flop are given in Chapter 6, and several examples showing design with each of the various flip-flop types are given, but the emphasis is upon the use of the J-K flip-flop type.

5.3.5. The D Flip-flop. One other type of flip-flop has found some acceptance and should be mentioned. It is the pulse-mode equivalent of the delay storage element and hence is called the *delay* or D flip-flop. Its basic configuration is shown in Fig. 5-30. When a clocking transition is

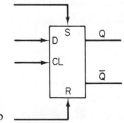

Figure 5-30. The D Flip-flop

received, the D flip-flop transfers the logical value that is present at its enabling input into the flip-flop.

The D flip-flop can be constructed from a J-K flip-flop and an inverter (see Problems 5-10 and 5-11), but it is not as closely related to the J-K flip-flop as are the S-R and T types. Thus, the D flip-flop is somewhat of an alternative to the J-K (rather than a relative) and should be considered as an alternative storage device. In some cases the use of the J-K will allow a cheaper system implementation, but in others, particularly storage registers, shift registers, and ring counters (sequential devices that are described in Chapter 6), the D type of flip-flop will allow a greater cost reduction.

5.3.6. Example Flip-flop Types: RTL and TTL.

In keeping with the practical orientation of this text, the example flip-flops that are used in the remaining chapters are typical of commercially available devices. The use of the single-clocked J-K flip-flop is of central concern, but techniques for using the S-R, T, and D flip-flops are also given.

The RTL and TTL J-K flip-flops are used as specific examples. For reference, the complete descriptions of the typical RTL and TTL flip-flops are given in Fig. 5-31. Note the mirrored input symbolism that is applicable

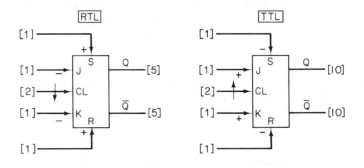

[Loading Factors shown in Brackets]

Figure 5-31. RTL and TTL Flip-flop Symbols

to the two types. The RTL J-K flip-flop responds to falling transitions; the TTL J-K flip-flop responds to rising transitions. Both types may have direct-coupled inputs for setting and resetting the flip-flop without using the transition-sensitive inputs. References to the RTL or the TTL flip-flop will imply the symbolism that is given in Fig. 5-31. The applicable input and output loading factors are also given in Fig. 5-31. The design techniques that are given hereafter can easily be extended to other types of flip-flops beside the RTL and TTL types.

It should be noted that some TTL J-K flip-flops are constructed in master-slave (rather than edge-triggered) form. In most applications the differences between these two forms are not really significant, other than the fact that most TTL master-slave flip-flops are sensitive to falling transitions, rather than to the rising transitions that activate their edge-triggered counterparts. The direct-coupling between the J-K inputs and the master flip-flop can cause trouble in some situations, however, especially if the *J* or *K* level inputs change state while the clocking input is TRUE. These changes violate the fundamental mode (in a sense) and may cause the subsequent behavior of the slave flip-flop to improperly reflect the values of its J-K control inputs. For these reasons, most manufacturers recommend that the clocking signals to master-slave flip-flops be kept TRUE for very brief periods. Level synchronization, as discussed in Sec. 5.6.2, can also be used to keep the *J* and *K* signals constant during the period when the clock input to a master-slave flip-flop is TRUE.

5.4 The One-shot

In addition to the flip-flop, other pulse-sensitive devices are used in many digital systems. One such device, the *monostable multivibrator* or *one-shot*, is useful for delaying pulses or transitions. A common symbol for the one-shot is shown in Fig. 5-32. The one-shot is controlled by a transition gate or, in some cases, by a direct-coupled input. When activated by a transition, the one-shot changes state. After a specified period of time it returns to its original quiescent state. The change in state is always such that the transition associated with the delayed return to the quiescent state is of the same polarity as the activating transition. Thus, a negative-sensitive one-shot will normally have low output voltage. When triggered, its output will go to the higher voltage for the delay period, *T*, and then return to the lower voltage, thereby producing a delayed negative transition. A timing diagram showing the behavior of a typical positive-sensitive one-shot is given in Fig. 5-33.

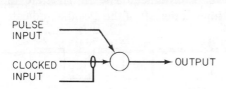
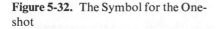
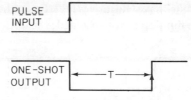
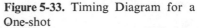

Figure 5-32. The Symbol for the One-shot

Figure 5-33. Timing Diagram for a One-shot

The period of the one-shot is set by its circuitry, usually by means of a resistance-capacitance charging circuit. Often, the period is made to be adjustable by means of an external resistance or capacitance. After being triggered, many types of one-shots must wait a specified *recovery time* before they can be retriggered.

One-shots are available in both discrete-component and integrated-circuit form. They are used for both delay and timing. They are generally considered to be asynchronous, since their delayed output pulse is not necessarily coincident with a system clock pulse. Applications of the one-shot will be shown in several examples that appear in later chapters. For further information on the design and use of the one-shot, see Ref. [4].

5.5 Timing Diagrams

In describing the action of flip-flops and other timing devices which are controlled by a clocking source, a diagram showing the time relationships among the various signals is often a most useful and descriptive design aid. The clocking signal is usually shown first, since it is the reference to which the activity of the devices is related.

The clock signal and other wave forms are normally shown in an idealized sense, with rise and fall times squared up and minor propagational delays omitted. Also, since a particular type of transition gate is sensitive only to either the rising or the falling transition of the clock signal, it is useful to mark this transition with an arrow so that it can be recognized. Signals can be shown with positive-TRUE or negative-TRUE symbolism. The symbolism and signal name should be included at the left of the timing diagram. The following examples illustrate this technique. Timing diagrams are used to describe sequential behavior in many of the examples in the remainder of this text.

EXAMPLE 5-11: The timing diagram in Fig. E5-11 shows the behavior of an S-R flip-flop under various combinations of control inputs. A flip-flop whose transition inputs are sensitive to falling transitions and whose S and R inputs have negative-TRUE symbolism is assumed. The initial state of the flip-flop is not specified; the dotted lines at the left of the Q signal indicate

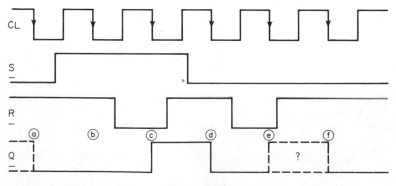

Figure E5-11

that Q can be either TRUE or FALSE. The S input is enabled at ⓐ, so the flip-flop is forced into the $Q = 1$ state. This condition is repeated at ⓓ. Both S and R are FALSE at ⓑ, so that the clock transition has no effect on the flip-flop. The R input is enabled prior to the clock transition at ⓒ, resulting in the $Q = 0$ state. Both S and R are enabled at ⓔ, so that the state of Q following that transition is indeterminate. The S input remains TRUE at ⓕ, however, while R is made FALSE. This assures $Q = 1$ following the clock transition at ⓕ.

EXAMPLE 5-12: The behavior of a J-K flip-flop under varying control conditions is shown in the timing diagram of Fig. E5-12. Again, negative-

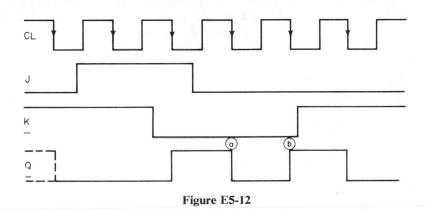

Figure E5-12

TRUE J and K inputs and a negative-sensitive transition input are assumed. Toggling action occurs at ⓐ and ⓑ, since both J and K are enabled.

5.6 Synchronization in Pulse-mode Systems

In most pulse-mode sequential systems, the primary inputs of both level and pulse type come from outside the system. As a result, these signals are not under the control of the system or its designer; they can change at any time and in any fashion. Section 5.2.3 shows how pulse generators can be used to enforce pulse-mode operation, and Sec. 5.3 introduces the common clocked flip-flop types which have internal pulse generators so that pulse-mode operation is assured. What of the fundamental-mode restriction, however? How can its satisfaction be guaranteed in a pulse-mode system?

Basically, the fundamental-mode restriction, when applied to a pulse-mode sequential system, requires that successive pulse signals cannot come too closely together. If pulses on two separate pulse inputs occur too closely or if successive pulses on the same pulse input occur without sufficient intervening null, then the sequential system may not reach stability between the pulses. This violates the fundamental-mode restriction.

Changes in level input signals may also cause erratic system operation, especially if any level input changes its logical value very close to the occurrence of a pulse input. In such a case, the change in the level input may or may not have propagated through all the delays in the sequential system. Some parts of the system may have recognized the new value when the pulse input occurs, while other portions of the system, due to longer delays, may respond as if the old input value were still present.

One solution to these problems is to place a restriction upon the user, making it his responsibility to see that the pulse and level input signals that he provides do not change so as to violate the fundamental mode. This is like placing speed limit signs in advance of a dangerous curve. Some may heed the warning; others may not, with disastrous results.

The second approach is to place protective circuitry between the inputs to the system and the system itself, so that the fundamental-mode restriction is enforced even if the actual input signals do not meet it. This approach corresponds to the erection of "speed breakers" (barriers or dips) that *force* drivers to slow down.

The interface devices that create fundamental-mode operation are commonly called *synchronizers*. They fall into two classes: synchronizers for pulse inputs and synchronizers for level inputs.

5.6.1. Pulse Synchronization.

The maximum rate at which a pulse-mode system can accept pulse inputs is determined by the worst-case settling time of the system. If pulse inputs arrive at an average rate that exceeds that maximum, then fundamental-mode operation cannot be enforced. If occasional pulses arrive in close proximity, however, but the *average* input pulse rate is slow enough to allow the system sufficient time to settle in between pulses, then synchronization techniques can be used to force a delay between successive pulses and hence to create the appearance of the fundamental mode.

Some synchronizing systems use one-shots (or some other delaying technique) to disable all pulse inputs for a sufficiently long period following the occurrence of a pulse. The disadvantage of this approach is that pulse inputs that occur during the delay period may be lost. A better synchronizing system uses flip-flops to store pulse inputs. Thus, if two or more pulses arrive in close proximity, the first pulse is passed on into the sequential system, while the others are held in storage. After a sufficient waiting period, the other pulses are allowed to enter, one at a time.

A common method for enforcing a delay period between pulse inputs is the use of one of the pulse inputs as a primary controlling signal, i.e., the system clock. Repetitive pulses are provided at this input, with the repetition rate set equal to or below the maximum allowed by the settling time of the sequential system. All other pulse inputs are stored temporarily and then entered in conjunction with the next clocking pulse. Thus, pulse inputs can never occur too closely together.

In a sense, this technique converts all pulse inputs except the clock pulse into level-type inputs by storing them prior to entry into the system. The clock pulses separate time into a sequence of intervals. One pulse is allowed to enter the system during each interval. If no pulse inputs are active, then the next interval passes without any activity taking place. The following example illustrates one technique for using a clock signal to enforce pulse separation.

EXAMPLE 5-13:   A pulse-mode, sequential system receives two pulse inputs, P_1 and P_2. The anticoincidence circuit shown in Fig. E5-13a makes use of

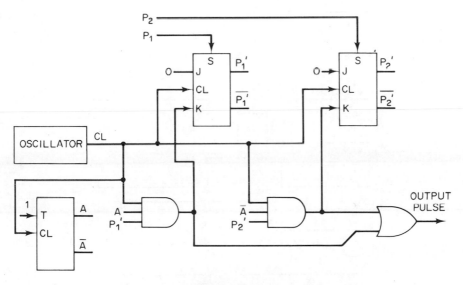

Figure E5-13a

a clocking oscillator that produces a sequence of narrow pulses, say *CL*, whose spacing is at least τ sec, where τ is the worst-case settling time of the sequential system. These pulses are used to form a series of "time slots" into which the input pulses are to be transferred. Since the input pulses may arrive at any time, they must be stored in flip-flops until their time slot occurs. Thus, the P_1 and P_2 pulses are used as direct setting inputs to two J-K (or S-R) flip-flops. The outputs from these flip-flops are labeled P_1' and P_2'.

Since there are two pulses, the clocking system must produce two separate time slots. A type T flip-flop (or a J-K flip-flop with its inputs joined) is used for this. When its output, say A, is TRUE, the P_1 time slot is enabled. When its output is FALSE, the P_2 time slot is enabled.

The separated output signals are generated by ANDing the stored pulse signals, the time-slot signals, and the clock pulses:

$$\text{Output Pulse} = (P_1' \cdot A \cdot CL) + (P_2' \cdot A \cdot CL)$$

The output pulses are also used to reset the flip-flops that store the original, unsynchronized pulses. This is accomplished through the K input to the flip-flop.

A timing diagram showing the separation of two nearly simultaneous P_1 and P_2 pulses in given in Fig. E5-13b. Negative-sensitive transition inputs are assumed.

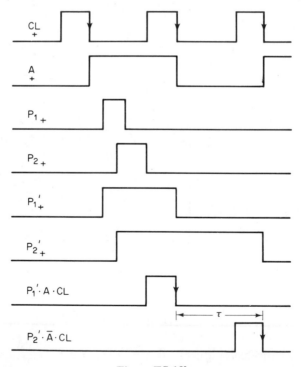

Figure E5-13b

A system similar to that of Ex. 5-13 can be used to separate a pair of pulses that appear at the same pulse input and are not sufficiently separated.

The modification of the anticoincidence circuit of Ex. 5-13 to suit that purpose is left to the problems in Sec. 5.11. A similar system can also be used to separate more than two pulses. It requires that more than two time slots be assigned.

5.6.2. Level Synchronization. As mentioned earlier, it is also essential that the level inputs to a pulse-mode, sequential system be synchronized so that they do not change in value just before or during the occurrence of a pulse input. If the pulse synchronization technique described in Sec. 5.6.1 is used, all pulse inputs are forced into synchronization with the clock, making it essential that level inputs do not change just ahead of a clock pulse.

Level synchronizers generally serve this purpose, keeping the level inputs that the sequential system actually sees constant during the period just ahead of each clock pulse. In fact, most level synchronization techniques cause all changes in the level inputs to occur just *after* a clock pulse, thereby allowing a full clock period for the level inputs to settle before the next clock pulse appears.

A type D flip-flop can be used for this purpose, since its output follows the logical value of its *D* control input. The unsynchronized level signal is used as the *D* input and the transition input to the flip-flop is tied to the system clock. Thus, the output from the flip-flop will follow the value of the *D* input, with its state changes occurring just *after* each clocking transition. The arrangement for using a D flip-flop as a level synchronizer is shown in Fig. 5-34, along with an example timing diagram.

One problem with the use of a single flip-flop as a level synchronizer is the appearance of a level input that becomes TRUE just before or just after a clock period and that has returned to FALSE before the next clock period, as shown at the right of Fig. 5-34. Such inputs are somewhat pulse-mode in nature, and a simple level synchronizer such as that of Fig. 5-34 may totally miss the presence of such an input.

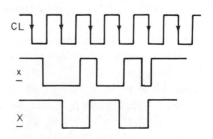

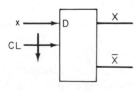

x = Unsynchronized Input
X = Synchronized Output

Figure 5-34. The D Flip-flop used as a Level Synchronizer

This type of signal can be "captured" and synchronized by the system shown in Fig. 5-35. The level input acts as a direct set to the latch. The output from the latch sets the Q flip-flop when the next clock pulse occurs. The latch remains SET as long as the level input is TRUE. When the level input goes FALSE, the latch is reset by the $Q \cdot \bar{L}$ logic. The $Q \cdot \bar{L}$ gating assures that the latch will not be reset until the synchronizing flip-flop has become SET.

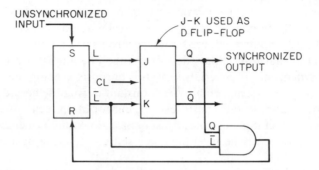

Figure 5-35. An Improved Level Synchronizer

Many other types of synchronization and signal conditioning circuits have been developed. Problem 5-8 shows how a latch can be used to remove the effects of switch "bounce." The *pulse-catcher*, a sequential subsystem that is useful for generating synchronous control signals from asynchronous switch or relay closures, is introduced in Sec. 7.3.

Other logical circuits of this genre are often presented in the "practical" literature. References [7] and [10] are also suggested as good sources for digital synchronization and signal-conditioning circuits. The major concern for the designer is that he recognizes when synchronization is needed and that he knows some good method (or methods) for providing it.

5.7   Clocking Signals in Pulse-Mode Systems

The timing pulses that control a pulse-mode sequential system must come from some source. Commonly, these pulses come from an oscillator that is called the *system clock*. This clock source is usually periodic. It provides a sequence of pulses that separate all time into discrete clock intervals. The clock signal need not be a square wave, although it often is. Each clock pulse may be very narrow. Some typical clock wave forms are shown in Fig. 5-36. In some systems *two* clock sources are present, each being 180° out of phase with the other. Such biphasic clock signals are also shown in Fig. 5-36. Some clocks have an enable control that allows or stops their oscillation. When

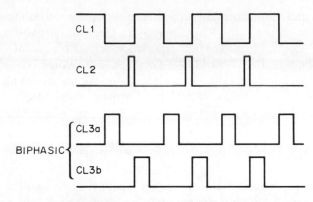

Figure 5-36. Typical Clock Waveforms

required, crystals or tuning forks can be incorporated into the clock oscillator to assure precise frequency control. Details of oscillator or clock circuit design are available in Refs. [10] or [12].

As a point of interest, two or more one-shots can be used to make a useful system clock oscillator, if precise frequency control is not required. Such a system is shown in Fig. 5-37. Once started, the delayed output from one-

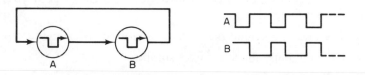

Figure 5-37. Two One-shots used as a Clocking Oscillator

shot *A* will trigger one-shot *B* and so on. The prime requirement for this type of clocking system is that the recovery times of each one-shot must be equal to or less than the duration of the delay of the other. Starting the system can be accomplished by an additional transition gate or a direct input to either one-shot.

5.8 Summary

Chapter 5 introduces sequential systems, i.e., digital systems that have storage or memory capability. The basic syntax that is applicable to sequential systems is defined, and state tables and diagrams are introduced as useful sequential system descriptors. Generalized sequential system specification involving present states having no specific state-variable assignment is also presented.

The instability problems that may arise in sequential systems are pointed out, along with the fundamental-mode and pulse-mode restrictions that help to eliminate these problems. The flip-flop (an elementary, two-state storage device) and its direct-coupled and transition-sensitive control inputs are described. The S-R, J-K, T, and D flip-flop types are defined.

Supporting comments concerning one-shots, the use of timing diagrams to describe sequential behavior, and synchronization techniques are introduced and illustrated by example.

5.9 Did You Learn?

1. The relationship between parallel (space-iterative) and serial (time-iterative) systems? Between present state and next state? Between present state and total present state?

2. The definitions of primary inputs and outputs? Secondary inputs and outputs? Next-state variables? Present-state variables? Excitation variables?

3. The function of the storage section of a sequential system?

4. The operation of the basic NAND/NOR latch?

5. The functions of direct control inputs? Of clocked (transition-sensitive) control inputs?

6. The meaning behind the fundamental-mode restriction?

7. How the pulse mode enforces stability?

8. How a pulse-narrowing circuit creates the pulse mode?

9. How to develop and use state tables and state diagrams?

10. How the transition gate combines a *level* and a *change in a level* in order to affect a flip-flop?

11. The definitions of the four flip-flop types?

12. The operation of the one-shot?

13. How to draw timing diagrams?

14. The voltage symbolism that is applicable to the example RTL and the TTL J-K flip-flops?

15. How a single-clock flip-flop differs from a separated-clock flip-flop? How a master-slave flip-flop differs from an edge-triggered flip-flop?

16. Why synchronization is needed in some pulse-mode systems?

17. How Mealy and Moore machines differ? How to generate both Mealy and Moore output signals?

5.10 References

[1] Givone, Donald D., *Introduction to Switching Circuit Theory*. New York: McGraw-Hill Book Company, 1970.

[2] Hill, Frederick J., and Gerald R. Peterson, *Introduction to Switching Theory and Logical Design*. New York: John Wiley & Sons, Inc., 1968.

[3] Huffman, D. A., "The Synthesis of Switching Circuits," *J. Franklin Inst.*, 257 (March 1954), 161–190, and (April 1954), 275–303.

[4] Hurley, R. B., *Transistor Logic Circuits*. New York: John Wiley & Sons, Inc., 1961.

[5] Kohavi, Zvi, *Switching and Finite Automata Theory*. New York: McGraw-Hill Book Company, 1970.

[6] Lindley, P. L., Letter to the Editor, *EDN* (August 1968).

[7] Maley, Gerald A., *Manual of Logic Circuits*. Englewood Cliffs, N.J.: Prentice-Hall, Inc., 1970.

[8] McCluskey, E. J., Jr., "Fundamental Mode and Pulse Mode Sequential Circuits," *Proceedings of the IFIP Congress 1962, International Conference on Information Processing*, Munich, August 27-September 1, 1962, Cicely M. Popplewell (ed.). Amsterdam: North-Holland Publishing Company, 1963, pp. 725–730.

[9] Mealy, G. H., "A Method for Synthesizing Sequential Circuits," *Bell Sys. Tech. Journal*, 34 (September 1955), 1045–1080.

[10] Millman, Jacob, and Herbert Taub, *Pulse, Digital, and Switching Waveforms*. New York: McGraw-Hill Book Company, 1965.

[11] Moore, E. F., "Gedanken Experiments on Sequential Machines," in *Automata Studies*. Princeton, N.J.: Princeton University Press, 1956, pp. 129–153.

[12] Sifferlen, Thomas, and Vartan Vartanian, *Digital Electronics with Engineering Applications*. Englewood Cliffs, N.J.: Prentice-Hall, Inc., 1970.

[13] Smith, E. Norbert, "The Not-so-famous Permistable Multivibrator," *EDN* (April 1968), 140.

5.11 Problems

5-1. A parallel arrangement for comparing the magnitudes of two binary numbers is described in Sec. 4.5.2. Describe the reorganization of that parallel comparator into a serial (time-iterative) comparison arrangement.

Illustrate the operation of the two comparators by example, using two six-bit binary numbers.

5-2. Convert each of the following state tables into state diagrams.

Present State	(a) Next State	(b) Next State	(c) Next State
0 0 0 0	1 1 1 1	X X X X	0 0 0 1
0 0 0 1	1 1 1 0	0 0 1 1	0 0 1 0
0 0 1 0	1 1 0 1	0 1 0 0	0 0 0 0
0 0 1 1	1 1 0 0	0 1 0 1	1 1 1 1
0 1 0 0	1 0 1 1	0 1 1 0	1 1 1 0
0 1 0 1	1 0 1 0	0 1 1 1	1 0 0 1
0 1 1 0	1 0 0 1	1 0 0 0	1 0 0 0
0 1 1 1	1 0 0 0	1 0 0 1	1 1 0 1
1 0 0 0	0 1 1 1	1 0 1 0	1 1 0 0
1 0 0 1	0 1 1 0	1 0 1 1	1 0 1 1
1 0 1 0	0 1 0 1	0 0 1 0	0 1 1 1
1 0 1 1	0 1 0 0	1 1 0 1	0 0 1 1
1 1 0 0	0 0 1 1	X X X X	0 1 0 1
1 1 0 1	0 0 1 0	1 1 1 1	0 1 0 0
1 1 1 0	0 0 0 1	X X X X	0 1 1 0
1 1 1 1	0 0 0 0	0 0 0 1	1 0 1 0

5-3. Convert each of the state diagrams in Fig. P5-3 into state tables.

5-4. A sequential system receives six primary inputs, generates ten primary outputs, and has five state variables. How many different present states can it take on? How many different primary input conditions can it receive? How many distinct total present states can it be in?

5-5. Circle all stable conditions on the following state table. Then convert the state table into a state diagram. Also, rearrange the state table into the *total* present-state form shown in Ex. 5-8.

Present State	Next States $I_1I_2 = 00$	01	10	11	Outputs, F_1 and F_2 $I_1I_2 = 00$	01	10	11
a	a	b	c	d	0, 0	0, 1	1, 1	1, 0
b	e	a	b	c	0, 1	1, 1	0, 0	0, 0
c	d	e	c	c	1, 0	0, 1	0, 1	0, 1
d	a	a	b	b	0, 0	0, 0	0, 0	0, 0
e	c	d	d	c	1, 1	1, 0	0, 1	0, 0

5-6. Describe the behavior of the logical system shown in Fig. P5-6 from a sequential systems viewpoint. Include both a state table and a state diagram.

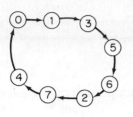

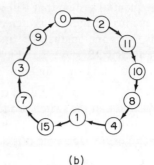

(a)   (b)

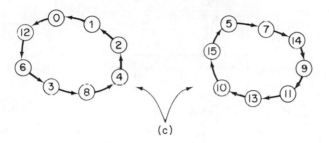

(c)

Figure P5-3

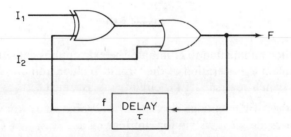

Figure P5-6

5-7.   An automatic system is to check a driver's sobriety prior to allowing him to start his car. To enable the ignition switch, the driver must perform the following sequence of actions with no errors:

    a. Turn on the radio, R.
    b. Push in the cigarette lighter, C.
    c. Set the turn indicator to LEFT, L.
    d. Shift into reverse, RV.
    e. Sound the horn twice, H.

Assuming that each of these actions generates a pulse signal, describe a general sequential system that will check his performance. Assume that inserting the key into the ignition switch clears the system, that the correct sequence of inputs starts the engine, *SE*, and that an incorrect sequence sets the system into an ERROR state, *E*, which forces the driver to walk home. Give both a state table and state diagram description of the system.

*5-8. A switch is to be used to provide clock pulses to a sequential system. It is desired that the switch produce only *one* clock pulse each time it is closed. Because of mechanical bouncing, the switch will produce more than one contact closure each time it is activated. The circuit shown in Fig. P5-8 uses

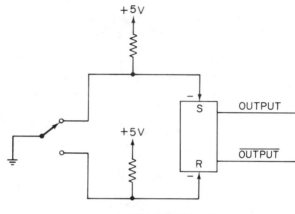

Figure P5-8

a SPDT switch and a latch to eliminate the extra pulses caused by the switch bounce. Explain the operation of the circuit. Include a timing diagram showing several switch reversals. TTL circuitry is assumed.

5-9. If the switch bounce described in Prob. 5-8 is known to last for 15 msec or less, design a system for eliminating its effects by using a one-shot.

*5-10. Show the input control logic that will make a J-K flip-flop act as a type D flip-flop. Also show how to make a J-K act as a type T flip-flop. Consider both RTL and TTL devices, showing complete logic diagrams for each case.

*5-11. Show the input control logic required to make a T flip-flop and a D flip-flop act as a J-K. Also show how to make a T flip-flop act as a D flip-flop and vice versa. Assume that both the T and D flip-flops require negative-TRUE level inputs and are sensitive to positive-going transitions. Use RTL gates.

5-12. The device shown in Fig. P5-12 is a proposed new flip-flop type, the X-Y flip-flop. It is made from an S-R flip-flop by adding the interconnec-

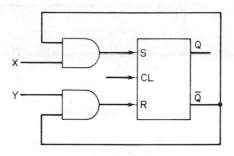

Figure P5-12

tions shown in Fig. P5-12. Derive the behavioral table for this device. Can it be used in place of any of the other flip-flop types?

5-13.   The device shown in Fig. P5-13 is a second new flip-flop type that has been proposed, the K-J flip-flop. Determine its behavioral table. What uses can this type of flip-flop have?

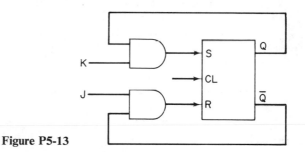

Figure P5-13

*5-14.   Two RTL J-K flip-flops are interconnected as shown in Fig. P5-14. Draw a timing diagram showing their operation through six complete clock cycles. Assume that $AB = 00$ initially.

*5-15.   Repeat the timing analysis of Prob. 5-9 for the interconnected T and D flip-flops shown in Fig. P5-15. Assume that $TD = 00$ to begin with.

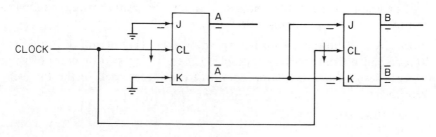

Figure P5-14

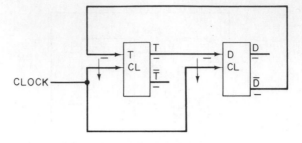

Figure P5-15

5-16. A partially completed timing diagram for a J-K flip-flop is shown in Fig. P5-16. Complete the diagram assuming that an RTL flip-flop is being used.

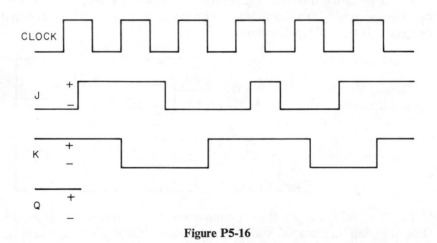

Figure P5-16

5-17. Rework Prob. 5-11 assuming that a TTL flip-flop is being used.

5-18. Show a modification of the pulse separation circuit of Ex. 5-13 that will separate pulses that appear on the *same* pulse input. Give a complete logic diagram for the system and also a timing diagram showing the separation of two closely occurring pulses.

*5-19. A particular one-shot has an adjustable delay period of from 10 to 20 μsec and a fixed recovery time of 15 μsec. Using two or more of these devices, design clock oscillators that will operate at each of the following frequencies.

    a. 10 KHz
    b. 20 KHz
    c. 25 KHz
    d. As fast as possible

Show the interconnections between one-shots, their delay settings, and timing diagrams.

*5-20.   Design an edge-triggered J-K flip-flop by using only 2- and 3-input RTL gates. Use the pulse-narrowing technique that is described in Ex. 5-9. Estimate the upper frequency limit for toggling this flip-flop.

*5-21.   Repeat Prob. 5-20 for TTL gates.

*5-22.   Modify the flip-flops of Probs. 5-20 and 5-21 to D and T form.

6 The Design of Sequential Systems

6.0 Introduction

Sequential system design is introduced in Chapter 5. It is now to be explained in detail and illustrated by example. The design process can be considered as a step-by-step procedure, just as combinational design can be separated into specification, minimization, and implementation. In fact, the combinational process serves as a portion of the larger design procedure that is used with sequential systems.

Figure 6-0 gives a fairly comprehensive outline of the sequential design process. Once a sequential system has been specified, the remainder of its design involves three distinct aspects: state reduction (to reduce the number of distinct present states and thereby possibly reduce the storage requirements); state assignment (to simplify the combinational requirements); and the design of the combinational logic that generates the primary output signals and controls the movement from present state to next state.

The combinational logic that generates the primary output signals is best designed by using normal combinational design techniques. The variables that define the total present state are used as inputs to this logic. The combinational expressions that are used to generate the output signals can be minimized by using the K-mapping or tabular techniques that are given in Chapter 4. As is the case for any combinational expression, the output logic can be implemented with any type of digital gating circuits by using the voltage symbolism technique.

The remainder of the combinational logic that is required by the sequential system is used to generate the secondary outputs (or excitation variables)

262

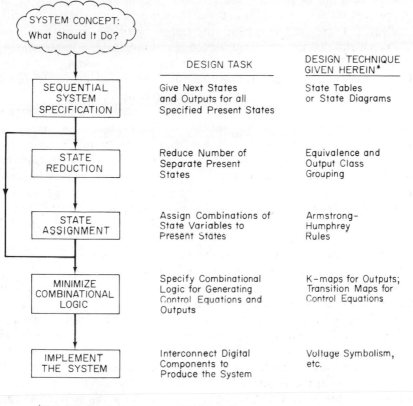

DESIGN TASK	DESIGN TECHNIQUE GIVEN HEREIN*

SYSTEM CONCEPT: What Should It Do?

SEQUENTIAL SYSTEM SPECIFICATION — Give Next States and Outputs for all Specified Present States — State Tables or State Diagrams

STATE REDUCTION — Reduce Number of Separate Present States — Equivalence and Output Class Grouping

STATE ASSIGNMENT — Assign Combinations of State Variables to Present States — Armstrong–Humphrey Rules

MINIMIZE COMBINATIONAL LOGIC — Specify Combinational Logic for Generating Control Equations and Outputs — K–maps for Outputs; Transition Maps for Control Equations

IMPLEMENT THE SYSTEM — Interconnect Digital Components to Produce the System — Voltage Symbolism, etc.

*Other Design Techniques are Cited in References

Figure 6-0. The Sequential Design Process

that control the movement from state to state. This portion of the system, the memory elements and their associated control logic, is often called a *counter*, since the movement from one state to the next is very much like the counting operation found in clocks, odometers, etc.

This chapter is concerned with the design of sequential systems that use flip-flops for memory. The transfer from the present state to the next state is controlled by the level inputs to these flip-flops. These input signals (J, K, S, R, etc.) are related to, but not necessarily equivalent to, the state variables themselves. Rather, with the exception of the type D flip-flop, the control inputs are excitation variables, as shown in Fig. 5-9. Since the output of the D flip-flop follows its control input directly, the D control input is not an excitation variable; it serves directly as the next-state signal, as shown in Fig. 5-8. The flip-flop outputs are the present-state variables in all cases.

The combinational design portion of the sequential design process relates each combination of present-state variables and primary inputs (i.e., each total present state) to the values of the control inputs (next-state or excitation variables) that are needed to cause the transfer into the correct next state. This combinational control logic should be minimized and implemented in accordance with the techniques given in Chapters 2, 3, and 4. This portion of the sequential design process can be called *counter design* (which should not be confused with terms such as counterespionage).

Chapter 6 begins by introducing *transition mapping*, a straightforward technique for converting the present-state/next-state information that describes the counter-like portion of a sequential system into the minimized control equations that make the system operate. In fact, the transition mapping technique converts the design of a counter from a sequential design problem into a combinational design problem. Transition mapping allows the control equations to be minimized in both SOP and POS form. It can be extended to tabular and computer-aided methods for problems involving more than five input variables.

The transition mapping technique also makes the generation of the control equations for different types of flip-flops quite easy, so that the costs involved in using each type can be compared. In practice, however, only the J-K and D flip-flop types need be compared, since the S-R flip-flop can never have cheaper control logic than the J-K, and the advantageous use of a T flip-flop will become apparent whenever the control logic for the *J* and *K* inputs to a given flip-flop is the same.

The emphasis in this text is upon pulse-mode sequential systems. For this class of systems, sequential behavior is controlled through the level inputs to the flip-flops. Clocking pulses actually *cause* the system to change states, however. The design techniques used with both single-clock systems (called *ripple-free* or *synchronous* systems herein) and systems that allow internal signals to act as secondary pulse inputs (called *ripple* systems or *asynchronous** systems herein) are considered.

The development of these techniques begins with the simplest type of sequential system (the single-mode counter), proceeds to systems having more complexity (multimode counters), and then proceeds to completely general sequential design techniques, with basic state reduction and state assignment techniques being given at that point.

Additional material dealing with special-purpose sequential systems and other supporting topics is included at the conclusion of this chapter.

---

*Many sources use the term *asynchronous* to describe non-pulse-mode systems. In this text, the term is used to describe pulse-mode systems having more than one pulse input, i.e., ripple systems. This choice of terminology contrasts an asynchronous system with a synchronous system, i.e., one having only *one* pulse input.

6.1   The Counter

The need for designing a counter can arise in a variety of ways. Many digital systems have functional subsystems that act as sequential counters. A digital time-of-day clock has several such counters; its seconds counter counts from 0 to 59 and resets, as does its minutes counter. Its hours counter repeatedly counts from 1 to 12 (or 0 to 23 for a 24-hour clock). Similarly, once a state assignment has been made for a general sequential system, its design is reduced to a counter design process, with the present-state/next-state sequence defining the counting operation.

In general, a counter may be thought of as any sequential device that uses flip-flops for arithmetic operations, for sequential signal generation, or, of course, for counting. The basic operational concept of a counter is that for any present state (defined by the present SET/RESET condition of each of its flip-flops) there is a well-defined next state. During the counting operation the counter moves from present state to next state in accordance with its specified counting sequence. The design of a counter centers around selecting the combinational control logic that interprets the present state and properly enables or disables the control inputs to each flip-flop so that, upon receipt of the next clock pulse, each flip-flop will be set, reset, or left unchanged, and the counter will transfer to the proper next state.

The input signals that a counter receives may be limited in number. A direct reset signal is usually provided to permit the counter to be forced into a known initial state. A counting signal, usually a pulse, is also required. In addition, control signals that vary the counting sequence may be provided, although in many applications no such signals are required.

The simplest of all types of sequential systems is the single-mode counter, shown in general form in Fig. 6-1. Note that this type of counter has *no* primary inputs. As a result, each present state is followed by only one next state. The counting sequence is generally well specified. Such a counter must exhibit at least one closed path in its state diagram. The length of the path must be 2^N states or less, where N is the number of flip-flops that are present in the memory portion of the counter. There may be more than one closed path.

The desired counting sequence can be specified by table, by diagram, or by an extension of the Karnaugh map, commonly called a *transition map*. The next two sections discuss these methods for describing the sequential operation of a counting device.

6.1.1.   Describing the Counting Sequence.
State-to-state maps and diagrams, two techniques for describing a sequential system, are introduced in Sec. 5.1.2. As an example of their use with a single-mode counter, consider

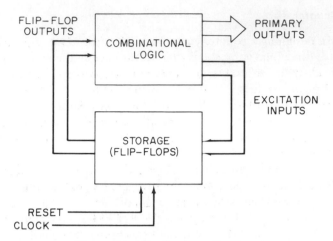

Figure 6-1. The Single-mode Counter

a three-flip-flop counter that counts upward in binary from 0 (000) to 7 (111). From 7 it resets to 0 and continues counting. The flip-flops may be labeled A (the 4 bit), B (the 2 bit), and C (the 1 bit). Figure 6-2a shows a state diagram for this counter. Note that the state of each flip-flop is shown in the circle representing the present state. Movement from state to state is indicated by arrows. A more compact form for the state diagram is shown in Fig. 6-2b. In the latter diagram the minterm number formed by the three flip-flops is used to define the states. This type of state diagram is used throughout this text.

The counting sequence can also be tabulated in a state table. The state table for the three-bit binary up-counter is presented in Table 6-0. It is equiva-

Table 6-0: State Table for a Three-Bit Binary Up-counter

Present State			Next State		
A	B	C	A_+	B_+	C_+
0	0	0	0	0	1
0	0	1	0	1	0
0	1	0	0	1	1
0	1	1	1	0	0
1	0	0	1	0	1
1	0	1	1	1	0
1	1	0	1	1	1
1	1	1	0	0	0

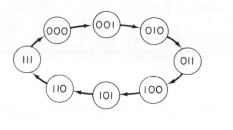

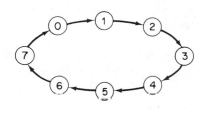

(a)

(b)

Figure 6-2. State-to-state Diagrams for a Three-bit Binary Up-counter

lent to the state diagrams of Fig. 6-2. Table 6-0 may be thought of as three separate tables that define the state of each of the three flip-flops following the clocking transition that signals the end of the present state. Note that since the counter is a pulse-mode system, each of its present states is stable between clocking pulses. The null input column is omitted from Table 6-0.

It is significant to note that the next state of any one of the flip-flops cannot be determined from knowledge of its present state alone. Rather, each next state is determined by considering the total present state of the entire counting system, i.e., by considering the present state of *all* the flip-flops that make up the counter. Therefore, the combinational control logic for each flip-flop must interpret the present states of all the other flip-flops in order to operate properly.

In many counters all possible present states are not present in the main counting sequence. Such counters are said to be *incompletely specified*, since no required next state is given for the unused states. The unspecified next states are useful as don't-care conditions in the design of the counter and generally lead to a simplification of the combinational control logic required by the counter. Such states are normally omitted from a state diagram and are marked by Xs in the next-state column of the state table.

6.1.2. Transition Mapping (Refs. [2], [3], [5], [14], [15], [16], [22]). A third approach to the specification of a counting system is the use of a set of maps that define the behavior of each particular flip-flop for each of the specified present states of the counter. For combinational functions the Karnaugh map with entries of 0, 1, and *X*, for TRUE, FALSE, and don't-care, was a sufficient descriptive tool. In counting systems, however, these three entries are not enough. The flip-flop has *two* stable present states, each of which can go to either of *two* possible next states upon receiving a clocking transition. This leads to *four* possible behavioral actions. Transition mapping extends the Karnaugh mapping technique to cover this type of activity.

The behavior of a flip-flop from one clock period to the next can be expressed symbolically as one of the following four conditions:

1. If the flip-flop is initially in the 0 state (RESET) and remains in the 0 state after being clocked, its behavior is *static* and can be represented by the symbol 0, indicating that the flip-flop is RESET in both the present and the next state.

2. If the flip-flop is in the 0 state and changes to a 1 state (SET), the behavior is *dynamic* and can be represented by the symbol α. [Of note is the fact that static behavior does not require *any* control stimulation to the flip-flop (it can be left unactivated), while dynamic behavior *requires* a control stimulus to cause the flip-flop to change state.]

3. If the flip-flop is in the 1 state and changes to the 0 state, the symbol β can be used. This, of course, is dynamic behavior.

4. If the flip-flop is in the 1 state initially and remains therein following the clock signal, the symbol 1 can be used.

Note that the symbol *X* can still be used to represent a present state with an unspecified next state. These definitions are tabulated in Table 6-1.

The behavioral activity of each flip-flop in a counter is determined by comparing the present states and the next states given in the state table.

Table 6-1: The Transition Mapping Symbols

Present State Q	Next State Q_+	Symbol on Map	Type of Behavior
0	0	0	Static
0	1	α	Dynamic
1	0	β	Dynamic
1	1	1	Static
0	X	X	—
1	X	X	—

The proper behavioral symbol is then entered into the transition map for each flip-flop, *with the location of the symbol being specified by the minterm formed by the present states of all the flip-flops that make up the counter*, i.e., the total present state. This follows from the necessity that the control logic for each flip-flop interpret the total present state in order to set up the flip-flop activity associated with transfer to the proper next state.

Consider, once again, the three-bit binary up-counter which was tabulated in Table 6-0. The first entry in its state table was

Present State			Next State		
A	B	C	A_+	B_+	C_+
0	0	0	0	0	1

Comparing A to A_+ and B to B_+ shows that both exhibit 0 behavior from present state m_0. The status of the C flip-flop changes from RESET to SET, however, which is designated as α behavior. Continuing this comparison process gives Table 6-2.

Table 6-2:   The Behavioral Table for the Three-bit Binary Up-Counter

Present State			Next State			Behavior		
A	B	C	A_+	B_+	C_+	A	B	C
0	0	0	0	0	1	0	0	α
0	0	1	0	1	0	0	α	β
0	1	0	0	1	1	0	1	α
0	1	1	1	0	0	α	β	β
1	0	0	1	0	1	1	0	α
1	0	1	1	1	0	1	α	β
1	1	0	1	1	1	1	1	α
1	1	1	0	0	0	β	β	β

Once the Behavior column is known, the Next-State column can be omitted, since the behavioral activity can be combined with the present state to obtain the next state.

EXAMPLE 6-0:   The behavior specified for present state m_3 in Table 6-2 is $\alpha\beta\beta$. Determine the next state.

Solution: The present state is

$$m_3 = \bar{A}BC$$

The $\alpha\beta\beta$ behavior specification requires that in the next state the A flip-flop change from RESET to SET and that the B and C flip-flops become RESET. Thus, the next state is

$$A\bar{B}\bar{C} = m_4$$

or, in state diagram form:

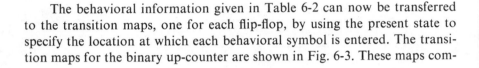

The behavioral information given in Table 6-2 can now be transferred to the transition maps, one for each flip-flop, by using the present state to specify the location at which each behavioral symbol is entered. The transition maps for the binary up-counter are shown in Fig. 6-3. These maps com-

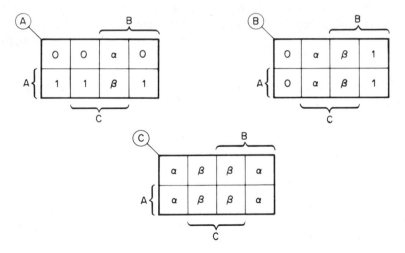

Figure 6-3. Transition Maps for the Three-bit Binary Up-counter

pletely describe the counting sequence; no additional information is needed. For each present state the behavioral symbols show the change (or lack of change) that each flip-flop will make in going to the next state. Thus, the state table and/or state diagram can be reconstructed from the transition maps. An example of inverse transition mapping follows.

EXAMPLE 6-1: The transition maps for a three-bit counter are shown in Fig. E6-1a. Develop a state table and state diagram for the counter.

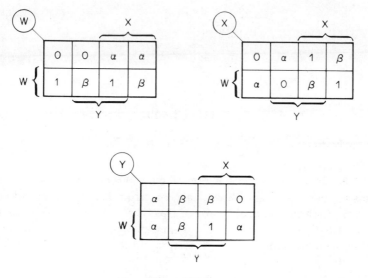

Figure E6-1a

Solution: The inverse transition mapping process can begin with present state $m_0 = 000$. The behavior from this present state is given as 00α (taken from the m_0 location on all three maps). Thus, the next state will be $m_1 = 001$, since W and X remain RESET and Y becomes SET. The behavior specified at m_1 is $0\alpha\beta$, making the next state be $m_2 = 010$ (W stays RESET, X becomes SET, and Y becomes RESET). Continuing this process through the remaining six states gives the following state table:

Present State			Next State		
W	X	Y	W_+	X_+	Y_+
0	0	0	0	0	1
0	0	1	0	1	0
0	1	0	1	0	0
0	1	1	1	1	0
1	0	0	1	1	1
1	0	1	0	0	0
1	1	0	0	1	1
1	1	1	1	0	1

The resultant state diagram is as shown in Fig. E6-1b, indicating that this counter has two separate closed counting paths.

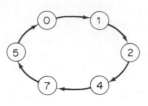

Figure E6-1b

6.1.3. Errors in Transition Mapping. While the transition mapping process is both simple and direct, careless errors are common, especially to beginners. Such errors are generally of two types: using the wrong behavioral symbol (usually mixing an α and a β) and placing the symbol in the wrong place on the map. Simple visual inspection of the final maps can detect many such errors.

Consider the transition map shown in Fig. 6-4, which describes the behavior of the A flip-flop in a four flip-flop counter. The transition map

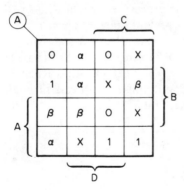

Figure 6-4. An Example Transition Map

may be considered as having two parts, one wherein A is FALSE in the present state (the upper half) and one wherein A is TRUE in the present state (the lower half). The known present state of A in each half of the map limits the type of behavior that the A flip-flop can have in that region. For example, if A is initially RESET, then its subsequent behavior can only be 0 (staying RESET), α (being SET), or, of course, X (indicating an unspecified next state). Thus, the only valid behavioral symbols that can appear in the $\bar{A}$ region of the transition map for the A flip-flop are 0, α, and X. The example map given in Fig. 6-4 contains at least two errors in its $\bar{A}$ region, the 1 at m_4 and the β at m_6.

Similarly, the only valid symbols that can appear in the A region of the A map are β, 1, and X, since the present state of the A flip-flop is known to be SET in that region. The example map given in Fig. 6-4 contains errors at m_8 and m_{15}.

This simple inspection should be applied to the transition map for each of the flip-flops that make up a counter, noting that the "halves" of the map will have different locations on the various transition maps, as shown in Fig. 6-5. This inspection process will not catch all possible errors, but it is very effective in detecting most careless errors. The best protection is accuracy, especially in locating the proper entry points on each map, remembering that entry position is specified by the *present state* of the counter.

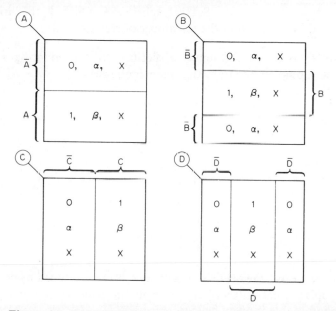

Figure 6-5 The Limitations on Behavioral Symbol Locations

EXAMPLE 6-2: The transition maps that define a four flip-flop counter are given in Fig. E6-2. List the locations that contain incorrect behavioral symbols.

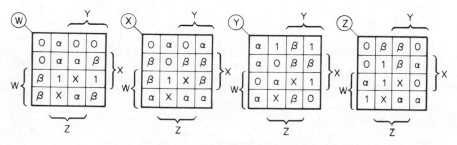

Figure E6-2

Solution: Consider the Z map. The Z region should contain only 1s, βs, or Xs. The α at m_{11} is an error. The $\bar{Z}$ region should contain only 0s, αs, and Xs. The 1 at m_8 is an error. Similarly, errors are present at

$$Y \text{ Map:} \quad m_1, m_{10}$$
$$X \text{ Map:} \quad m_5$$
$$W \text{ Map:} \quad m_6, m_{11}$$

6.1.4. More Examples.   Several more examples of the transition mapping process are given in the remainder of this section.

EXAMPLE 6-3:   A four-bit counter is specified by the counting sequence given below. Complete the transition mapping process.

$$A \quad B \quad C \quad D$$

$$
\begin{bmatrix}
0 & 0 & 0 & 0 \\
0 & 0 & 1 & 0 \\
1 & 0 & 0 & 1 \\
1 & 1 & 0 & 0 \\
0 & 1 & 1 & 0 \\
1 & 0 & 0 & 0 \\
0 & 1 & 1 & 1 \\
0 & 1 & 0 & 1 \\
1 & 1 & 1 & 0 \\
0 & 1 & 0 & 0
\end{bmatrix}
$$

Solution: First, the behavioral activity table is developed by comparing each flip-flop's present state with its next state (the state below it in the table). The behavioral table is given below:

Present State				Minterm Number	Behavior			
A	B	C	D	m_i	A	B	C	D
0	0	0	0	0	0	0	α	0
0	0	1	0	2	α	0	β	α
1	0	0	1	9	1	α	0	β
1	1	0	0	12	β	1	α	0
0	1	1	0	6	α	β	β	0
1	0	0	0	8	β	α	α	α
0	1	1	1	7	0	1	β	1
0	1	0	1	5	α	1	α	β
1	1	1	0	14	β	1	β	0
0	1	0	0	4	0	β	0	0

Note that the minterm description of each present state has been included as an aid to locating the correct entry points on the transition maps. The behavior from the six unspecified states (m_1, m_3, m_{10}, m_{11}, m_{13}, and m_{15}) is plotted as an X.

The completed transition maps are given in Fig. E6-3.

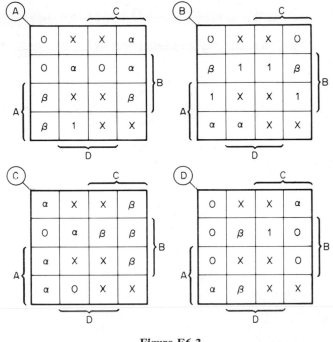

Figure E6-3

EXAMPLE 6-4: The counting sequence and final transition maps for a second counter, the 8421 BCD down-counter are shown in Fig. E6-4.

6.2 Designing the Counter

Once the transition maps that specify the behavior of the counting portion of a sequential system have been completed, the type of flip-flop that will be used to make the counter must be selected. The specified behavior of each flip-flop must then be related to its particular set of control inputs. This process converts the design of the sequential system into a combinational

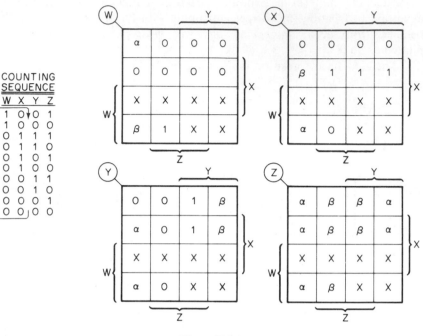

COUNTING SEQUENCE

W	X	Y	Z
1	0	0	1
1	0	0	0
0	1	1	1
0	1	1	0
0	1	0	1
0	1	0	0
0	0	1	1
0	0	1	0
0	0	0	1
0	0	0	0

Figure E6-4

design problem involving the present states of the flip-flops as inputs and the control signals to each flip-flop as outputs. Combinational minimization can then be used to simplify the control equations. These equations are commonly called *excitation equations*. The complexity of the excitation equations will vary depending on the type of flip-flop that is used to construct the counter.

The remainder of Sec. 6.2 describes a systematic design process that converts the counter design process into one mapping-grouping operation. This technique is most useful for counters with five flip-flops or less. For larger counters, a tabular minimization is usually necessary; the tabular design process is discussed in Sec. 6.2.6. Fully synchronous design techniques are described in Secs. 6.2.2–6.2.4. Asynchronous design is covered in Sec. 6.2.5. Other topics that relate to counter design are covered in the remainder of Sec. 6.2.

6.2.1. Excitation Tables. It was shown earlier that a flip-flop, in moving from a known present state to a specified next state as part of a counting sequence, can only take *five* behavioral actions: 0, 1, α, β, or X. For a particular type of flip-flop, each of these actions (or lack of action) will require a certain set of control inputs. Static behavior may require that all control inputs be disabled, while dynamic behavior may require that all inputs be enabled. Redundant input conditions may arise. The next several paragraphs discuss this behavior/control relationship for the four flip-flop types defined in Chapter 5.

The S-R Flip-flop.

The excitation table for the S-R flip-flop is given in Table 6-3. The first row shows that if the flip-flop is RESET in the present state and is to remain RESET in the next state, corresponding to 0 behavior, then its S input *must*

Table 6-3: The Excitation Table for the S-R Slip-flop

Behavior	S Input	R Input
0	0	X
1	X	0
α	1	0
β	0	1
X	X	X

not be enabled. The R input can be used as a don't-care, since the flip-flop will remain RESET whether R is enabled or not, so long as it is assured that the S input is disabled.

The second row of Table 6-3 shows that 1 behavior requires that the R input be disabled and allows the S input to be used as a redundancy. For α behavior the flip-flop must become SET; hence the S input must be enabled. Then, since S is enabled, R *must be* disabled (if both S and R are enabled, the next state of the flip-flop is undefined). The last rows show that β behavior requires that *only* R be enabled, while X behavior allows both control inputs to be used as redundancies.

The J-K Flip-flop.

The J-K excitation table is shown in Table 6-4, which is the same as the S-R excitation table except for the α and β entries. For the S-R only one control input could be enabled at a time. For the J-K flip-flop, however, the toggling action that results when both J and K are enabled introduces two additional redundancies into the excitation table.

Consider α behavior. Since the flip-flop must be changed from RESET to SET, the J input *must* be enabled. If the K input is also enabled, the flip-flop action is the same, since either a SET or a TOGGLE will move the flip-flop to the proper next state. Thus, whether K is enabled or not, the flip-flop will exhibit α behavior if J is enabled.

Table 6-4: **The Excitation Table for the J-K Flip-flop**

Behavior	J Input	K Input
0	0	X
1	X	0
α	1	X
β	X	1
X	X	X

The don't-care condition of the K input is actually a result of the internal connection of the J-K flip-flop. Whenever the J-K flip-flop is RESET, its K input is internally disabled by the $Q = 0$ condition (see Fig. 5-27). Thus, the external condition at the K input cannot have any effect on the behavior of the flip-flop whenever the present state is the RESET condition. The excitation table shows this, in that the K entries for 0 and α (the behavioral actions that begin with $Q = 0$) are both don't-cares.

Similarly, the J input of the flip-flop is internally disabled whenever the flip-flop is SET. Thus, the J input is a don't-care for both 1 and β behavior. The K input must be enabled in order to produce β (SET to RESET) behavior. Note that the additional redundancies present in the J-K excitation table will usually lead to a simplification of J-K control logic relative to S-R control logic.

The D Flip-flop.

The excitation table for the D flip-flop is shown in Table 6-5. Clearly, the D input must be enabled whenever the next state is to be $Q = 1$ and disabled whenever the flip-flop must be RESET in the next state.

Table 6-5: **The Excitation Table for the D Flip-flop**

Behavior	D Input
0	0
1	1
α	1
β	0
X	X

The T Flip-flop.

The excitation table for the T flip-flop is shown in Table 6-6. The T input must be enabled for dynamic behavior (α or β) and disabled otherwise to prevent change of state.

**Table 6-6:   The Excitation Table
for the T Flip-flop**

Behavior	T Input
0	0
1	0
α	1
β	1
X	X

6.2.2. Synchronous Excitation Mapping: Indirect.   The excitation tables for the four flip-flop types described in Sec. 6.2.1 permit sequential flip-flop behavior to be transformed into combinational form. To illustrate this process, the transition map for the B flip-flop in the binary up-counter, Fig. 6-3, is repeated in Fig. 6-6. As pointed out earlier, the transition map is independent of the type of flip-flop that is used to implement the counter.

If S-R flip-flops are used, the transition map shown in Fig. 6-6 can be converted to two other maps, one for the S input to the B flip-flop (S_B) and one for its R input (R_B). These are normal Karnaugh maps describing the combinational signals S_B and R_B. They are called *excitation maps*. The conversion process follows from the S-R excitation table, Table 6-3. According to the table, the K-map for S_B must contain a 0 at locations that contain either a 0 or a β on the transition map. TRUE entries on the S_B excitation map will appear at locations where an α appears on the transition map. Either X or 1 behavior will transfer as an X.

The S_B excitation map that results from this conversion process is shown in Fig. 6-7. This is a *combinational* map showing the conditions at the S

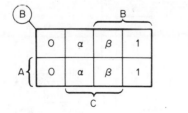

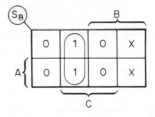

Figure 6-6. The Two-bit of a Binary
Up-counter

Figure 6-7. The S_B Excitation Map

input to the B flip-flop that are required by the specified counting sequence. Grouping the 1s on the S_B map gives the S_B control logic:

$$S_B = \bar{B}C \cdot CL \qquad\qquad (6\text{-}0)$$

where the *CL* term indicates that the $\bar{B}C$ logic at the S_B input is ANDed with the system clock (the counting pulse) by the internal transition gating circuitry of the flip-flop.

The R_B excitation map is given in Fig. 6-8. It is produced by converting the α and 1 entries on the *B* transition map to 0s, the β entries to 1s, and the 0

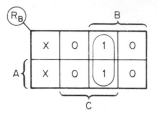

Figure 6-8. The R_B Excitation Map

and *X* entries to *X*s. The combinational logic for R_B is obtained by grouping the 1s on the K-map given in Fig. 6-8, giving

$$R_B = BC \cdot CL \tag{6-1}$$

The S_B logic shows that whenever the 1 bit of the binary counter (*C* flip-flop) is SET and the 2 bit (*B* flip-flop) is RESET, the *B* flip-flop will be set by the next clock pulse. The R_B logic specifies that whenever *B* and *C* are both SET, the *B* flip-flop will be reset by the next clock pulse. This agrees with the behavior specified for the *B* flip-flop in counting upward in the binary sequence, as given in Table 6-3.

A logic diagram for the *B* flip-flop and its control logic is given in Fig. 6-9. RTL circuits are shown. Since a clocked S-R flip-flop is not available in

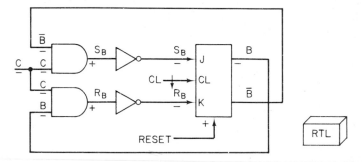

Figure 6-9. Logic Diagram for the *B* Flip-flop

RTL form, an RTL J-K flip-flop is used in its place. The control logic needs never exceed two levels of gating (unless loading requires the use of buffers), since the flip-flops provide their output signals in both TRUE and complemented form. Higher levels of gating may be used to save on cost, but this will introduce delay and lower the maximum rate at which the counter can operate.

Next, consider the design of the same counter with J-K flip-flops used in place of S-R. The J and K excitation maps are found from application of their excitation tables to the B transition map. The results are shown in Fig. 6-10. Note the additional don't-care conditions which appear. The control logic is

$$J_B = K_B = C \cdot CL \qquad (6\text{-}2)$$

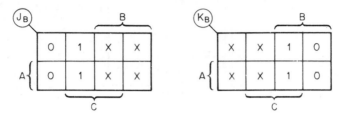

Figure 6-10. The J_B and K_B Excitation Maps

showing that both J_B and K_B are enabled whenever C is SET. This will cause the B flip-flop to change state following each clock period wherein the C flip-flop is SET.

Similarly, the D_B and T_B excitation maps are given in Fig. 6-11. Their control equations are

$$D_B = (B\bar{C} + \bar{B}C) \cdot CL$$
$$= (B \oplus C) \cdot CL \qquad (6\text{-}3)$$

and

$$T_B = C \cdot CL \qquad (6\text{-}4)$$

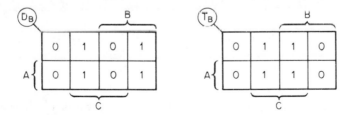

Figure 6-11. The D_B and T_B Excitation Maps

Comparison of the T_B and the J_B, K_B logic verifies that the J-K will act as a T flip-flop if its J and K terminals are connected together and used as a T input.

EXAMPLE 6-5: Show the complete excitation mapping process for the Y flip-flop in the 8421 BCD down-counter described in Ex. 6-4. Consider S-R, J-K, D, and T flip-flops.

Solution: The Y transition map is repeated in Fig. E6-5a. The S-R excitation maps and the simplified control equations for S_Y and R_Y are given in Fig. E6-5b and c. Similarly, the J-K excitation maps and control equations are as shown in Fig. E6-5d and e. The D and T results are shown in Fig. E6-5f and g. Only SOP forms are shown.

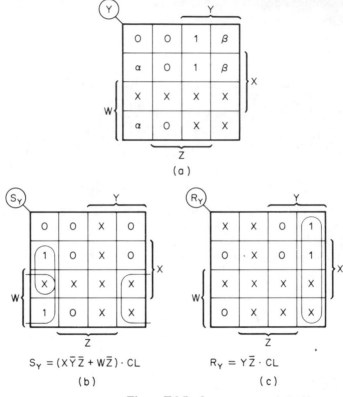

$$S_Y = (X\overline{Y}\overline{Z} + W\overline{Z}) \cdot CL$$

$$R_Y = Y\overline{Z} \cdot CL$$

(b)   (c)

Figure E6-5a, b, c,

Excitation maps are useful in determining the combinational control logic for a particular counter. Since a different map must be plotted for each type of flip-flop, however, the excitation mapping process requires extra effort and introduces opportunity for error beyond the transition map. For this reason, it is generally advantageous to obtain the control equations directly from the transition map.

6.2.3. Synchronous Excitation Mapping: Direct.   The process of obtaining the control equations without actually plotting separate excitation maps may be called *direct excitation mapping* or *transition mapping*. The

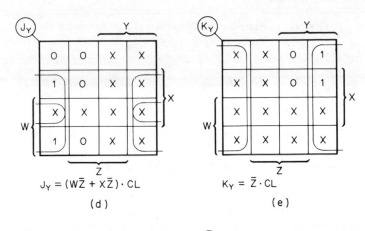

$J_Y = (W\bar{Z} + X\bar{Z}) \cdot CL$

(d)

$K_Y = \bar{Z} \cdot CL$

(e)

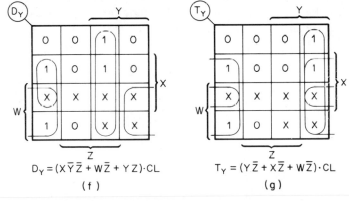

$D_Y = (X\bar{Y}\bar{Z} + W\bar{Z} + YZ) \cdot CL$

(f)

$T_Y = (Y\bar{Z} + X\bar{Z} + W\bar{Z}) \cdot CL$

(g)

Figure E6-5d, e, f, g

basic philosophy of this method is to consider the transition map as containing three types of entries:

1. *Must Include:* those transition symbols which would plot as a 1 on an excitation map and hence must be included in the groupings which give the logical control equations.

2. *Must Avoid:* those transition symbols which would plot as a 0 on an excitation map and hence must be avoided in the combinational groupings to find the control logic.

3. *Redundancies:* all other entries on the transition map.

The symbols which fall into each category will differ depending on the type of flip-flop control input that is being considered. Table 6-7, derived from the excitation tables given earlier, summarizes the necessary information for

Table 6-7: **Must-Include/Must-Avoid Conditions for the Various Flip-flop Types**

Flip-Flop Control Input	Must Include	Must Avoid	Redundancies
S	α	$0, \beta$	$1, X$
R	β	$1, \alpha$	$0, X$
J	α	0	$1, \beta, X$
K	β	1	$0, \alpha, X$
D	$1, \alpha$	$0, \beta$	X
T	α, β	$0, 1$	X

the four common flip-flop types. Clearly, the J-K flip-flop offers the most grouping flexibility because of its large amount of redundancy.

To illustrate the direct transition mapping process, consider the transition map given in Fig. 6-12. This map describes the behavior of the *C* flip-

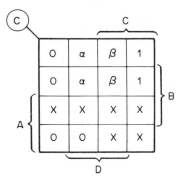

Figure 6-12. The Two-bit of an 8421 BCD Up-counter

flop (the 2 bit) in an 8421 BCD up-counter. Using the transition mapping process, the *S* input to an S-R flip-flop becomes

$$S_C = \bar{A}\bar{C}D \cdot CL \tag{6-5}$$

as shown by the groupings in Fig. 6-13a. Note that the specified grouping covers *all* the αs on the map and does not cover any 0s or βs.

For the *R* input, the grouping shown in Fig. 6-13b yields

$$R_C = CD \cdot CL \tag{6-6}$$

which covers all the βs and avoids both 1s and αs.

If a J-K flip-flop is used, the single grouping shown in Fig. 6-13c covers all the αs and βs, so that the control equations become

$$J_C = K_C = \bar{A}D \cdot CL \tag{6-7}$$

Note that the $\bar{A}D$ group covers all the αs and βs and no 0s or 1s. It can therefore be used for both J_C and K_C.

The transition maps for D and T flip-flops are shown in Figs. 6-13d and 6-13e. Their control equations are

$$D_C = (\bar{A}\bar{C}D + C\bar{D}) \cdot CL \tag{6-8}$$

and

$$T_C = \bar{A}D \cdot CL \qquad \text{[compare with Eq. (6-7)]} \tag{6-9}$$

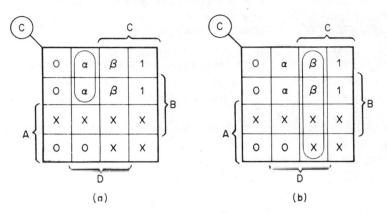

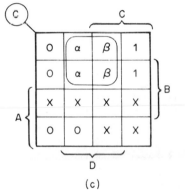
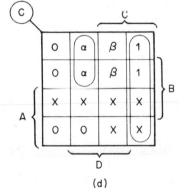

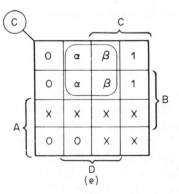

Figure 6-13. Transition Mapping for Various Types of Flip-flops

It should be noted, for error-detecting purposes, that the control logic for a J-K flip-flop, if properly minimized by the transition mapping process, should never contain the output from that flip-flop, either in TRUE or complemented form. In a four-flip-flop counter labeled A, B, C, and D, for example, J_A and K_A should depend only on B, C, or D. The logic for J_B and K_B should not contain either B or $\bar{B}$, but depend only on A, C, or D. Similar restrictions apply to the C and D flip-flops. This restriction arises from the internal cross-connections of the J-K flip-flop.

The transition mapping process will result in the same combinational groupings as the indirect excitation mapping process. The former process is easier to use, however, and allows the designer to compare the control logic required by each type of flip-flop without plotting separate excitation maps.

EXAMPLE 6-6:   As an example of the complete transition mapping process, consider the design of a 4221 BCD down-counter. The state table and behavior listing are given below:

Decimal Value	Present State* 4 2 2 1 W X Y Z	Minterm m_i	Behavior W X Y Z
9	1 1 1 1	15	1 1 1 β
8	1 1 1 0	14	1 1 β α
7	1 1 0 1	13	1 1 0 β
6	1 1 0 0	12	β 1 α α
5	0 1 1 1	7	α β β β
4	1 0 0 0	8	β 0 α α
3	0 0 1 1	3	0 0 1 β
2	0 0 1 0	2	0 0 β α
1	0 0 0 1	1	0 0 0 β
0	0 0 0 0	0	α α α α

*Next state is listed below present state; the last state resets to the top of the listing.

The transition maps for the counter are shown in Fig. E6-6a. The S-R groupings and control equations for the W and X flip-flops are shown in Fig. E6-6b. Note that both S_X and R_X were previously generated as part of the control logic for the W flip-flop and hence introduce no additional cost so long as loading requirements are not exceeded. The process is continued in Fig. E6-6c. The S_Z and R_Z control equations indicate toggling action. The Z flip flop is actually connected as a T flip-flop. The groupings and control equations for J-K flip-flops are shown in Fig. E6-6d and e. The 1s shown for the level in puts to J_Z and K_Z indicate that both J_Z and K_Z should be connect-

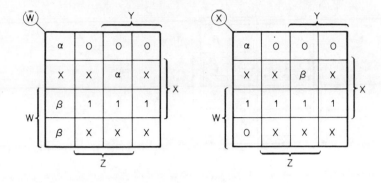

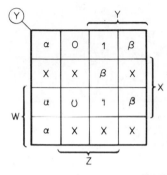

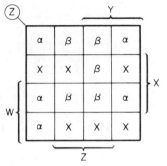

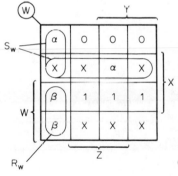

Figure E6-6a

$S_w = (\overline{W}X + \overline{W}Y\overline{Z}) \cdot CL$

$R_w = W\overline{Y}\overline{Z} \cdot CL$

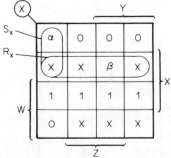

$S_x = \overline{W}Y\overline{Z} \cdot CL$

$R_x = \overline{W}X \cdot CL$

Figure E6-6b

287

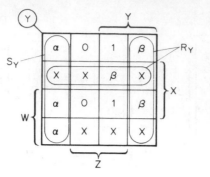

$$S_Y = \overline{Y}\overline{Z} \cdot CL$$

$$R_Y = (Y\overline{Z} + \overline{W}X) \cdot CL$$

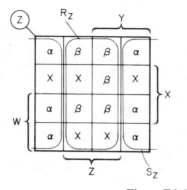

$$S_Z = \overline{Z} \cdot CL$$

$$R_Z = Z \cdot CL$$

Figure E6-6c

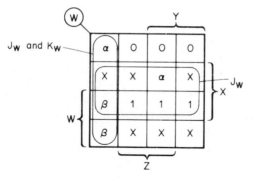

$$J_W = (\overline{Y}\overline{Z} + X) \cdot CL$$

$$K_W = \overline{Y}\,\overline{Z} \cdot CL$$

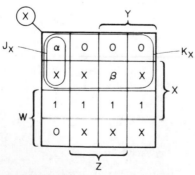

$$J_X = \overline{W}Y\overline{Z} \cdot CL$$

$$K_X = \overline{W} \cdot CL$$

Figure E6-6d

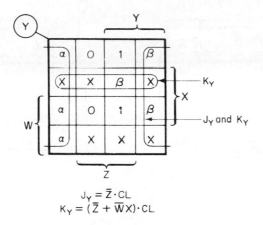

$$J_Y = \overline{Z} \cdot CL$$
$$K_Y = (\overline{Z} + \overline{W}X) \cdot CL$$

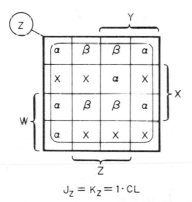

$$J_Z = K_Z = 1 \cdot CL$$

Figure E6-6e

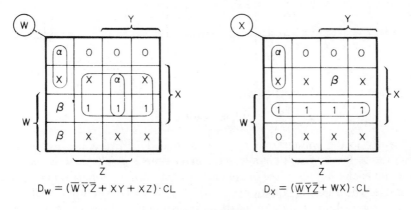

$$D_W = (\overline{W}\,\overline{Y}\,\overline{Z} + XY + XZ) \cdot CL$$

$$D_X = (\overline{W}\,\overline{Y}\,\overline{Z} + WX) \cdot CL$$

Figure E6-6f

289

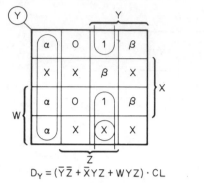

$$D_Y = (\overline{Y}\overline{Z} + \overline{X}YZ + WYZ) \cdot CL$$

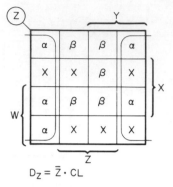

$$D_Z = \overline{Z} \cdot CL$$

Figure E6-6g

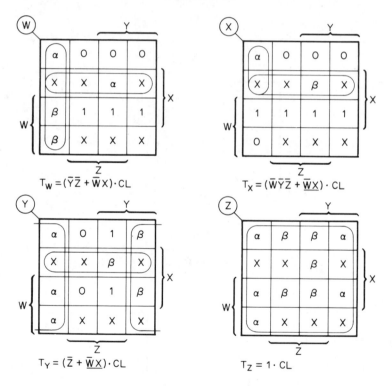

$$T_W = (\overline{Y}\overline{Z} + \overline{W}X) \cdot CL$$

$$T_X = (\overline{W}\overline{Y}\overline{Z} + \overline{W}X) \cdot CL$$

$$T_Y = (\overline{Z} + \overline{W}X) \cdot CL$$

$$T_Z = 1 \cdot CL$$

Figure E6-6h

ed directly to the voltage that represents their TRUE (enabled) condition. Thus, the Z flip-flop will change state each time a clock pulse is received. Note that each flip-flop is free of dependence on itself. Comparison of the J-K equations with the S-R equations shows this characteristic quite well.

For the D flip-flop the results are as shown in Fig. E6-6f and g. For the T flip-flop, see Fig. E6-6h. An RTL implementation and a timing diagram for this counter (in J-K form) are shown in Fig. E6-6i and j. A RESET signal

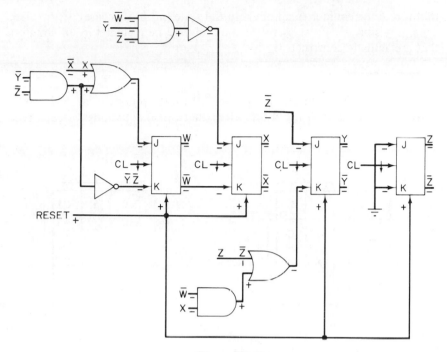

Figure E6-6i

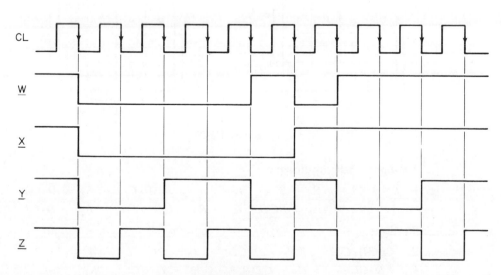

Figure E6-6j

that forces the counter to a known initial state (m_0 in this case) is provided. This provision is a most useful one and should be included in all counters that are actually constructed.

EXAMPLE 6-7: The transition maps for all four flip-flops in an 8421 BCD up-counter are given in Fig. E6-7. Find the control equations for each type of flip-flop.

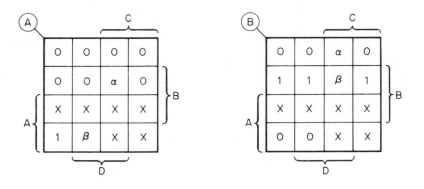

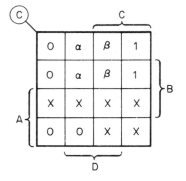

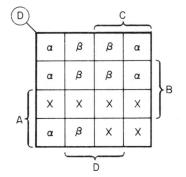

Figure E6-7

Solution: S-R flip-flops:

$$S_A = BCD \cdot CL \qquad S_B = \bar{B}CD \cdot CL \qquad S_C = \bar{A}\bar{C}D \cdot CL \qquad S_D = \bar{D} \cdot CL$$
$$R_A = AD \cdot CL \qquad R_B = \underline{BCD \cdot CL} \qquad R_C = CD \cdot CL \qquad R_D = D \cdot CL$$

J-K flip-flops:

$$J_A = BCD \cdot CL \qquad J_B = CD \cdot CL \qquad J_C = \bar{A}D \cdot CL \qquad J_D = 1 \cdot CL$$
$$K_A = D \cdot CL \qquad K_B = \underline{CD} \cdot CL \qquad K_C = D \cdot CL \qquad K_D = 1 \cdot CL$$

T flip-flops:

$$T_A = (AD + BCD) \cdot CL \qquad T_B = CD \cdot CL \qquad T_C = \bar{A}D \cdot CL \qquad T_D = 1 \cdot CL$$

D flip-flops:

$$D_A = (A\bar{D} + BCD) \cdot CL \qquad D_B = (\bar{B}CD + B\bar{C} + B\bar{D}) \cdot CL$$
$$D_C = (\bar{A}\bar{C}D + C\bar{D}) \cdot CL \qquad D_D = \bar{D} \cdot CL$$

6.2.4. POS Forms for Control Equations. The transition mapping process can easily be extended to obtain control equations in the product-of-sums form, which can then be compared with the SOP control equations in order to determine the best control logic to use.

To find the POS control equations, the grouping process is carried out with the Must Include and Must Avoid columns given in Table 6-7 reversed. This gives the SOP form for the complement of each control signal. De Morgan's theorem is then used to give the desired control equations in POS form.

EXAMPLE 6-8: Consider, once again, the W flip-flop of the 4221 BCD down-counter of Ex. 6-6. With an S-R flip-flop the groupings for $\bar{S}_W$ and $\bar{R}_W$ are as shown in Fig. E6-8a. The $\bar{S}_W$ groupings must cover 0s and βs while avoiding

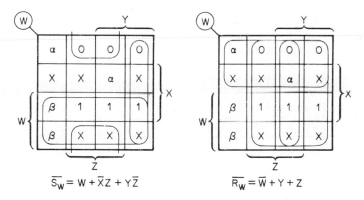

$$\bar{S}_W = W + \bar{X}Z + Y\bar{Z} \qquad\qquad \bar{R}_W = \bar{W} + Y + Z$$

Figure E6-8a

αs. The $\bar{R}_W$ groupings cover all 1s and αs and avoid βs. Applying De Morgan's theorem gives the desired results:

$$S_W = [\bar{W}(X + \bar{Z})(\bar{Y} + Z)] \cdot CL$$
$$R_W = W\,\bar{Y}\bar{Z} \cdot CL$$

For the J input to a J-K flip-flop the 0s must be included in the $\bar{J}$ groups; only αs must be avoided. The $\bar{K}$ groups must include all the 1s and avoid all βs. The POS logic for the J-K flip-flop is as shown in Fig. E6-8b. Both these results are the same as the results of the SOP grouping process.

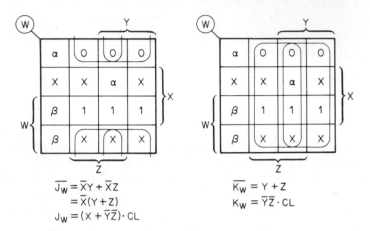

$$J_W = \overline{X}Y + \overline{X}Z$$
$$= \overline{X}(Y + Z)$$
$$J_W = (X + \overline{Y}\overline{Z}) \cdot CL$$

$$\overline{K_W} = Y + Z$$
$$K_W = \overline{Y}\overline{Z} \cdot CL$$

Figure E6-8b

For the D flip-flop, 0s and βs must be covered and 1s and αs must be avoided in obtaining the POS control equations. The W transition map yields the type D control logic shown in Fig. E6-8c. Similarly, the T flip-flop

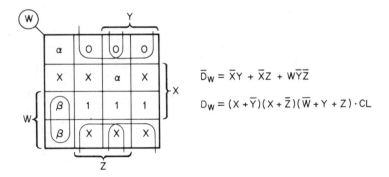

$$\overline{D}_W = \overline{X}Y + \overline{X}Z + W\overline{Y}\overline{Z}$$

$$D_W = (X + \overline{Y})(X + \overline{Z})(\overline{W} + Y + Z) \cdot CL$$

Figure E6-8c

POS groupings must cover all 0s and 1s and avoid αs and βs, giving, for the W transition map, the control equation shown in Fig. E6-8d.

These examples illustrate the POS grouping technique. The selection of the SOP or POS form for each control equation can be made by comparison of their relative costs.

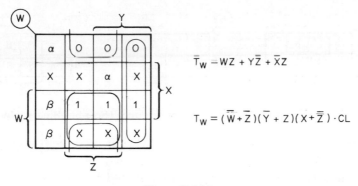

$$\overline{T}_W = WZ + Y\overline{Z} + \overline{X}Z$$

$$T_W = (\overline{\overline{W}+\overline{Z}})(\overline{Y}+Z)(X+\overline{\overline{Z}})\cdot CL$$

Figure E6-8d

6.2.5. Asynchronous Transition Mapping.

In some counting sequences, particularly those that are closely related to the binary counting sequence, a simplification of the combinational control logic required to implement the counter can be obtained by using transitions from the outputs of one flip-flop, rather than the system clock, as secondary pulse inputs that initiate changes in one or more of the other flip-flops that make up the counter.

At least one of the flip-flops must be activated by the system clock. Thereafter, the state changes of that flip-flop can be used as clocking signals for other flip-flops, should their use lead to a savings in gating cost relative to fully synchronous control logic.

The propagational delay inherent in each flip-flop will cause such secondary changes to take place *after* the system clock, however, making the operation of such a counter asynchronous. Delays may accumulate, limiting an asynchronous counter to a slower maximum counting rate than its fully synchronous equivalent. This effect is discussed in Sec. 6.2.9.

In addition to the reduced counting rate, an asynchronous counter will momentarily assume states that are different from its initial state and its eventual next state, since all flip-flops will not change state at once. These *transition states* may cause problems in some counting systems.

Transition mapping can be extended to the design of asynchronous counters. This technique is best introduced by example. The three-bit binary up-counter is described by Table 6-2 and its transition maps are given in Fig. 6-3 and are repeated in Fig. 6-14. Transition mapping yields the following synchronous J-K control equations:

$$\begin{aligned}
J_A &= K_A = BC \cdot CL \\
J_B &= K_B = C \cdot CL \\
J_C &= K_C = 1 \cdot CL
\end{aligned} \qquad (6\text{-}10)$$

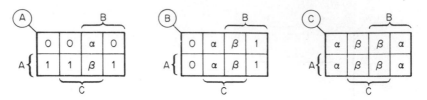

Figure 6-14. The Transition Maps for the Three-bit Binary Up-counter

Since one of the flip-flops must be operated synchronously, the one with the simplest (cheapest) synchronous logic should be selected as a starting point for the asynchronous design process. This is clearly flip-flop C.

Once the C flip-flop is operational, two clocking signals are available in addition to the system clock. These asynchronous clocking signals may be labeled as α_C and β_C. α_C occurs only when the C flip-flop makes a RESET $\longrightarrow$ SET transition, e.g., following present states m_0, m_2, m_4, and m_6 in the example. β_C occurs when C makes a SET $\longrightarrow$ RESET transition, as at m_1, m_3, m_5, and m_7. Excitation maps showing these three clocking signals are given in Fig. 6-15.

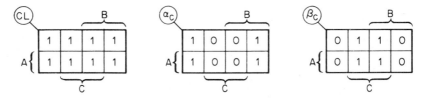

Figure 6-15. The Excitation Maps for CL, α_C, and β_C

The usefulness of asynchronous design is illustrated by these maps. For synchronous logic, each flip-flop is clocked following *every present state*. Thus, its level inputs must always be enabled or disabled in accordance with the desired behavior. If α_C is used as a clocking source, however, the states of the level controls at m_1, m_3, m_5, or m_7 become don't-cares, since no α_C transition occurs following these present states. The additional don't-care conditions that asynchronous clocking signals allow may be used to simplify the combinational logic that controls the level inputs for other flip-flops in the counter.

It is necessary that a clocking source cover *all* the must-include entries on a given transition map. Its points of occurrence may also cover any redundancies. If the clocking source covers any must-avoid entries, however, then the level inputs must be disabled when in those present states just as in the synchronous case. If the control logic that results from asychronous design is less costly than the fully synchronous control logic, asynchronous counting may be chosen.

In the binary up-counter the α_C transitions do not cover the must-include entries on either of the other maps. The β_C entries completely cover the α and β entries on the B map, however, allowing asynchronous control for the B flip-flop:

$$J_B = 1 \cdot \beta_C$$
$$K_B = 1 \cdot \beta_C \qquad \qquad (6\text{-}11)$$

indicating that the B flip-flop is toggled each time the C flip-flop makes a SET $\longrightarrow$ RESET transition.

Once the B flip-flop is operational, *five* potential clocking signals are available. They include CL, α_C, β_C, α_B, and β_B. Intermap comparison shows that β_B covers the must-include entries for both J and K on the A map, giving

$$J_A = 1 \cdot \beta_B$$
$$K_A = 1 \cdot \beta_B \qquad \qquad (6\text{-}12)$$

The asynchronism in this counter is three levels deep. The system clock will always cause C to change state. If the C flip-flop changes from SET to RESET, then B will receive a clocking transition, causing it, too, to change state. If B makes a β change, then A will also change state.

Consider behavior following present state $m_3 - 011$. The system clock will cause C to toggle, giving a next state of 010. Since C changed from 1 to 0, a β_C transition is generated. This β_C change will cause B to toggle, giving a second "next state" of 000 and producing a β_B transition. Finally, the β_B transition will toggle the A flip-flop, giving the desired next state of 100 $= m_4$. Transition states such as these may be shown within parentheses in a state sequence or *without* circles on a state diagram, as in Fig. 6-16. The complete

Figure 6-16. The Representation of Transition States

state sequence and state diagram for the asynchronous binary up-counter designed above are given in Fig. 6-17.

State Sequence

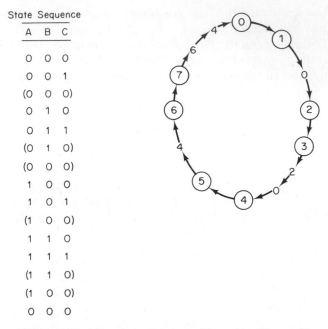

A	B	C
0	0	0
0	0	1
(0	0	0)
0	1	0
0	1	1
(0	1	0)
(0	0	0)
1	0	0
1	0	1
(1	0	0)
1	1	0
1	1	1
(1	1	0)
(1	0	0)
0	0	0

Figure 6-17. The Asynchronous Three-bit Binary Up-counter

To illustrate the combined use of the level and transition inputs in asynchronous design, consider again the A and C transition maps for the binary up-counter, Fig. 6-14. Clearly, α_C is of no use in clocking the A flip-flop. β_C, however, covers both the must-include entries on the A transition map. But control logic such as

$$J_A = 1 \cdot \beta_C$$
$$K_A = 1 \cdot \beta_C \qquad (6\text{-}13)$$

will not give correct operation of the A flip-flop. The J_A logic shown in Eq. (6-13) would activate the J control input to the A flip-flop following present states m_1, m_3, m_5, and m_7. The enabling of the J input at m_3 is a necessity, since the α at m_3 is a must-include condition for the J input. The m_5 and m_7 inputs are don't-care conditions for the J input, so their being enabled and clocked is also suitable. The action at m_1 is incorrect, however, since the 0 entry at that location on the A transition map is a must-avoid condition for the J input. Thus, the J input to the A flip-flop must be disabled at m_1 if β_C is to be used as a clocking signal to A.

A suitable selection for the control logic would be

$$J_A = \overline{m_1} \cdot \beta_C$$
$$= (A + B + \bar{C}) \cdot \beta_C \qquad (6\text{-}14)$$

This logic activates the J_A input at m_3, m_5, and m_7. It requires a three-input gate, however, which is more than the cost of the synchronous control logic that is given in Eq. (6-10).

The asynchronous control logic of Eq. (6-14) can be simplified, though, by remembering that m_3 is the only present state that *must be* enabled. Thus, the J_A input can be disabled by the largest group that avoids m_1 and still covers m_3. Consideration of the transition map for the A flip-flop (Fig. 6-14) shows that the *simplest* J_A logic is therefore

$$J_A = B \cdot \beta_C \qquad (6\text{-}15)$$

This may be represented by the maps of Fig. 6-18, showing that the final J_A signal occurs only at m_3 and m_7.

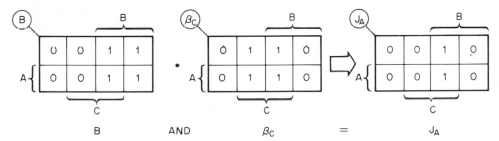

Figure 6-18. Asynchronous Generation of the J_A Control Signal

Similarly, the best control logic for K_A can be found to be

$$K_A = B \cdot \beta_C \qquad (6\text{-}16)$$

if β_C is to be used as the clocking signal. The interaction between the asynchronous signal and the conditionally enabled level input is used to give the proper excitation to the A flip-flop.

In implementing an asynchronous counter, the asynchronous transition signals must be brought from the outputs of one flip-flop to the transition inputs on the other flip-flop(s) with the proper voltage symbolism. For example, if RTL J-K flip-flops are used, the control logic for B that is given in Eq. (6-11) requires that the clock input to the B flip-flop be given a negative-going voltage transition when the C flip-flop changes from SET to RESET. The question is, does this require connection to C with positive-TRUE or with negative-TRUE symbolism?

To select the proper polarity for C, a simple table may be made, relating the logical change symbolized by β_C to the change in voltage that is required to activate the transition input to the flip-flop. Table 6-8 gives this relationship for β_C and an RTL flip-flop. The "Logical" row shows that β_C behavior requires a TRUE present state (1) and a FALSE next state (0). The "Voltage"

Table 6-8: Tabulation of β_C Behavior for an RTL Flip-flop

β_C	Before	After
Logical	1	0
Voltage	+	−

row shows that a negative voltage transition (RTL) requires that the higher voltage be present prior to the transition and the lower voltage be present afterward. Since the 1 and the + are in the same column, the asynchronous clocking signal should be C with positive symbolism, i.e., the C signal that is HIGH when C is TRUE and goes LOW when the C flip-flop is RESET. Similarly, the connection for J_A and K_A uses β_B and hence requires that B be brought to the CL input of the A flip-flop with positive-TRUE symbolism. The logical diagram for the asynchronous binary up-counter is shown in Fig. 6-19.

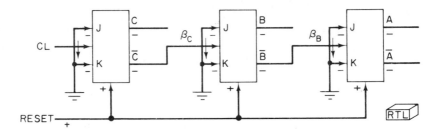

Figure 6-19. The Asynchronous Three-bit Binary Up-counter

EXAMPLE 6-9: As a second example of the asynchronous design process, consider the design of an asynchronous 4221 BCD up-counter. J-K flip-flops are to be used. The state/behavior table is shown below:

Decimal Value	Present State				Minterm Number m_i	Next State				Behavior			
	A	B	C	D		A_+	B_+	C_+	D_+	A	B	C	D
0	0	0	0	0	0	0	0	0	1	0	0	0	α
1	0	0	0	1	1	0	0	1	0	0	0	α	β
2	0	0	1	0	2	0	0	1	1	0	0	1	α
3	0	0	1	1	3	0	1	1	0	0	α	1	β
4	0	1	1	0	6	1	0	0	1	α	β	β	α
5	1	0	0	1	9	1	1	0	0	1	α	0	β
6	1	1	0	0	12	1	1	0	1	1	1	0	α
7	1	1	0	1	13	1	1	1	0	1	1	α	β
8	1	1	1	0	14	1	1	1	1	1	1	1	α
9	1	1	1	1	15	0	0	0	0	β	β	β	β

The transition maps for the four flip-flops are shown in Fig. E6-9. The synchronous control logic for the counter is

$$J_A = B \cdot CL \qquad\qquad K_A = CD \cdot CL$$
$$J_B = (A + \overline{CD}) \cdot CL \qquad K_B = (\overline{A} + CD) \cdot CL$$
$$J_C = (\overline{A}D + BD) \cdot CL \qquad K_C = (\overline{A}B + \overline{BD}) \cdot CL$$
$$J_D = 1 \cdot CL \qquad\qquad K_D = 1 \cdot CL$$

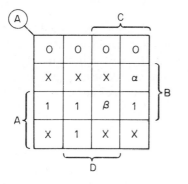

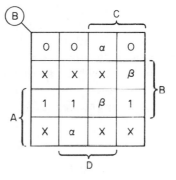

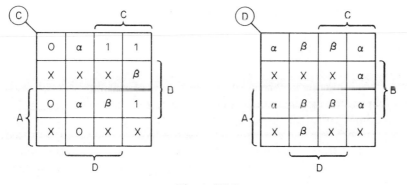

Figure E6-9

The D flip-flop is an obvious choice for synchronous operation because of the simplicity of its control logic.

This allows α_D and β_D transitions to be used as clocking signals for A, B, or C. The α_D signal is of no use, since it does not cover all the must-include terms for either J or K on any of the other flip-flops. β_D can be used to clock J_C, however, since it covers all the α entries (must include for J) on the C map. Since the β_D entry at m_9 falls on a 0 on the C map, the level control must be used to "strip off" m_9. Thus, if β_D is used as a transition source, the best choice for J_C is

$$J_C = (\bar{A} + B) \cdot \beta_D$$

The $\bar{A} + B$ logic strips off the bottom row, thereby preventing the β_D transition at m_9 from activating J_C.

The K_C logic must be completed synchronously:

$$K_C = (\bar{A}B + BD) \cdot CL$$

Note that the asynchronous logic for J_C and K_C that is given above cannot be used with a single-clock J-K flip-flop, since that type of flip-flop requires that J and K receive the same transition signal. Thus, if typical RTL or TTL integrated circuit flip-flops are used to build the given counter, the fully synchronous logic for the C flip-flop that was shown earlier must be used.

The β_D signal is also useful as a clocking source for J_B, provided that separated-clock flip-flops are available. The separated-clock, asynchronous logic for B becomes

$$J_B = (A + C) \cdot \beta_D$$
$$K_B = 1 \cdot \beta_C$$

Note that β_C completely covers the β entries on the B map. With the single-clock flip-flop, however, only synchronous operation of the B flip-flop is allowable.

The β_C and β_B overlap illustrates a potential trap in the asynchronous design process. A careless designer might arrive at the following logic for K_B and K_C:

$$K_B = 1 \cdot \beta_C$$
$$K_C = 1 \cdot \beta_B$$

While this logic seems correct, it will not work. Each flip-flop will wait on the other to change state, and, in fact, neither will do so. This phenomenon is sometimes called the "I'll follow you" situation, because each flip-flop expects the other to make the initial transition. A systematic design procedure of proceeding from one flip-flop to the next in selecting the gating logic and never reversing the direction of the design process will prevent this type of design flaw from occurring.

Once B, C, and D are operational, the A flip-flop has seven possible clocking sources. Two are most useful:

$$J_A = K_A = 1 \cdot \beta_B$$

or

$$J_A = K_A = 1 \cdot \beta_C$$

Either is correct, with the choice being made so as to balance the loading on the B and C flip-flops.

In summary, asynchronous design can often be used to reduce the amount of control logic required to construct a counter, particularly if the counting sequence is close to the natural binary counting sequence. The disadvantages of asynchronous operation include a reduction in maximum counting rate and the presence of transition states between stable states. The asynchronous design procedure can be summarized as follows:

1. Make a complete synchronous design for the counter.

2. Select the flip-flop that has the simplest synchronous design as a beginning.

3. Using the system clock and the transitions from any flip-flops that are operational, select the control logic for each flip-flop. Use the enabling inputs to strip off transitions occurring in the wrong places.

4. Proceed from one flip-flop to the next; never retrace steps.

The behavior of an asynchronous counter from its redundant states can be investigated just as can be done for a synchronous counter, as shown in Sec. 6.2.7, although it is a more complex process since both level and transition conditions are involved. Asychronous design techniques also apply to S-R, D, and T flip-flops.

6.2.6. Transition Mapping and Tabular Minimization.

Just as the utility of the Karnaugh map as a Boolean reduction tool diminishes for combinational problems involving six or more variables, transition mapping is difficult to use with counters containing more than five flip-flops or multimode counters having more than five flip-flops and mode-control inputs combined. For problems of this genre a tabular minimization is usually necessary to find the minimal control equations.

The entries to a Quine-McCluskey table are easy to find via the transition mapping process, however. As usual, the state table for the counter is used to generate the behavioral table that describes the counting sequence. Then, by using Table 6-7, the TRUE (must-include) and don't-care entries into column 1 of the tabulation are derived from the behavioral tabulation for each control input to each flip-flop. The particular type of flip-flop that is to be used enters into the design process at this stage, since its particular must-include and don't-care entries are used in converting the behavioral listing into the column 1 starting tabulation. The must-include entries are the *essential minterms* in the grouping process.

One such tabular reduction must be carried out for *each* control input to *each* flip-flop. The tabulation leads to the simplified SOP control equations. As shown in Sec. 4.3.5, however, the POS forms for the control equations can be obtained by entering the must-avoid terms (rather than the must-include terms) as the essential minterms and applying De Morgan's theorem to the minimized equations which result from the tabular process. The redundancies are the same in both cases. The SOP and POS forms for the control equations can be compared to select the better form of implementation. The tabular process is tedious, however, and large problems of this type are best suited to computer solution.

6.2.7. Behavior from Redundant States.

Whenever states for which no next state is specified are used as don't-care terms in designing a counter, the question of "What will the counter do if it accidentally gets into one of these present states?" arises. It should be noted that once a set of control equations for the counter is selected, those equations specify the *complete* behavior of the counter, not only from the present states that are in the basic counting sequence but also from the redundant states.

To complete the counter design, an investigation of the next-state behavior from these redundant states is often desirable, in order to determine any potentially troublesome behavior. It is highly desirable that the counter, when started in any of its redundant present states, will return to the main counting sequence. Situations that should be avoided include a "dead" state (next state same as present state) in which the counter may become trapped and any closed paths totally external to the basic counting sequence. Counters which avoid these conditions are said to be *self-correcting*. The state diagram for such a counter is said to be a *bush* if all states external to the main sequence eventually join back into it. The term *bush* relates to the "branches" that extend from the don't-care states back into the main "trunk" or counting sequence.

The analysis of the behavior from redundant states is a reverse of the usual combinational design process. Here, the logical control equations are known, and by using each of the redundant present states as an input to these control equations, the logical condition at each flip-flop control input is determined. Once these control signals are known, the next state of each flip-flop can be established, thereby predicting the next state of the counter.

EXAMPLE 6-10: Consider again the 4221 down-counter designed in Ex. 6-6. Its don't-care present states are m_4, m_5, m_6, m_9, m_{10}, and m_{11}. Its main counting diagram is repeated in Fig. E6-10a. The control logic for use with S-R flip-flops was found to be

$$S_W = (\bar{W}X + \bar{W}\bar{Y}\bar{Z}) \cdot CL \qquad R_W = W\bar{Y}\bar{Z} \cdot CL$$
$$S_X = \bar{W}\bar{Y}\bar{Z} \cdot CL \qquad R_X = \bar{W}X \cdot CL$$
$$S_Y = \bar{Y}\bar{Z} \cdot CL \qquad R_Y = (\bar{W}X + Y\bar{Z}) \cdot CL$$
$$S_Z = \bar{Z} \cdot CL \qquad R_Z = Z \cdot CL$$

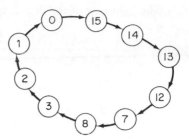

Figure E6-10a

Consider behavior from $m_5 = \bar{W}X\bar{Y}Z$. For this particular present state it can be seen that S_W, R_X, and R_Y will be TRUE, since $\bar{W}X$ is TRUE in the given present state. R_Z will be TRUE since Z is TRUE, as well. Thus, the next clock pulse will cause W to become SET and X, Y, and Z to become RESET. Any flip-flop whose control inputs were disabled would have stayed in the same state in which it began. As a result, the next state following m_5 is $W\bar{X}\bar{Y}\bar{Z} = m_8$, which is in the main counting sequence.

With S-R flip-flops a problem may arise when investigating behavior from don't-care states. Consider the action of the counter from present state $m_4 = \bar{W}X\bar{Y}\bar{Z}$. This present state will enable S_W, S_X, R_X, S_Y, R_Y, and S_Z. Since *both* S_X and R_X are enabled, the next state of the X flip-flop is indeterminant, as is the next state for the Y flip-flop. This indicates that the redundancy at m_4 was included in the grouping for both S_X and R_X as well as S_Y and R_Y. Because of this, the behavior of the counter cannot be fully described. The W flip-flop may or may not change states following present state m_4.

This indeterminant condition arises only with S-R flip-flops and should be avoided. To eliminate it, the groupings for the S and R inputs to each flip-flop should not share the same redundancies. One possible modification of the S-R control logic for the 4221 down-counter so as to avoid indeterminant behavior results in the following Boolean equations:

$$S_W = (\bar{W}X + \bar{W}\bar{Y}\bar{Z})\cdot CL \qquad R_W = W\bar{Y}\bar{Z}\cdot CL$$
$$S_X = \bar{W}\bar{Y}\bar{Z}\cdot CL \qquad R_X = \bar{W}Y\cdot CL$$
$$S_Y = \bar{Y}\bar{Z}\cdot CL \qquad R_Y = (Y\bar{Z} + \bar{W}XY)\cdot CL$$
$$S_Z = \bar{Z}\cdot CL \qquad R_Z = Z\cdot CL$$

The new logic is 5 c.u. more expensive than the original logic, since the R_Y logic is changed from $\bar{W}X + Y\bar{Z}$ to $\bar{W}XY + Y\bar{Z}$ to avoid "sharing" the redundancy at m_4 with S_Y. The new R_X logic is also more expensive, since the new term must be generated separately. The original logic for R_X shared a term, $\bar{W}X$, with S_W. The increase in cost is offset by the elimination of indeterminant behavior.

The behavior of the new counter from the redundant states is tabulated •
below:

Present State	Control Inputs								Next State
	S_W	R_W	S_X	R_X	S_Y	R_Y	S_Z	R_Z	
$m_4 = \bar{W}X\bar{Y}\bar{Z}$	1	0	1	0	1	0	1	0	$WXYZ = m_{15}$
$m_5 = \bar{W}X\bar{Y}Z$	1	0	0	0	0	0	0	1	$WX\bar{Y}\bar{Z} = m_{12}$
$m_6 = \bar{W}XY\bar{Z}$	1	0	0	1	0	1	1	0	$W\bar{X}\bar{Y}Z = m_9$
$m_9 = W\bar{X}\bar{Y}Z$	0	0	0	0	0	0	0	1	$W\bar{X}\bar{Y}\bar{Z} = m_8$
$m_{10} = W\bar{X}Y\bar{Z}$	0	0	0	0	0	1	1	0	$W\bar{X}\bar{Y}Z = m_9$
$m_{11} = W\bar{X}YZ$	0	0	0	0	0	0	0	1	$W\bar{X}Y\bar{Z} = m_{10}$

The complete state diagram is given in Fig. E6-10b. This is clearly a bush,
since the counter will always return to its main sequence in three clock cycles
or less, no matter what initial state it is in.

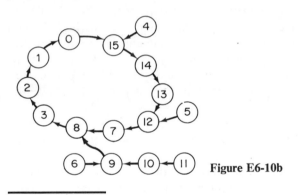

Figure E6-10b

For counters designed with J-K flip-flops, the determination of behavior
from redundant states is easier than for S-R counters. This is a result of
the internal cross-connections of the J-K flip-flop. Recall that when a J-K
flip-flop is SET, its *J* input is disabled. When RESET, the *K* input is similarly
nonfunctional. Thus, for a J-K flip-flop that is SET in the present state, only
the condition of its *K* input needs to be determined in order to predict its
next state; its *J* input, whether enabled or disabled, can have no effect on the
next state. Similarly, when a flip-flop's present state is RESET, only its *J*
input needs to be considered. This reduces the logical considerations of determining behavior from redundant states by one-half.

A tabular approach to this problem is useful. The present state is entered
at the left, and the *J* entry for each flip-flop that is SET is canceled out, as is

the K input for all flip-flops that are RESET in the present state. The states of the remaining J and K entries are then determined from the logical conditions of the present state. The next state can then be predicted.

EXAMPLE 6-11: Consider the J-K control equations for the 4221 BCD down-counter of Ex. 6-6:

$$J_W = (X + \bar{Y}\bar{Z}) \cdot CL \qquad K_W = \bar{Y}\bar{Z} \cdot CL$$
$$J_X = \bar{W}\bar{Y}\bar{Z} \cdot CL \qquad K_X = \bar{W} \cdot CL$$
$$J_Y = \bar{Z} \cdot CL \qquad K_Y = (\bar{Z} + \bar{W}X) \cdot CL$$
$$J_Z = 1 \cdot CL \qquad K_Z = 1 \cdot CL$$

The tabular entries for the m_4 present state begin as shown below:

Present State	Control Inputs								Next State
	J_W	K_W	J_X	K_X	J_Y	K_Y	J_Z	K_Z	
$m_4 = \bar{W}X\bar{Y}\bar{Z}$		X	X			X		X	?

The K_W, K_Y, and K_Z entries are crossed out since they cannot have any effect on their respective flip-flops. The J_X input is also of no useful concern.

Considering the J_W logic, it is seen to be enabled both by X and by $\bar{Y}\bar{Z}$ when the present state is m_4. K_X is enabled by $\bar{W}$, J_Y is enabled by $\bar{Z}$, and J_Z is always enabled. Thus, all four flip-flops will change state, making the complete tabular entry for present state m_4 become

Present State	Control Inputs								Next State
	J_W	K_W	J_X	K_X	J_Y	K_Y	J_Z	K_Z	
$m_4 = \bar{W}X\bar{Y}\bar{Z}$	1	X	X	1	1	X	1	X	$W\bar{X}YZ = m_{11}$

Completing the table for the remaining redundant states gives

Present State	Control Inputs								Next State
	J_W	K_W	J_X	K_X	J_Y	K_Y	J_Z	K_Z	
$m_5 = \bar{W}X\bar{Y}Z$	1	X	X	1	0	X	X	1	$W\bar{X}\bar{Y}\bar{Z} = m_8$
$m_6 = \bar{W}XY\bar{Z}$	1	X	X	1	X	1	1	X	$W\bar{X}\bar{Y}Z = m_9$
$m_9 = W\bar{X}\bar{Y}Z$	X	0	0	X	0	X	X	1	$W\bar{X}\bar{Y}\bar{Z} = m_8$
$m_{10} = W\bar{X}Y\bar{Z}$	X	0	0	X	X	1	1	X	$W\bar{X}\bar{Y}Z = m_9$
$m_{11} = W\bar{X}YZ$	X	0	0	X	X	0	X	1	$W\bar{X}Y\bar{Z} = m_{10}$

The complete state diagram for the J-K counter is shown in Fig. E6-11. The diagram is a bush; the counter is self-correcting within four clock pulses.

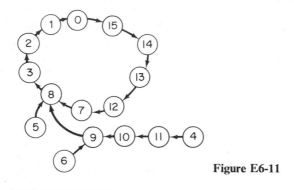

Figure E6-11

EXAMPLE 6-12: Show the complete state diagram for the J-K version of the 8421 BCD up-counter of Ex. 6-7.

Solution: The don't-care present states are $m_{10}, m_{11}, m_{12}, m_{13}, m_{14}$, and m_{15}. Using the J-K control equations given in Ex. 6-7, the next-state tabulation is

Present State	Control Inputs								Next State
	J_A	K_A	J_B	K_B	J_C	K_C	J_D	K_D	
$m_{10} = A\bar{B}C\bar{D}$	X	0	0	X	X	0	1	X	$A\bar{B}CD = m_{11}$
$m_{11} = A\bar{B}CD$	X	1	1	X	X	1	X	1	$\bar{A}B\bar{C}\bar{D} = m_4$
$m_{12} = AB\bar{C}\bar{D}$	X	0	X	0	0	X	1	X	$AB\bar{C}D = m_{13}$
$m_{13} = AB\bar{C}D$	X	1	X	0	0	X	X	1	$\bar{A}B\bar{C}\bar{D} = m_4$
$m_{14} = ABC\bar{D}$	X	0	X	0	X	0	1	X	$ABCD = m_{15}$
$m_{15} = ABCD$	X	1	X	1	X	1	X	1	$\bar{A}\bar{B}\bar{C}\bar{D} = m_0$

The complete state diagram for the J-K 8421 BCD up-counter is shown in Fig. E6-12.

Similar next-state analysis procedures can be carried out for counters designed with D or T flip-flops.

6.2.8. The Generation of Output Signals.   The primary outputs from a sequential system are generated by combinational logic that uses the present-state variables and the primary input variables as arguments. A single-mode counter has only present-state variables, making it fall most

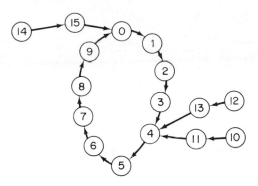

Figure E6-12

naturally into the Moore machine class, Fig. 5-22a. These output signals are derived directly from the values of the flip-flops that store the present state. Karnaugh mapping can be used to minimize the output logic. All unused present states appear as redundancies.

EXAMPLE 6-13: The counting sequence for the 8421 BCD down-counter is given in Ex. 6-4. An output signal, F, that indicates whenever the present state of the counter represents a decimal value of 6 or more is to be generated.

Solution: The output signal is determined from the *present* state of the counter. The following table lists the value of the output for each of the assigned present states:

Present State				Output F
W	X	Y	Z	
0	0	0	0	0
0	0	0	1	0
0	0	1	0	0
0	0	1	1	0
0	1	0	0	0
0	1	0	1	0
0	1	1	0	1
0	1	1	1	1
1	0	0	0	1
1	0	0	1	1

The control logic for the counter is designed in Ex. 6-5. The K-map for the F output signal is given in Fig. E6-13. The unassigned present states are entered as redundancies. The simplified expression for F is

$$F = W + XY$$

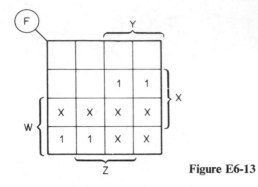

Figure E6-13

This logic generates a Moore-type output (a level signal) that depends *only* on the present-state variables.

EXAMPLE 6-14:   The state sequence and control equations for a 4221 BCD down-counter are given in Ex. 6-6. Ten output signals, $A, B, \ldots, I, J$, are to be generated from the counter, corresponding to the decimal values from 0 to 9. The following table relates the output signals to the present states of the counter:

Decimal Value	Present State 4 2 2 1 $W\ X\ Y\ Z$	Minterm Number m_i	Output that is TRUE
0	0 0 0 0	0	A
1	0 0 0 1	1	B
2	0 0 1 0	2	C
3	0 0 1 1	3	D
4	1 0 0 0	8	E
5	0 1 1 1	7	F
6	1 1 0 0	12	G
7	1 1 0 1	13	H
8	1 1 1 0	14	I
9	1 1 1 1	15	J

The K-map in Fig. E6-14 gives the TRUE/FALSE information for the A output function. The unspecified present states are used as redundancies. The grouping shown on the map gives

$$A = \bar{W}\bar{Y}\bar{Z}$$

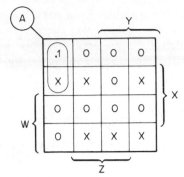

Figure E6-14

Similarly,

$$B = \bar{W}\bar{Y}Z, \qquad C = \bar{W}Y\bar{Z}, \qquad D = \bar{X}YZ, \qquad E = W\bar{X}, \qquad F = \bar{W}X,$$
$$G = X\bar{Y}\bar{Z}, \qquad H = X\bar{Y}Z, \qquad I = XY\bar{Z}, \qquad J = WYZ$$

The H and I output functions can also be implemented as

$$H = W\bar{Y}Z, \qquad I = WY\bar{Z}$$

All these expressions correspond to the Moore type of output.

Problems can arise during the generation of Moore output signals, however, due to differences in the switching speeds and propagational or ripple delays of the flip-flops that store the present-state variables. The presence of transitional states between the stable states of an asynchronous counter can also cause Moore-type output signals to appear at the wrong times. For example, Fig. 6-17 shows that state m_0 appears three times during the counting sequence of an asynchronous three-bit binary up-counter. Thus, a Moore-type output signal that detects $m_0 = \bar{A}\bar{B}\bar{C}$ would produce two extra TRUE conditions. Also, even though all the flip-flops in a synchronous counter receive the same clocking signal, differences in propagational delays may cause the flip-flops that change state to move to their new states at slightly different times. Thus, brief transition states can appear, even for a fully synchronous counter.

Consider a synchronous three-bit binary up-counter that is counting from 011 to 100. If the high-order flip-flop becomes SET more rapidly than the other two flip-flops become RESET, then the counter will move from 011 to 111 (momentarily) and then on to 100. Other momentary state sequences are possible, as well.

Generally, the differences between the setting and resetting delays of a group of flip-flops are small enough to make the momentary states that are present between the stable states of a synchronous counter be of very short duration. Even so, if simple combinational gating is used to generate output signals from the present state and input variables, the gating may detect the presence of states that produce output signals even though the states appear only momentarily. This condition can produce spurious output signals.

To prevent such spurious outputs, it is common to include the clocking signal in the AND logic that generates Moore-type output signals from a counter. The polarity of the clock pulse that is ANDed with the outputs is chosen so that the output signals are enabled during the period prior to the clocking transition that activates the flip-flop and are disabled during the period when the flip-flops are changing state just after each clock pulse. This technique will eliminate spurious output pulses. It is very similar to the generation of Mealy-type outputs, as discussed in Sec. 6.3.

EXAMPLE 6-15:   A four-bit binary up-counter is to be built with RTL flip-flops. An output signal is to be generated that indicates whenever the counter is in state m_4. What is the best combinational logic for generating the output signal?

Solution: Let the flip-flops be called A, B, C, and D, in order of decreasing weight. The basic output logic is

$$F = \bar{A}B\bar{C}\bar{D}$$

If the counter is synchronous, no problems may arise. If the counter is asynchronous, however, the presence of m_4 as a transition state in between stable states m_7 and m_8 will produce an extra output signal. To eliminate any possibility of such spurious output signals, the logic for the F output should include the clock signal, CL.

RTL flip-flops trigger on the falling edge of the CL pulses (see Fig. 5-31). Thus, the CL signal is at the higher voltage just before the active clocking transition. Since an RTL gate requires negative-TRUE input symbolism in order to act as an AND gate, the CL signal itself will be at the wrong voltage level (for TRUE input to the AND gate) prior to the falling transition. Thus, the complement of the CL signal, $\overline{CL}$, should be included in the logic for F:

$$F = \bar{A}B\bar{C}\bar{D} \cdot \overline{CL}$$

A logic diagram for this gating is shown in Fig. E6-15, along with an illustrative timing diagram.

━━━━━━━━━

6.2.9. Estimating Allowable Counting Rates   All flip-flops have an upper frequency limit above which they cannot reliably change state. In

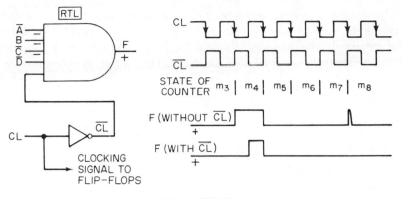

Figure E6-15

counting systems the control logic that is used to generate the desired counting sequence adds delays that limit the overall counting rate to something
less than that upper limit. In fact, especially in asynchronous counters,
the allowable counting rate may be much lower than the maximum counting
rate for a single flip-flop.

Exact limits cannot be established without building a counter and making measurements, but the typical device characteristics provided by manufacturers can be used to estimate such limits. Parameters of concern are

1. Propagational delay through a gate, t_g.

2. Propagational delay through a flip-flop, t_f.

3. Setup time for control inputs to a flip-flop, t_s.

For reference, typical propagation delays for RTL and TTL gates are 15 and
5 nsec, respectively. A typical RTL flip flop has a propagation delay of 80
nsec. The corresponding figure for a TTL flip-flop is 25 nsec. Setup times
are negligible for most integrated-circuit flip-flops.

In synchronous counters all flip-flops change state together, so that only
one flip-flop delay need be allowed for. To this must be added the net propagational delay through the longest gating path present in the control logic—
normally two levels of gating, or $2\,t_g$. Thus, the minimum period (related to
maximum frequency) for a typical synchronous counter is

$$P_{\min} = 2t_g + t_f + t_s \qquad (6\text{-}17)$$

The setup time, if significant, must be included in order to permit the control
inputs to the flip-flops to recognize their new values. If multilevel control
gating is used, the number of gate delays must be increased in keeping with
the longest logical path.

In an asynchronous counter the flip-flops may follow one another, changing state in domino-like fashion. Thus, additional propagational delays must be added. For example, in a four-bit asynchronous binary up-counter the worst-case delay occurs during the $m_{15} \rightarrow m_0$ transition, when four levels of asynchronous change are present. Even though there is no gating delay in this particular counter, the minimum period is

$$P_{\min} = 4t_f \qquad (6\text{-}18)$$

For a long asynchronous binary counter this delay may become prohibitively restrictive. For an interesting discussion of the effects of exceeding this rate limitation, see Ref. [12].

The values given by Eqs. (6-17) and (6-18) are at best only approximate, but such an analysis is useful during design phases to assure that a counting system will operate as desired.

EXAMPLE 6-16:   The asynchronous control logic for the 4221 BCD up-counter is given in Ex. 6-9. If this counter is constructed with RTL components, what is the approximate upper limit on its counting rate?

Solution: The single-clock nature of the RTL J-K flip-flop restricts the asynchronous control logic for the four flip-flops to

$$J_D = K_D = 1 \cdot CL$$
$$J_C = (\bar{A}D + BD) \cdot CL \qquad K_C = (\bar{A}B + \underline{BD}) \cdot CL$$
$$J_B = (A + CD) \cdot CL \qquad K_B = (\bar{A} + \underline{CD}) \cdot CL$$
$$J_A = K_A = 1 \cdot \beta_B$$

Except for the control logic for the A flip-flop, the counter is fully synchronous. The minimum period allowed during a fully synchronous counting cycle (whenever the A flip-flop does not change state) is

$$P_{\min} = t_f + 2t_g$$
$$= 80 + 2(15)$$
$$= 110 \text{ nsec}$$

with t_s being neglected.

When the A flip-flop is involved in the state change of the counter, the β_B logic delays the change in A until the B flip-flop has changed states. This asynchronism adds to the previous value, as:

$$P_{\min} = 2t_f + 2t_g$$
$$= 190 \text{ nsec}$$

The larger period must be chosen. It restricts the counting rate to

$$F_{max} = \frac{1}{P_{min}}$$
$$= \frac{1}{190} \times 10^{-9}$$
$$= 5.3 \text{ mHz}$$

Additional delays are introduced by wiring and contact capacitance and other nonideal factors. It is unlikely that the RTL counter will operate above 5 mHz, but the number is useful as an estimate of the upper counting rate.

6.3 Multimode Counter Design and Mealy Outputs

The single-mode counters described in Sec. 6.2 do not receive any primary input signals other than their counting pulse, *CL*. A more general situation arises when a counter receives one or more primary level inputs in addition to the counting signal. These inputs, often called mode-control signals, are used to change the counting sequence. Each different combination of these inputs can be used to define a different counting sequence. The pulse input is still used to cause the counting activity to take place.

A multimode counter more closely represents a general sequential system than does the single-mode counter. Each present state of a multimode counter will have more than one next state, with the choice of next state depending on the *total* present state, i.e., the present values of the mode-control inputs *and* the flip-flops. There will only be one next state following each total present state.

Transition mapping can be used to design multimode counters. The primary inputs are simply included as part of the present-state description. Usually, the state tables that describe each mode of operation of the counter are combined into one state table that lists the next state following each total present state, as shown previously in Ex. 5-8. This table is used to initiate the transition mapping process.

As an example, consider the design of an 8421 BCD counter that is to count upward when a mode-control signal, *U*, is TRUE, and downward when *U* is FALSE. The composite state and behavioral table for this up/down counter is given in Table 6-9. All unlisted total present states have unspecified next states. Note that the control signal *U* is used as part of the total present-state identification. It thus affects the location of the behavioral entries on the transition maps. There will always be as many transition maps as there are flip-flops (four in the example under consideration). Each transi-

Table 6-9: State Table for the 8421 BCD Up/Down Counter

Total Present State					Next State				Behavior			
U	W	X	Y	Z	W_+	X_+	Y_+	Z_+	W	X	Y	Z
0	0	0	0	0	1	0	0	1	α	0	0	α
0	0	0	0	1	0	0	0	0	0	0	0	β
0	0	0	1	0	0	0	0	1	0	0	β	α
0	0	0	1	1	0	0	1	0	0	0	1	β
0	0	1	0	0	0	0	1	1	0	β	α	α
0	0	1	0	1	0	1	0	0	0	1	0	β
0	0	1	1	0	0	1	0	1	0	1	β	α
0	0	1	1	1	0	1	1	0	0	1	1	β
0	1	0	0	0	0	1	1	1	β	α	α	α
0	1	0	0	1	1	0	0	0	1	0	0	β
1	0	0	0	0	0	0	0	1	0	0	0	α
1	0	0	0	1	0	0	1	0	0	0	α	β
1	0	0	1	0	0	0	1	1	0	0	1	α
1	0	0	1	1	0	1	0	0	0	α	β	β
1	0	1	0	0	0	1	0	1	0	1	0	α
1	0	1	0	1	0	1	1	0	0	1	α	β
1	0	1	1	0	0	1	1	1	0	1	1	α
1	0	1	1	1	1	0	0	0	α	β	β	β
1	1	0	0	0	1	0	0	1	1	0	0	α
1	1	0	0	1	0	0	0	0	β	0	0	β

tion map will have as many variables as there are flip-flops *and* mode-control signals (five in this example).

The five-variable transition maps for the up/down counter can be plotted from Table 6-9. They are shown in Fig. 6-20. SOP grouping on these transition maps gives the following J-K control equations:

$$J_W = (UXYZ + \bar{U}\bar{X}\bar{Y}Z)\cdot CL$$
$$J_X = (UYZ + \bar{U}W\bar{Z})\cdot CL$$
$$J_Y = (U\bar{W}Z + \bar{U}X\bar{Z} + \underline{\bar{U}W\bar{Z}})\cdot CL$$
$$J_Z = 1\cdot CL$$
$$K_W = (UZ + \bar{U}\bar{Z})\cdot CL = (U \odot Z)\cdot CL$$
$$K_X = (\underline{UYZ} + \bar{U}\,\bar{Y}\bar{Z})\cdot CL$$
$$K_Y = (\underline{UZ} + \bar{U}\bar{Z})\cdot CL$$
$$K_Z = 1\cdot CL$$

(6-19)

The transition mapping process leads most naturally to the minimized SOP or POS control equations, with all Boolean reductions being clearly evident on the transition maps. The analysis can be extended to S-R, T, and D flip-flops, as usual.

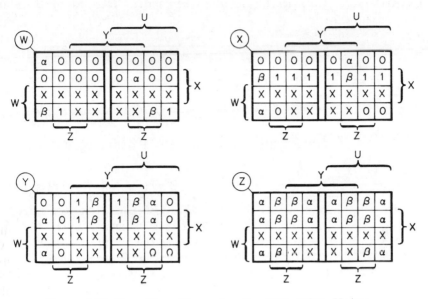

Figure 6-20. Transition Maps for the 8421 BCD Up/Down Counter

For illustration, the logical diagram for the control of the W flip-flop is shown in Fig. 6-21. RTL symbolism is assumed. The control signal U changes the control logic such that each flip-flop behaves properly in either the up or down mode.

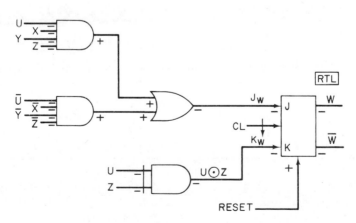

Figure 6-21. RTL Implementation of the W Flip-flop Control Logic

EXAMPLE 6-17: Consider the multimode counter described by the state table given below:

Present State		Next States							
		$XY = 00$		$XY = 01$		$XY = 10$		$XY = 11$	
A	B	A_+	B_+	A_+	B_+	A_+	B_+	A_+	B_+
0	0	0	1	1	1	1	1	0	0
0	1	1	0	0	0	1	0	0	1
1	0	1	1	0	1	0	1	1	0
1	1	0	0	1	0	0	0	1	1

There are two control inputs, X and Y. When both are FALSE the counter acts as a binary up-counter. It acts as a binary down-counter when $XY = 01$. When $XY = 10$ the counter complements itself on each clock pulse, and when X and Y are both TRUE, it remains in its present state.

Combining these four state tables into a single state/behavioral table gives

Total Present State				Next State		Behavior	
X	Y	A	B	A_+	B_+	A	B
0	0	0	0	0	1	0	α
0	0	0	1	1	0	α	β
0	0	1	0	1	1	1	α
0	0	1	1	0	0	β	β
0	1	0	0	1	1	α	α
0	1	0	1	0	0	0	β
0	1	1	0	0	1	β	α
0	1	1	1	1	0	1	β
1	0	0	0	1	1	α	α
1	0	0	1	1	0	α	β
1	0	1	0	0	1	β	α
1	0	1	1	0	0	β	β
1	1	0	0	0	0	0	0
1	1	0	1	0	1	0	1
1	1	1	0	1	0	1	0
1	1	1	1	1	1	1	1

The transition maps for the counter are shown in Fig. E6-17. From the maps the J-K control equations may be found to be

$$J_A = K_A = (X\bar{Y} + B\bar{Y} + \bar{X}YB) \cdot CL$$
$$J_B = K_B = (\bar{X} + \bar{Y}) \cdot CL$$

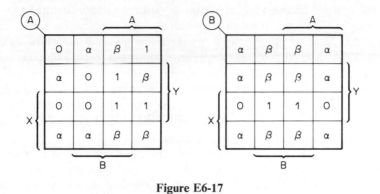

Figure E6-17

The presence of the mode-control signals in a multimode counter intro-duces the possible need for synchronization between the counting pulses and the control signals. If not assured by the user, synchronization must be pro-vided to prevent any of the mode-control signals from changing just prior to or during a clocking transition. Changes in the primary inputs at these critical times may lead to unpredictable or erratic behavior, since some of the control inputs to the flip-flops may respond to the new control signals, while others, due to longer delays in their control logic, may respond to the control signals that were in existence before the change.

Synchronization of the mode-control inputs can be provided by using D flip-flops, as shown in Sec. 5.5. As an example, Fig. 6-22 shows how a D

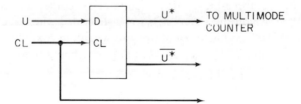

Figure 6-22. Synchronizing a Mode-control Signal

flip-flop can be used to synchronize the U signal that controls the BCD up/down counter that is described earlier in this section. If U changes just prior to a CL transition, the synchronized signal, U^*, may or may not change. If not, it will "catch up" on the next CL pulse. In either case, the U^* signal will change state only just *after* a CL transition, giving the control logic an entire clock period to settle to its new values.

If two or more mode-control inputs are present, then additional synchronizers must be provided. If the mode-control signals change in the near vicinity of a clocking transition, then some of the synchronizers may change to new values, while others may delay their change until the following clocking signal. As a result, the set of synchronized input signals that are seen by the counter may change from the original combination to an intermediate combination and then to the correct combination. This condition cannot really be eliminated, since the fact that the unsynchronized inputs changed close to the clock violated the fundamental mode, and synchronization cannot completely eliminate the consequences of this violation.

The presence of one or more primary input signals in a multimode counter or a more general sequential system allows output signals to be generated in either Moore or Mealy form. Moore output signals do not include these input signals in their combinational logic. They are determined from the present-state variables (flip-flop outputs) alone, as discussed in Sec. 6.2.8.

Mealy-type output signals, on the other hand, allow the inclusion of the primary input signals in their determination. As a result, Mealy output signals must be ANDed with the controlling pulse input, as shown in Fig. 5-22b. Otherwise, a change in one of the primary inputs that occurs between pulses will change the total present state, even though the present state (as represented by the states of the flip-flops) does not change, thereby allowing an asynchronous change in the output signals. The level inputs must be synchronized so as to remain constant for the period just prior to and during a pulse input, so that ANDing the Mealy-type outputs with the clocking pulse allows them to become TRUE only at a time when the primary inputs are unchanging.

EXAMPLE 6-18: A three-bit up/down counter is controlled by a mode-control signal, U. An output signal, F, is to be generated whenever the counter is in state $m_5 = A\bar{B}C$ and is to count upward when the next clock pulse, CL, appears. Assume that TTL gates and flip-flops are being used.

Solution: The basic Mealy output logic is

$$F = A\bar{B}CU$$

This logic includes the U signal, so that it must be enabled only when U is unchanging. The U signal is assumed to be stable just prior to and during the CL transition that activates the counter. Since TTL flip-flops respond to rising transitions, the CL signal will be low just before its active transition. Since a TTL gate requires positive-TRUE symbolism in order to perform the AND function, the CL signal will not be at the proper value prior to the transition. Thus, the complement of the CL signal, $\overline{CL}$, must be used as the enabling signal:

$$F = A\bar{B}CU \cdot \overline{CL}$$

A logic diagram for this signal appears in Fig. E6-18.

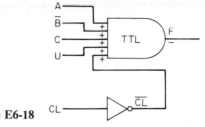

Figure E6-18

6.4 General Sequential Systems Design

The counter design techniques that are given in Secs. 6.1, 6.2, and 6.3 assume that the combinations of state variables that represent each present and next state are well known. In many applications, however, the sequence of states that represents the behavior of a sequential machine is not so explicitly defined. Rather, as introduced in Sec. 5.1.3, a generalized state sequence is used, with each present state represented by a name, say "state a" or "state b," rather than by a particular combination of state variables.

In these cases, no transitional behavior can be specified until a state assignment is made, thereby relating each of the generalized present states to a specific combination of state variables. Once the state assignment is made, however, the sequential system can be designed by using the techniques given in Secs. 6.1–6.3.

The state variables in a digital system are stored in flip-flops. The number of state variables (and hence the number of flip-flops) that are required for a sequential system having N different present states is lower-bounded by K, where

$$K \geq \log_2(N) \qquad \text{(5-5) (repeated)}$$

and K must be an integer. More than K flip-flops may be used in some situations.

One consequence of Eq. (5-5) is that a reduction in the number of present states may permit a reduction in the number of flip-flops that are required to implement the sequential system. Thus, state reduction is generally attempted prior to the state assignment process.

6.4.1. State Reduction. State reduction techniques center around the identification of two or more states that are *equivalent*, meaning (roughly) that the operations performed during the states can be performed simultaneously and the events that follow their control intervals are the same. Equivalence is not always easy to identify, especially when unspecified next states are present. For detailed discussion of the state reduction topic the reader should see Refs. [4], [10], [13], [18], [20], and [23]. This section considers only a simplified state reduction approach that is applicable to the reduction of fully specified sequential systems.

Prior to investigating state reduction possibilities, the complete state table that describes the system must be developed. Each output signal that the system must produce should be included in the table. As an example, consider the generalized sequential system described by Table 6-10. Pulse-mode operation is assumed.

Table 6-10:   State Table for an Example System

Present State	Next State		Outputs	
	$I = 0$	$I = 1$	F_1	F_2
a	b	c	0	1
b	c	d	1	0
c	d	e	1	1
d	b	c	0	1
e	e	e	0	0

The system generates two Moore-type output signals, F_1 and F_2. These outputs are independent of the state of the single control input, I.

Examination of Table 6-10 shows that the output signals that are produced during states a and d are the same. Also, both states a and d transfer to state b when $I = 0$ and to state c when $I = 1$. This corresponds to equivalence, since

two states are equivalent if and only if their specified next states and specified output signals are correspondingly the same for all combinations of input signals.

Since states a and d are equivalent, they may be combined into a single state, say state f, with no effect on the operation of the sequential system. The reduced state table is given in Table 6-11. The reduction permits the system to be implemented by using only two flip-flops (four states) rather than three flip-flops as originally required.

Whenever two or more states are recognized as being equivalent, they may be combined into a single state which acts in place of all the originally separate states. The next two examples further illustrate this principle.

Table 6-11 :   State Table after States *a* and *d* are Combined

Present State	Next State I = 0	Next State I = 1	Outputs F_1	Outputs F_2
f	*b*	*c*	0	1
b	*c*	*f*	1	0
c	*f*	*e*	1	1
e	*e*	*e*	0	0

EXAMPLE 6-19 :   A digital system is to monitor the output of a binary receiver that receives a string of TRUE/FALSE values from a distant transmitter. The digital system is to extract the bits in groups of three and is to examine each successive group of three bits to determine if the combination 011 (most recent value on the right) is present. The presence of this combination is to generate an output signal, *F*.

Solution: One approach to the design of this system combines the decision-making and data storage operations into one sequential operation. The system begins at state P_0. During the first time interval the system observes the value of the first bit and if $I = 0$ moves to state P_1. If $I = 1$, it moves to state P_2. Since the three-bit sequence is not yet complete, no output signal is given in either case.

From P_1 the system moves to P_3 if $I = 0$ and to P_4 if $I = 1$. Thus, the system can only be in P_3 if a 00 sequence has been received. Similar behavior follows P_2, with states P_5 and P_6 corresponding to received sequences of 10 and 11, respectively. From any of the last four states the system returns to P_0, in order to begin the examination of the next three bits. The output signal, $F = 1$, is generated whenever the system is in state P_4 (01 received thus far) and the $I = 1$ condition appears.

The complete state table for this system is given below:

Present State	Next State I = 0	Next State I = 1	Output, F I = 0	Output, F I = 1	Information Received
P_0	P_1	P_2	0	0	XXX
P_1	P_3	P_4	0	0	0XX
P_2	P_5	P_6	0	0	1XX
P_3	P_0	P_0	0	0	00X
P_4	P_0	P_0	0	1	01X
P_5	P_0	P_0	0	0	10X
P_6	P_0	P_0	0	0	11X

This system is clearly a sequential machine, moving from state to state during the identification process. The state diagram for this system is given in Fig. E6-19.

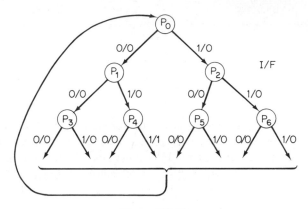

Figure E6-19

The state table shows that the next states and output signals for states P_3, P_5, and P_6 are the same when $I = 0$ and when $I = 1$. Thus, P_3, P_5, and P_6 are equivalent and can be combined into a single state, say P_3. This combination is called *initial state reduction*. The initially reduced state table for the system is

Present State	Next State		Output, F	
	$I = 0$	$I = 1$	$I = 0$	$I = 1$
P_0	P_1	P_2	0	0
P_1	P_3	P_4	0	0
P_2	P_3	P_3	0	0
P_3	P_0	P_0	0	0
P_4	P_0	P_0	0	1

No further state equivalence is present. Note that states P_3 and P_4 would be equivalent except for the differences in their output behavior.

State P_3 is now a "catch-all" state corresponding to input sequences 00, 10, and 11, all of which cannot precede the special sequence, 011. State P_4 is kept distinct from P_3. It is reached only following the 01 sequence, via $P_0 \rightarrow P_1 \rightarrow P_4$. If the third bit is a 1, then the combination

$$F = P_4 \cdot I$$

will generate the correct output signal. Since there are five states even after reduction, at least three flip-flops (and their control logic) will be required.

EXAMPLE 6-20:   Consider a system similar to that of Ex. 6-19 except that each group of *four* input values is to be compared to the valid code words in the excess-three BCD code, and any erroneous combinations are to be indicated by the output signal, E. The errors will correspond to receiving $m_0, m_1,$ $m_2, m_{13}, m_{14},$ or m_{15}. The state table for such a system is

Present State	Next State $I = 0$	Next State $I = 1$	Output, E $I = 0$	Output, E $I = 1$	Inputs Received
a	b	c	0	0	XXXX
b	d	e	0	0	0XXX
c	f	g	0	0	1XXX
d	h	i	0	0	00XX
e	j	k	0	0	01XX
f	l	m	0	0	10XX
g	n	o	0	0	11XX
h	a	a	1	1	000X
i	a	a	1	0	001X
j	a	a	0	0	010X
k	a	a	0	0	011X
l	a	a	0	0	100X
m	a	a	0	0	101X
n	a	a	0	1	110X
o	a	a	1	1	111X

Initially, states h and o and states $j, k, l,$ and m may be recognized as equivalent groups. Combinations may be made as

Let h and o become h

Let $j, k, l,$ and m become j

The initially reduced state table is

Present State	Next State $I = 0$	Next State $I = 1$	Output, E $I = 0$	Output, E $I = 1$
a	b	c	0	0
b	d	e	0	0
c	f	g	0	0
d	h	i	0	0
e	j	j	0	0
f	j	j	0	0
g	n	h	0	0
h	a	a	1	1
i	a	a	1	0
j	a	a	0	0
n	a	a	0	1

At this point, a *secondary reduction*, the combination of states *e* and *f* into a single state, *e*, is apparent. This simplifies the state table to

Present State	Next State $I = 0$	$I = 1$	Output, E $I = 0$	$I = 1$
a	*b*	*c*	0	0
b	*d*	*e*	0	0
c	*e*	*g*	0	0
d	*h*	*i*	0	0
e	*j*	*j*	0	0
g	*n*	*h*	0	0
h	*a*	*a*	1	1
i	*a*	*a*	1	0
j	*a*	*a*	0	0
n	*a*	*a*	0	1

The system has been reduced from 15 states to 10 states. Since no additional combinations are available, the last state table is the *finally reduced state table* for the system.

The output signal, *E*, can be obtained by using combinational logic of the form

$$E = h \cdot \bar{I} + h \cdot I + i \cdot \bar{I} + n \cdot I$$
$$= h + i\bar{I} + nI$$

indicating that an erroneous minterm has been received when the system reaches state *h*, or when the system is in state *i* and the last input value is FALSE, or when it is in state *n* and the last value is TRUE.

The systems that are reduced in Exs. 6-19 and 6-20 are both *reset systems*, meaning that they always return to some particular state (commonly called the *home state*) after going through a fixed number of other states. The reduction of reset systems is relatively straightforward. Next-state and output behavior is compared and states having the same entries are equivalent and can be combined.

A broader class of sequential systems exhibit nonreset behavior, meaning that they do not necessarily return to some home state at periodic intervals. With such *nonreset systems* an additional state reduction process is required. This method is called *Moore reduction* [19] or *output class grouping* [13] and is used to identify any equivalent states that are not apparent when next states and outputs are compared.

Output class grouping can be used with both reset and nonreset systems, but it will never yield any state reduction beyond next-state and output comparison if the system is a reset system. Thus, identification of the reset property will save going through the effort of the output class grouping procedure.

However, the reset property is not always easy to identify. Observation of the state table or state diagram of the system is helpful, since nonreset conditions may be apparent. If two closed state sequences of different lengths are present, then the system must be of the nonreset type. Also, if any state closes on itself for *any* input condition, as shown in Fig. 6-23, then the system is nonreset.

Figure 6-23. A Non-reset Condition

Thus, the identification of the reset property will save work, but failing to recognize the property, if it is present, will not invalidate the state reduction effort. It will only cause extra work (the use of output class grouping) with no gain. As a guideline, therefore, it is best to reduce all systems by next-state and output comparison first. Then, assume the nonreset property and apply output class grouping *unless the reset property is apparent.*

The output class grouping process begins by placing all states that have similar output behavior in the same class, usually indicated by a Roman numeral. The members of a given class have *potential equivalence*, since their output behavior is in agreement.

A present-class/next-class table is then generated, showing the next classes (rather than next states) that follow from each present state. Members of the same class that have different next-class behavior must be separated, forming new classes. The tabulation and separation process continues until all members of each class have similar next-class behavior. The members of each class can then be combined into a single state.

EXAMPLE 6-21:   Reduce the sequential system described by the state table at the top of the page 328. The system has one control input, K, and generates one output signal, F.

Solution: Note that at least two closed paths from state a back to state a are possible: $a \longrightarrow e \longrightarrow a$ and $a \longrightarrow b \longrightarrow d \longrightarrow c \longrightarrow f \longrightarrow a$. Since these are of different length, the controller is nonreset in nature. Next-state and output comparison gives no reduction, since no two states have corresponding next states.

Present State	Next State		Output, E	
	$K = 0$	$K = 1$	$K = 0$	$K = 1$
a	e	b	0	0
b	a	d	0	1
c	f	d	0	0
d	c	b	0	1
e	c	a	0	0
f	a	c	0	0

States *a*, *c*, *e*, and *f* fall into the (0,0) output class; this may be called class I. States *b* and *d* may be grouped into class II, corresponding to outputs of (0,1).

The state table, when grouped by classes, becomes

Present Class		Next Class	
		$K = 0$	$K = 1$
I	a	I	II
	c	I	II
	e	I	I
	f	I	I
II	b	I	II
	d	I	II

No output column needs to be given, since all members of a class are known to have the same output behavior.

Clearly, class I must be broken into two classes. States *a* and *c* can remain in class I, but states *e* and *f* must be placed into a new class, say III, because of their different next-class behavior. Remember that states *a* and *c* cannot be combined with states *b* and *d* even though their next classes are similar, since their outputs are dissimilar. Cross-class combinations are not allowed.

The new state table becomes

Present Class		Next Class	
		$K = 0$	$K = 1$
I	a	III	II
	c	III	II
II	b	I	II
	d	I	II
III	e	I	I
	f	I	I

This final table shows that states a and c, states b and d, and states e and f are equivalent, since the next-class behavior is the same for each pair.

Letting states a and c become state a, states b and d become state b, and states e and f become state f gives

Present State	Next State		Output, F	
	$K = 0$	$K = 1$	$K = 0$	$K = 1$
a	e	b	0	0
b	a	b	0	1
e	a	a	0	0

The original six states have been reduced to three.

EXAMPLE 6-22: A sequential system is to receive a sequence of TRUE/FALSE values through its lone primary input signal, I. The system is to indicate whenever the previous three values and the current input value form $m_{10}, m_{11}, m_{12}, m_{13}$, or m_{14} (assuming current input value on the left and oldest value on the right). Describe this system in general terms and reduce it to its minimum number of states.

Solution: The system must be able to "store" the three input values. In general, this will require that the system have eight possible present states, one for each of the possible combinations of the last three inputs. In tabular form, this yields

Last Three Input Values	Present State	Next State		Output	
		$I = 0$	$I = 1$	$I = 0$	$I = 1$
000	a	a	e	0	0
001	b	a	e	0	0
010	c	b	f	0	1
011	d	b	f	0	1
100	e	c	g	0	1
101	f	c	g	0	1
110	g	d	h	0	1
111	h	d	h	0	0

As the current value of I replaces the oldest of the three stored input values and the other two input values move to the right, the present-state/next-state sequence given above evolves. For example, state d represents values 011 having been received. If $I = 0$, then the next state is 001, or state b. If $I = 1$, the next state is 101, state f.

This is clearly a nonreset system, as evidenced by the appearance of state *a* in its own next-state row. Even so, states having equivalent next-state and output entries can be combined prior to the application of output class grouping. Examination of the state table gives the following initial reductions:

$$a, b \longrightarrow a; \quad c, d \longrightarrow c; \quad e, f \longrightarrow e$$

yielding the following initially reduced state table:

Present State	Next States $I = 0$	$I = 1$	Output $I = 0$	$I = 1$
a	*a*	*e*	0	0
c	*a*	*e*	0	1
e	*c*	*g*	0	1
g	*c*	*h*	0	1
h	*c*	*h*	0	0

Output class grouping gives

Present Class		Next Class $I = 0$	$I = 1$
I	*a*	I	II
	h	II	I
II	*c*	I	II
	e	II	II
	g	II	I

showing that all five states must be kept separate.

———

For more information on output class grouping, the reduction of systems having unspecified next states, and state reduction/assignment in general, Kohavi [13] is recommended.

As shown in this section, state reduction techniques can be used to reduce the number of states that are required by a system, but it should be remembered that design clarity and operational simplicity are of major importance. Often, the presence of potentially reducible states in a sequential system is an asset, provided that those states lead to more understandable system performance. Systems should always be designed with the user and the repairman in mind.

6.4.2. State Assignment. Once the number of states of a generalized sequential system have been reduced to an acceptable minimum, the number of state variables that are required can be determined. Equation (5-5) gives the *minimum* number of state variables, although in some applications more state variables than this minimum are used. Examples of the latter situation include the ring counter that is described in Sec. 6.5.3 and the ring-counter controller described in Chapter 7.

If M state variables are to be used, then 2^M different combinations of these variables are available for representing the present states of a general sequential machine. The design problem centers around the assignment of the flip-flop combinations to the states. Since the sequential system will, in essence, act as a counting device, combinational control logic must be designed to activate the flip-flops in the desired sequence. Given the state sequence, it is to be expected that the complexity and cost of the control logic will vary for differing assignments of flip-flop combinations to allowed states. The designer should seek a state assignment that minimizes the cost of the combinational logic required to control the sequential system. To illustrate the effect of differing state assignments on the cost of the system, a simplified example follows.

EXAMPLE 6-23: Consider the design of the following sequential system:

Present State	Next State		Output, F	
	$I = 0$	$I = 1$	$I = 0$	$I = 1$
a	a	b	0	0
b	c	b	0	1
c	a	d	1	0
d	c	b	1	1

There can be no state reduction since each state is in a different output class. Since there are four states, two flip-flops, say W and X, will suffice for storing the state variables. One possible state assignment is

Generalized State	Assigned Combination of State Variables	
	W	X
a	0	0
b	1	0
c	0	1
d	1	1

which may be shown on a map, as in Fig. E6-23a. Once this assignment is made, the state table shown earlier may be replaced by a present-state/next-

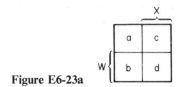

Figure E6-23a

state table summarizing the behavior required of the W and X flip-flops. In tabular form this is

Present State			Next State						Output, F	
			$I = 0$			$I = 1$				
W	X		W_+	X_+		W_+	X_+		$I = 0$	$I = 1$
0	0	(a)	0	0	(a)	1	0	(b)	0	0
1	0	(b)	0	1	(c)	1	0	(b)	0	1
0	1	(c)	0	0	(a)	1	1	(d)	1	0
1	1	(d)	0	1	(c)	1	0	(b)	1	1

Here, the assigned values of W and X have replaced the original state names.

The above table can be rewritten to show the behavior of W and X as a two-flip-flop counter. The input variable, I, is included in the Total Present State column, as shown in the multimode counter design process of Sec. 6.3. This new table is

Total Present State				Next State			Output	Behavior	
I	W	X		W_+	X_+		F	W	X
0	0	0	(a)	0	0	(a)	0	0	0
0	0	1	(c)	0	0	(a)	1	0	β
0	1	0	(b)	0	1	(c)	0	β	α
0	1	1	(d)	0	1	(c)	1	β	1
1	0	0	(a)	1	0	(b)	0	α	0
1	0	1	(c)	1	1	(d)	0	α	1
1	1	0	(b)	1	0	(b)	1	1	0
1	1	1	(d)	1	0	(b)	1	1	β

Note that the $I = 0$ entries appear at the top of this table and the $I = 1$ entries follow. Also, the present-state sequence has been rearranged into ascending minterm order. The first row shows that when the present state is $\bar{W}\bar{X} = 00$ (state a) and $I = 0$, the specified next state is $WX = 00$ (state a again).

The behavioral column results from comparing the present states and next states of W and X. The transition maps for these two flip-flops are shown in Fig. E6-23b. Using J-K flip-flops, the control logic becomes

$$J_W = I \cdot CL \qquad K_W = \bar{I} \cdot CL$$
$$J_X = \bar{I}W \cdot CL \qquad K_X = (\bar{I}\bar{W} + IW) \cdot CL$$

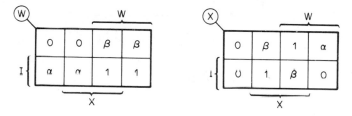

Figure E6-23b

The Karnaugh map for the output function, F, is shown in Fig. E6-23c, yielding

$$F = IW + \bar{I}X$$

where the IW term is cost-free. This is a Mealy-type output.

If it is desired to know the present state in terms of the original-state description, the following logic must be included:

$$a = \bar{W}\bar{X}, \qquad b = W\bar{X}, \qquad c = \bar{W}X, \qquad d = WX$$

along with four inverters, if both the state signals and their complements are needed.

As an alternative, consider the state assignment shown in Fig. E6-23d.

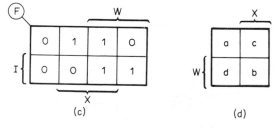

(c)   (d)

Figure E6-23c, d

The resultant state table, in rearranged form, is

Total Present State				Next State			Output	Behavior	
I	W	X		W_+	X_+		F	W	X
0	0	0	(a)	0	0	(a)	0	0	0
0	0	1	(c)	0	0	(a)	1	0	β
0	1	0	(d)	0	1	(c)	1	β	α
0	1	1	(b)	0	1	(c)	0	β	1
1	0	0	(a)	1	1	(b)	0	α	α
1	0	1	(c)	1	0	(d)	0	α	β
1	1	0	(d)	1	1	(b)	1	1	α
1	1	1	(b)	1	1	(b)	1	1	1

The W and X transition maps are shown in Fig. E6-23e, yielding

$$J_W = I \cdot CL \qquad\qquad K_W = \bar{I} \cdot CL$$
$$J_X = (I + W) \cdot CL \qquad K_X = \bar{W} \cdot CL$$

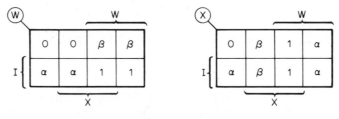

Figure E6-23e

which costs about three less cost units to implement. The output function is shown in Fig. E6-23f, which is more expensive than the output gating associated with the other assignment, however.

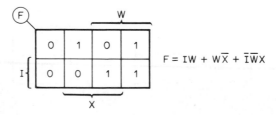

$$F = IW + W\bar{X} + \bar{I}\bar{W}X$$

Figure E6-23f

Example 6-23 illustrates the effect that state assignment has upon the design of the state counter. How, then, can a designer select a "good" assignment and hence minimize the cost of his sequential system?

The identification of good state assignments has been the subject of a considerable amount of research. References [4], [7], [9], [13], and [21] point interested readers to published results. In particular, the work of Armstrong [1] and Humphrey [11] has pointed to two suggestions concerning state assignment that, in keeping with the scope of this text, are easy to apply to systems having a relatively small number of states.

These state assignment suggestions are formulated into two "rules," which are

1. Two or more states which have the same next state should be given adjacent assignments.

2. Two or more states which are the next states of the same state should be given adjacent assignments.

The term *adjacent assignments* means that the states appear next to each other on the mapped representation of the state assignment. In other words, the combinations of flip-flops that represent the two states differ in only one variable. For example,

$$a = WX$$
$$b = W\bar{X}$$

	X	
W	b	a

are adjacent, while

$$a = WX$$
$$b = \bar{W}\bar{X}$$

	X	
	b	
W		a

are not. The groupings suggested by rule 1 should be given precedence over rule 2 groupings.

These two types of state pairings are illustrated in state diagram form in Fig. 6-24. The rule 1 pairings are those states whose state transitions terminate on the same next state. Rule 2 identifies states that have a common previous state.

Rule 1 Pairing   Rule 2 Pairing

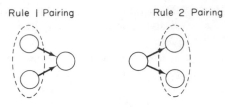

Figure 6-24. The Armstrong-Humphrey Rules in State Diagram Form

These two heuristic state assignment suggestions are by no means the state-of-the-art in state assignment techniques. Other state assignment approaches, such as partitioning [12], are available and are covered in the references. These two rules are presented here because of the simplicity of their application and their usefulness as state assignment guidelines for simple sequential systems. It should be noted that these rules apply best to systems that are implemented with J-K flip-flops and that they make no attempt to minimize the logic that is required to generate primary output signals.

To apply rule 1, a "reverse" state table can be made, showing the previous states associated with each present state. Giving these states adjacent assignments will satisfy rule 1. For the system of Ex. 6-23 this table is

Present State	Previous States
a	a, c
b	a, b, d
c	b, d
d	c

The (a,b,d) grouping should be interpreted as three pairings, (a,b), (b,d), and (a,d), of which only two can be satisfied (three states cannot all be adjacent to each other). This shows that (a,c), (a,b), (a,d), and (b,d) should be given adjacent assignments, if possible. The (b,d) grouping appears twice.

The groupings for rule 2 appear in the next-state column of the original state table. These suggested groupings for the example system are (a,b), (b,c), and (a,d), with (b,c) appearing twice.

From these suggestions, the state assignment,

$$a = \bar{W}\bar{X}, \qquad b = WX, \qquad c = \bar{W}X, \qquad d = W\bar{X} \qquad (6\text{-}20)$$

can be seen to be of merit since it satisfies all adjacencies suggested by rules 1 and 2 except for (a,b). Both of the twice-suggested groupings are satisfied. As predicted, Ex. 6-23 shows that the assignment given in Eq. (6-20) results in a simplification in the control logic relative to the alternative assignment presented at the beginning of that example.

The use of the pairings suggested by these rules leads to a "good" state assignment, but the best assignment cannot be obtained without exhaustive trial. The Armstrong and Humphrey suggestions are easy to apply, however, and lead to a reasonably minimal sequential system design.

EXAMPLE 6-24: Make a state assignment for the sequential system of Ex. 6-22 and complete the sequential system design.

Solution: The final, reduced state table given in Ex. 6-22 is

Present State	Next States		Output	
	$I = 0$	$I = 1$	$I = 0$	$I = 1$
a	a	e	0	0
c	a	e	0	1
e	c	g	0	1
g	c	h	0	1
h	c	h	0	0

The reversed state table for this system is

Present State	Previous State	
a	a, c	$\begin{cases} e, g \end{cases}$
c	e, g, h ⟶	e, h
e	a, c	g, h
g	e	
h	g, h	

The rule 1 and 2 groupings are

Rule 1	Rule 2
a, c (2)	a, e (2)
g, h (2)	c, h (2)
e, g	c, g
e, h	

Three state variables, say W, X, and Y, must be used. The state assignment shown in Fig. E6-24 is one of several that satisfy as many of the suggested adjacencies as possible. Note that all the doubly suggested adjacencies are satisfied. The state assignment can now be used to design the sequential system. Inserting the assignment into the original state table and rearranging the table to list the total present states in ascending minterm order gives the table that follows Fig. E6-24.

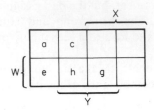

Figure E6-24

Total Present State				Next State			Output	Behavior		
I	W	X	Y	W_+	X_+	Y_+	F	W	X	Y
0	0	0	0 (a)	0	0	0	0	0	0	0
0	0	0	1 (c)	0	0	0	0	0	0	β
0	0	1	0	X	X	X	X	X	X	X
0	0	1	1	X	X	X	X	X	X	X
0	1	0	0 (e)	0	0	1	0	β	0	α
0	1	0	1 (h)	0	0	1	0	β	0	1
0	1	1	0	X	X	X	X	X	X	X
0	1	1	1 (g)	0	0	1	0	β	β	1
1	0	0	0	1	0	0	0	α	0	0
1	0	0	1	1	0	0	1	α	0	β
1	0	1	0	X	X	X	X	X	X	X
1	0	1	1	X	X	X	X	X	X	X
1	1	0	0	1	1	1	1	1	α	α
1	1	0	1	1	0	1	0	1	0	1
1	1	1	0	X	X	X	X	X	X	X
1	1	1	1	1	0	1	1	1	β	1

Transition mapping gives the following control equations for W, X, and Y:

$$J_W = I \cdot CL \qquad J_X = IW\,\bar{Y} \cdot CL \qquad J_Y = W \cdot CL$$
$$K_W = \bar{I} \cdot CL \qquad K_X = 1 \cdot CL \qquad K_Y = \bar{W} \cdot CL$$

Karnaugh mapping of the output function gives

$$F = \underline{IW\bar{Y}} + I\bar{W}Y + IX$$

Other examples of state reduction, state assignment, and the subsequent sequential system design are given in Chapter 7.

6.5 Registers and Related Sequential Devices

In many applications flip-flops are used for functions other than counting or sequencing. A major application of this type is the use of groups of flip-flops to store data for subsequent use by a sequential system. Such devices are called *registers*, a carry-over from computer applications wherein flip-flops are often used for the temporary storage (registration) of operands during the computational process. The design of registers centers around the logic that is required to transfer the information into and out of the flip-flops.

Some types of registers provide for simple modifications of the operands that they are storing. Also, certain types of registers may be used as counters, although they exhibit very restricted state sequences. The design and use of registers and several closely related digital devices are covered in this section.

6.5.1. Entering Data into a Register. As mentioned above, registers are used for the temporary storage of information. Normally, the information to be stored is available as a group of Boolean signals which are to be sampled at some specific time and the resultant set of TRUE and FALSE values stored in the register. There are several ways in which this can be done.

Direct-Entry Registers.

If flip-flops with *both* direct-coupled SET and RESET inputs are available, data can be transferred into a register in one operation. The hardware arrangement for a single flip-flop in this type of register is shown in Fig. 5-18 and is repeated in Fig. 6-25 (RTL symbolism shown). The input signal, I_i,

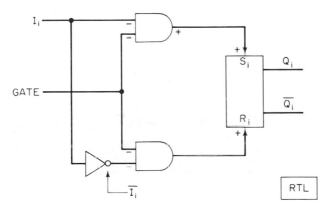

Figure 6-25. A Direct-coupled Register (Single-rail Input)

is ANDed with the *GATE* signal to set the flip-flop. Thus, if I_i is TRUE when *GATE* is made TRUE, the flip-flop will be set and will remain so when *GATE* is made FALSE. In the same manner, the I_i signal is inverted, ANDed with *GATE*, and used to reset the flip-flop. The *GATE* signal forces the flip-flop into the proper state because of the presence of either I_i or $\bar{I}_i$. Since these two signals can never both be TRUE at the same time, the flip-flop can never receive simultaneous S and R inputs. If I_i changes its logical value during the period when *GATE* is TRUE, the flip-flop will follow the changes until *GATE* is made FALSE. The flip-flop will then be left in the state the I_i was in just before *GATE* became FALSE.

The register shown in Fig. 6-25 may be called a *direct-coupled, single-rail* register. *Single-rail* implies that only one data input is required, with an inverter being used to provide the $\bar{I}_i$ signal. If *both* I_i and $\bar{I}_i$ are provided to the register, the term *double-rail* is applicable, and the inverter shown in Fig. 6-25 can be eliminated.

This type of register entry system leaves the clocked inputs to the flip-flop free for use in subsequent modification of the stored data. This is of great utility in many applications, as will be shown in later examples in this chapter.

Common-Reset, Direct-Entry Registers.

To eliminate the need for inversion and still require only single-rail inputs, the previously described register organization can be modified as shown in Fig. 6-26. Prior to data entry, the entire register is cleared by

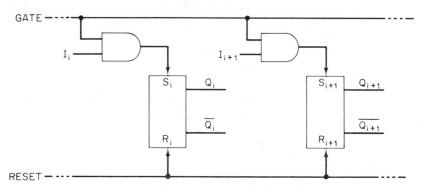

Figure 6-26. A Common-reset, Direct-entry Register

the *RESET* signal. Then, since all the flip-flops are known to contain a FALSE value, only those flip-flops that must store a TRUE value need to be changed. The gating need only set these flip-flops. Conversely, it is possible to set all the flip-flops initially and then gate in only those inputs that are FALSE.

This type of register is not truly parallel in nature, since a *sequence* of controlling signals is required to store the data. The system that controls the register must provide the *RESET* and the *GATE* signals in the proper order, so that the register will be cleared first and then filled with the necessary TRUE values. The savings in hardware (one gate per bit of the register) is offset by an increase in the complexity of the register control logic. The savings is sufficient, especially in registers that contain many flip-flops, to merit the use of this type of register entry logic in many applications.

Clocked Registers.

Just as the direct control inputs to the flip-flops that make up a register can be used for data entry, the clocked control inputs can serve the same purpose. Figure 6-27 shows how the J-K, S-R, D, and T flip-flop types can

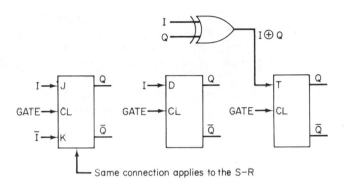

Figure 6-27. Various Flip-flop Types used for Clocked Register Entry

be used for clocked data entry and storage. The D flip-flop is clearly the best choice for this application, since it requires only single-rail inputs. The J-K and S-R flip-flops require either double-rail inputs or the use of one inverter per flip-flop to convert single-rail inputs to double-rail form. The T flip-flop requires the most complicated gating and is a very poor choice for register applications. The proof of the correctness of the EXCLUSIVE-OR gating used with the type T flip-flop is part of Prob. 5-11.

One advantage of using the clocked inputs for data entry is that only one change in state will take place no matter how long the *GATE* signal is kept TRUE. The internal pulse-narrowing circuitry of the flip-flops assures this. Only when the *GATE* signal makes a voltage change in the direction that activates the clock inputs to the flip-flops will the stored values be changed (or left unchanged) in accordance with the states of their control inputs. Thus, the pulse-mode *GATE* signal acts as a sampling strobe of narrow duration.

Input changes that occur just prior to or just after a gating transition violate the fundamental-mode restriction and may lead to trouble due to setup and holdover effects. Consider a four-bit register that is to store a binary number, as shown in Fig. 6-28 (D flip-flops shown). If the control inputs change from m_7 (0111) to m_8 (1000) just prior to a gating transition,

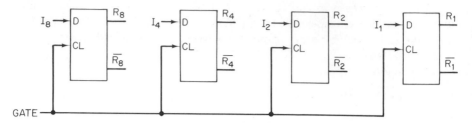

Figure 6-28. A Four-bit Clocked-entry Register

the following may happen. The D inputs to flip-flops R_4, R_2, and R_1 will continue to be enabled for the duration of their holdover time. At the same time, if flip-flop R_8 has a short setup period, it may recognize the presence of the TRUE signal at its I_8 input before the control inputs to the other flip-flops have sensed their change from TRUE to FALSE. Thus, the gating signal may set all four flip-flops, storing 15 (1111) rather than 8. Other permutations of the "before" and "after" input conditions may appear in the register as well. Clearly, asynchronous changes in input conditions just prior to a gating transition may lead to trouble.

Mixed-Mode Registers.

A third type of register entry technique, *mixed mode* data entry, uses both the direct *and* the clocked capabilities of the flip-flops. This technique is most useful with either J-K or S-R flip-flops that have only a single direct-coupled input. This type of flip-flop is common to RTL, TTL, and other product lines when two flip-flops are packaged together in one 14-pin dual in-line package. The limited number of input/output connections forces this reduction in control capability.

Two closely related techniques for using this type of flip-flop are shown in Figs. 6-29 and 6-30. The first method, Fig. 6-29, is useful when the clocked capabilities of the flip-flop are not needed for any use other than for data entry. Initially, the register is reset by using the direct inputs. Then, a gating transition is used to transfer any TRUE input signals into their flip-flops via

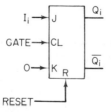

Figure 6-29. A Mixed-mode Register Entry Technique

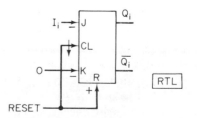

Figure 6-30. A Second Mixed-mode Register Entry Technique

the clocked J (or S) inputs. This is the mixed-mode equivalent of the common-reset entry method described earlier. The AC SET input takes the place of the DC SET used in that system. The mixed-mode system is useful even if a direct SET input is available, since it does not require any outside gating to enter single-rail data.

Figure 6-30 shows a simplification of this type of register whereby only a single clocking signal (rather than a *RESET* followed by a *GATE*) can be used. An RTL J-K flip-flop is shown. Its R input is enabled by a high level. The *RESET* signal is also applied to the clock input so that all the clock inputs are made high when *RESET* is TRUE. When the *RESET* signal goes FALSE (low), its trailing edge will give a falling transition to the clock inputs, thereby setting the flip-flop if I_i is TRUE.

This second entry technique simplifies the control system required to operate the register, since only one entry pulse is required. Other flip-flops, e.g., those with a different voltage symbolism at their DC input or different transition sensitivity, may require a single inverter between the *RESET* and the *CLOCK* input. The requirements are as follows: The proper *RESET* polarity must be present *during* the pulse, and the proper transition must be given to the *CLOCK* input on the *trailing* edge of the *RESET* pulse.

6.5.2. The Shift Register.

Once data have been entered into a register, it is often desirable to have the register perform simple arithmetic or logical operations upon its stored contents. The simplest of these types of data-modification registers is the *shift register.*

A shift register is a linear assemblage of flip-flops that will store a group of Boolean values and, upon receiving a clocking signal, will transfer or rotate its contents to the left or right. As an example, consider the generalized four-bit shift register shown in Fig. 6-31. If this is a *right* shift register, the *SHIFT* pulse will cause the logical value of the input signal to be stored

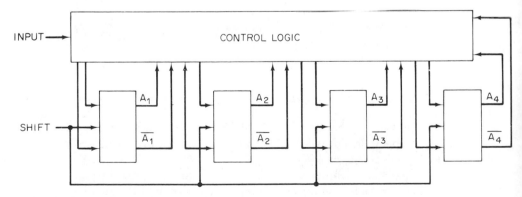

Figure 6-31. A Generalized Shift Register

into *A1*, the value of *A1* to transfer to *A2*, *A2* to *A3*, *A3* to *A4*, and the state of *A4* to be lost, unless it is returned to *A1* in a rotational fashion. A *left* shift register would transfer into the opposite direction.

Consideration of the *A2* flip-flop in a right shift register shows that its next state follows the present state of *A1*. This operation is shown in Table 6-12. The behavior of *A2* is transition-mapped in Fig. 6-32. If *A2* is imple-

Table 6-12: The *A2* Flip-flop in a Right Shift Register

Present State		Next State	Behavior
A1	*A2*	*A2*₊	*A2*
0	0	0	0
0	1	0	β
1	0	1	α
1	1	1	1

mented with the various flip-flop types, the control equations are

$$S_{A2} = A1 \cdot SHIFT \qquad\qquad R_{A2} = \overline{A1} \cdot SHIFT$$
$$J_{A2} = A1 \cdot SHIFT \qquad\qquad K_{A2} = \overline{A1} \cdot SHIFT \qquad (6\text{-}21)$$
$$T_{A2} = (A1 \oplus A2) \cdot SHIFT \qquad D_{A2} = A1 \cdot SHIFT$$

Clearly, the D flip-flop is the best choice for shift register implementation, with the J-K or S-R as next-best choices. A left shift register connects *A3* to *A2* in the same manner shown above. Complete logic diagrams for a four-bit right shift register constructed with J-K, S-R, and D flip-flops are shown in Fig. 6-33. Some special shift registers transfer more than one place to the left or right, but their design follows from the control equations given above.

Figure 6-32. Transition Map for *A2* in a Right Shift Register

The shifting process can be related to multiplication and division of binary operands. For more details on the arithmetic properties of shift operations, see Refs. [6] and [8].

A common symbol for a shift register unit is shown in Fig. 6-34. The output of the shift register will lag the state of the input by four clock periods, with the last four states of the input being held in *A1–A4*. Thus, the shift register is ideal for use as a serial/parallel conversion device.

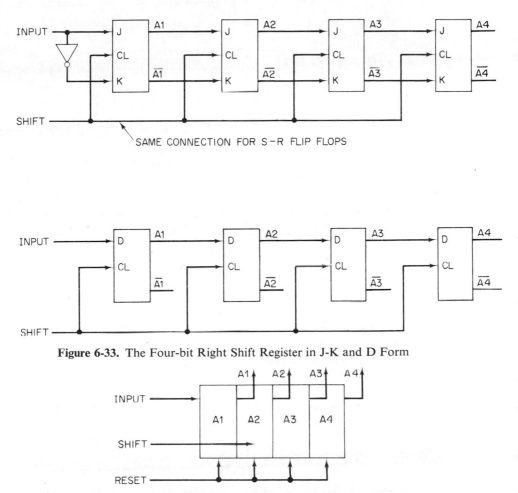

Figure 6-33. The Four-bit Right Shift Register in J-K and D Form

Figure 6-34. A Generalized Symbol for the Shift Register

EXAMPLE 6-25:   Consider the simple application of sending a six-bit code word over a single pair of wires such as a telephone line. At the source end a six-bit word is placed in a shift register by using the direct S and R inputs to the flip-flops that make up the register. Shift pulses then move the word, one bit at a time, to the output of the register, as shown in Fig. E6-25.

At the receiving end, the line is sampled when each *SHIFT* pulse occurs, and the TRUE/FALSE value of the line is shifted into receiving shift register. After all six bits have been received, the transmitted word is available for parallel output. Problems can arise in synchronizing the *SHIFT* pulses at the transmitter and the receiver, possibly requiring very stable clock oscillators or the provision of a second transmission path to carry synchronizing information.

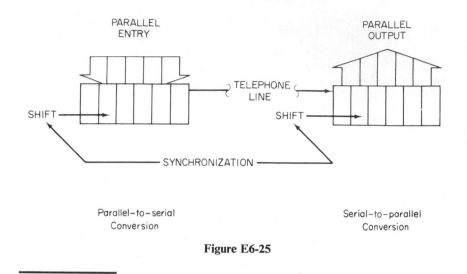

Parallel-to-serial
Conversion

Serial-to-parallel
Conversion

Figure E6-25

Many types of shift registers are available from logic circuit manufacturers in preassembled (MSI) form. Typical configurations include linear shift registers having from four to eight bits, multimode shift registers having both left and/or right shift capability, and large shift registers of 64–1024 bits (or more) in length.

6.5.3. Ring Counters.   While the shift register is normally used for data transfer or modification, it may also be used as a counting/timing device. If its output is returned to its input, a circular shift register or *ring counter* is formed, as shown in Fig. 6-35. The normal ring-counter sequence begins

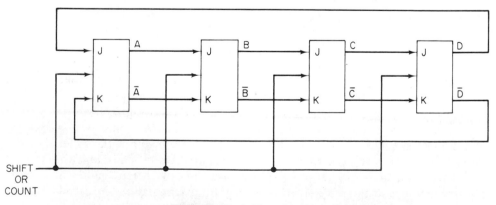

Figure 6-35. A Four-bit Ring Counter

with A being SET and B, C, and D being RESET. The counter can then be said to be in "state A," indicating that only the A flip-flop is SET. This corresponds to state m_8, since $ABCD = 1000$. The first shift pulse resets A and sets B, moving the counter to state B (m_4). States C (m_2) and D (m_1) then follow, with the fourth clock pulse returning the counter to state A. A timing diagram for the counter is shown in Fig. 6-36.

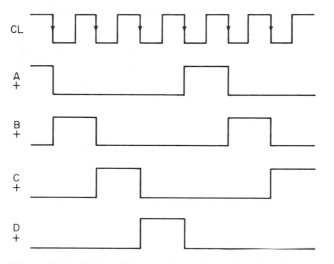

Figure 6-36. Timing Diagram for a Four-bit Ring Counter

This basic ring-counter counting sequence includes only 4 of the 16 states that are available to the four flip-flops. A complete state diagram for the ring counter can be determined by following the analysis procedures given in Sec. 6.2.7. The results are shown in Fig. 6-37. Clearly, this is not a bush sequence.

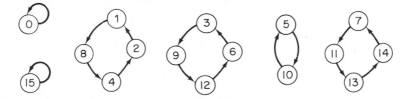

Figure 6-37. The Complete State Diagram for a Four-bit Ring Counter

To provide for bush behavior, the basic ring counter can be modified as shown in Fig. 6-38. Thus, the A flip-flop will be RESET whenever it is

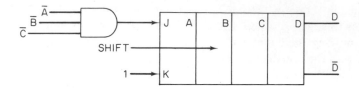

Figure 6-38. The Self-correcting Ring Counter

SET in its present state and will be SET again only when A, B, and C are RESET. The state diagram for this, the *self-correcting ring counter*, is shown in Fig. 6-39. This counter will always return to the main counting sequence

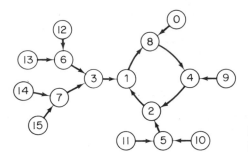

Figure 6-39. State Diagram for the Self-correcting Ring Counter

within three clock periods. The input logic for longer ring counters is simply an extension of the $\bar{A}\bar{B}\bar{C}$ logic that is used to SET the A flip-flop.

Obviously, the ring counter makes very inefficient use of its possible states. For n flip-flops it uses only n of its 2^n possible states, leaving $2^n - n$ states unused. For large n this loss becomes quite significant. Thus, ring counters are not often used when *many* states (more than ten, for example) are required.

The advantages of a ring counter include the simplicity of its control logic and the ease of determining what state it is in, since the A flip-flop is SET *only* when the counter is in state A, etc. For counters with relatively few states these advantages make the ring counter a good choice.

A modification of the ring counter leads to the *switch-tail ring counter*, also called the Lippel or Johnson counter or a *walking ring counter*. The return connection is made with logical inversion, as shown in Fig. 6-40. A timing diagram for this counter is given in Fig. 6-41. This counter has a period of $2n$ rather than n. Its basic state diagram is shown in Fig. 6-42.

These eight states of the four-bit, switch-tail ring counter may be labeled as states $A, B, ..., H$. State A may be represented as

$$A = m_8 = W\bar{X}\bar{Y}\bar{Z} \tag{6-22}$$

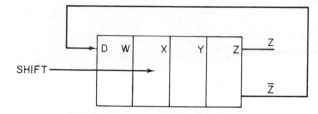

Figure 6-40. The Switch-tail Ring Counter

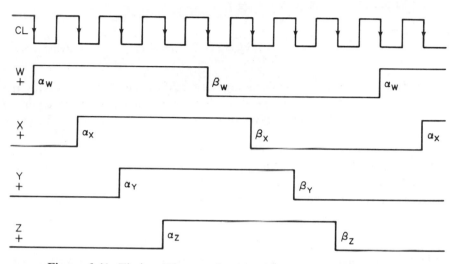

Figure 6-41. Timing Diagram for the Four-bit Switch-tail Ring Counter

Figure 6-42. State Diagram for the Four-bit Switch-tail Ring Counter

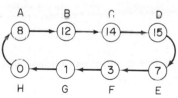

This expression can be simplified, however, by using the states external to the basic counting sequence as don't-cares. The Karnaugh map for the state A logic with the can't-happen states used as redundancies is given in Fig. 6-43. The grouping shows that the state A logic can be reduced to

$$A = W\bar{X} \tag{6-23}$$

Similarly,

$$B = X\bar{Y}, \quad C = Y\bar{Z}, \quad D = WZ, \quad E = \bar{W}X,$$
$$F = \bar{X}Y, \quad G = \bar{Y}Z, \quad H = \bar{W}\bar{Z} \tag{6-24}$$

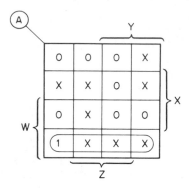

Figure 6-43. The K-map for Identifying State A

The simplicity of this state-decoding logic and of the control logic for the counter makes the switch-tail ring counter a good choice for many counting applications. The fact that it gives twice as many states as an equivalent ring counter is also of merit.

Note that if a sequence of timing signals is needed and only transitions (not levels) are required, no gating logic is required to use a switch-tail ring counter having n flip-flops to generate $2n$ sequential pulses. The timing diagram of Fig. 6-41 shows the following sequence of nonoverlapping transitions:

$$\alpha_W, \ \alpha_X, \ \alpha_Y, \ \alpha_Z, \ \beta_W, \ \beta_X, \ \beta_Y, \ \beta_Z$$

Thus the counter can provide eight timing transitions without any external logic.

The *complete* state diagram for this type of counter is shown in Fig. 6-44. The figure shows that a closed path external to the desired sequence exists. Changing the control logic for the W flip-flop to

$$J_W = \bar{Y} K_W = Z\bar{Z} \tag{6-25}$$

will make the counter self-correcting. Proof of this is left to the problems at the end of this chapter.

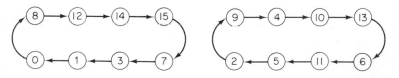

Figure 6-44. The Complete State Diagram for the Four bit Switch-tail Ring Counter

6.5.4. Shift Register Counters.

A shift register can be made to count through a sequence of states just as a normal counter does, with the exception that its present-state/next-state behavior is somewhat restricted. Consider the shift register shown in Fig. 6-45. The control logic interprets the present state of the shift register and determines the TRUE/FALSE condition of the input to the *A* flip-flop. Unlike a normal counter, however, the next states of *B*, *C*, and *D* are not affected by the control logic. Rather, they follow from the present states of *A*, *B*, and *C* in normal shift register fashion.

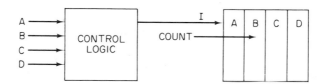

Figure 6-45. The Shift Register Counter

Thus, each present state has only *two* possible next states, one with $A = 1$ and one with $A = 0$. Consider present state m_0. Since A, B, and C are all RESET in the present state, B, C, and D will be RESET in the next state. If the control input is FALSE, A will remain RESET and the next state will be m_0. If the control input is TRUE, the next state will be m_8, as shown in Fig. 6-46. Continuing this sequence leads to the complete state diagram for

Figure 6-46. Conditional Next State Behavior for a Shift Register Counter

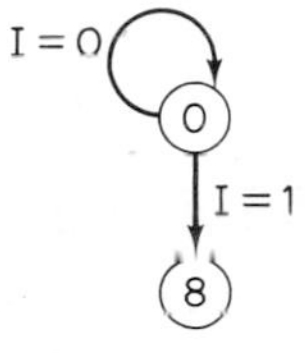

a four-bit shift register counter that is given in Fig. 6-47. Many closed sequences exist within this diagram. For example, the basic ring-counter sequence appears on the left-hand side.

By trial and error, sequences of various lengths can be identified. The maximum length sequence for an *n*-bit shift register counter will include $2^n = 16$ states. One such sequence is shown in Fig. 6-48.

The sequential operation of the shift register counter is determined by the control logic that enables or disables the input to the first stage of the register. The design of this logic can be considered as either a combinational *or* as a transition mapping problem. Considering the maximum length sequence given in Fig. 6-44, a tabulation of the required type D input condition,

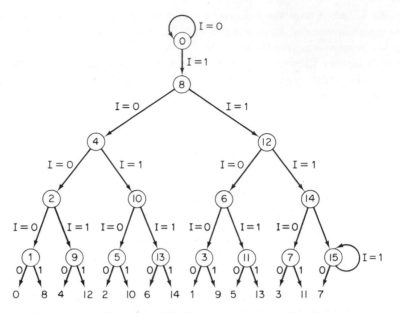

Figure 6-47. The Allowable Shift Register Counting Sequences

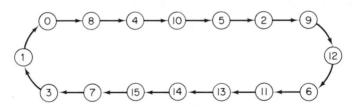

Figure 6-48. A Maximum Length Shift Register Counting Sequence

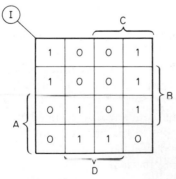

Figure 6-49. The K-map of the Shift Register Input

I, can be made. The *I* signal must be TRUE during any present state whose next state has the *A* flip-flop SET. Table 6-13 presents this information in tabular form. The Karnaugh mapping for the *I* function is given in Fig. 6-49. Its groups yield

$$I = \bar{A}\bar{D} + BC\bar{D} + A\bar{C}D + A\bar{B}D \qquad (6\text{-}26)$$

Table 6-13: **Tabulation of the Shift Register Counting Sequence Shown in Fig. 6-48**

Present State				Next State				Input
A	*B*	*C*	*D*	*A*₊	*B*₊	*C*₊	*D*₊	*I*
0	0	0	0	1	0	0	0	1
0	0	0	1	0	0	0	0	0
0	0	1	0	1	0	0	1	1
0	0	1	1	0	0	0	1	0
0	1	0	0	1	0	1	0	1
0	1	0	1	0	0	1	0	0
0	1	1	0	1	0	1	1	1
0	1	1	1	0	0	1	1	0
1	0	0	0	0	1	0	0	0
1	0	0	1	1	1	0	0	1
1	0	1	0	0	1	0	1	0
1	0	1	1	1	1	0	1	1
1	1	0	0	0	1	1	0	0
1	1	0	1	1	1	1	0	1
1	1	1	0	1	1	1	1	1
1	1	1	1	0	1	1	1	0

The logic diagram for this counter is shown in Fig. 6-50, with type D flip-flops being used.

Alternatively, if the control of the *A* flip-flop is considered as a transition mapping problem, a present-state/next-state tabulation is required. Only the next state of the *A* flip-flop needs to be considered, however, since *B*, *C*, and *D* act in shift register fashion. This behavior is tabulated in Table 6-14. The behavior of the *A* flip-flop is transition-mapped in Fig. 6-51, leading to the following control equations:

$$J_A = \bar{D} \cdot CL \qquad K_A = (\bar{C}\bar{D} + \bar{B}\bar{D} + BCD) \cdot CL \qquad (6\text{-}27)$$

These are somewhat cheaper than Eq. (6-26). With type D flip-flops the two sets of control equations will be the same. With J-K or S-R flip-flops, the transitional approach usually gives the cheaper solution.

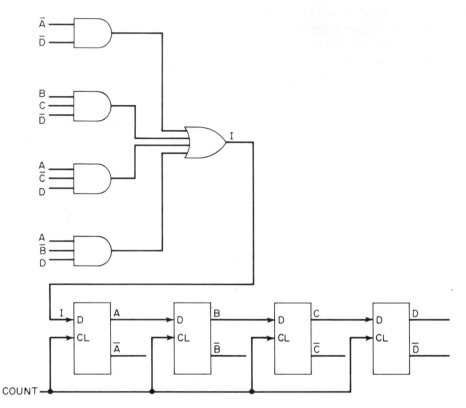

Figure 6-50. Logic Diagram for an Example Shift-register Counter

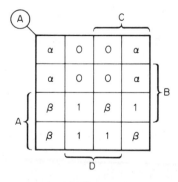

Figure 6-51. Transition Map of the *A* Flip-flop in an Example Shift-register Counter

Shift register counters offer ease of design, synchronous operation, and low cost for those applications wherein only the *number* of states in a counting sequence is specified. However, the restricted state sequence which they produce does limit their applicability.

**Table 6-14: Transitional Behavior of the Shift
Register Counter**

Present State				Next State	Behavior
A	B	C	D	A_+	A
0	0	0	0	1	α
0	0	0	1	0	0
0	0	1	0	1	α
0	0	1	1	0	0
0	1	0	0	1	α
0	1	0	1	0	0
0	1	1	0	1	α
0	1	1	1	0	0
1	0	0	0	0	β
1	0	0	1	1	1
1	0	1	0	0	β
1	0	1	1	1	1
1	1	0	0	0	β
1	1	0	1	1	1
1	1	1	0	1	1
1	1	1	1	0	β

6.6 Design with Sequential MSI Devices

Just as the development of MSI technology has prompted the production of many function-level combinational devices in prepackaged form, function-level sequential devices are now available as MSI "components." Typical sequential devices include binary counters, decade counters, and shift registers. These types of devices are produced in a wide variety of forms. Both synchronous and asynchronous binary counters are available, for example, as are decade counters having controllable presetting capability. Multimode devices are also available; up/down counters and left/right shift registers are examples of these.

The use of these predesigned devices whenever possible is strongly recommended, since their use generally saves cost, design expense, and space. Voltage symbolism can be used to describe the logical/electrical relationships at all level-sensitive inputs to devices of this sort, and the arrow-symbols can be used to describe transition-sensitive inputs. Thereafter, the use of these devices follows the techniques that are presented in this chapter.

The real problem is in keeping abreast of the new devices that become available.

6.7   Summary

Just as Chapters 2, 3, and 4 combine to present combinational design, Chapters 5 and 6 present sequential systems design. Chapter 5 introduces the concepts and devices that apply to sequential systems. Chapter 6 shows how to design the systems themselves.

The general sequential design technique is given in Fig. 6-0. The process resolves into four parts: system specification, state reduction, state assignment, and counter design. System specification is covered in Chapter 5. Simple techniques for state reduction and state assignment are presented in Chapter 6. Once a state assignment has been made, the design of a sequential system reduces to a counter design problem. Transition mapping is introduced as a counter design tool. It allows sequential behavior to be described with brevity, the counter control logic to be minimized in both SOP and POS forms for any type of flip-flop, the easy extension to tabular techniques for larger problems, and the design of both synchronous and asynchronous counting devices.

The general sequential design procedure given in Chapter 6 can easily be augmented by learning more advanced state reduction and state assignment techniques (such as partitioning, compatibility classes, etc.), but the usefulness of the transition mapping technique as a design tool during the final design step remains unaffected.

Chapter 6 includes several supporting topics. The behavior of a counter from its unspecified states (if any) is analyzed. Shift registers and related counters are described, as are techniques for estimating allowable counting rates.

6.8   Did you Learn?

1. The relationships between counting sequences, state tables, state diagrams, behavioral tables, and transition maps?

2. The reasoning behind the *must include* and *must avoid* transition map entries for the four flip-flop types?

3. How a transition map is plotted? How it describes sequential behavior? How to detect certain types of errors that may arise during the transition mapping process?

4. How to obtain simplified SOP control expressions for each type of flip-flop by using transition mapping? Simplified POS expressions?

5. How to convert a large sequential design problem into a tabular reduction problem?

6. How to use transition maps for asynchronous design? How transition states arise in an asynchronous counter?

7. How Moore and Mealy output signals differ? How to generate each type?

8. How to determine counter behavior from unspecified states?

9. How to estimate allowable counting rates?

10. The general sequential design process?

11. How to reduce the number of states of both reset and nonreset sequential systems?

12. How to use the Armstrong-Humphrey rules for making a state assignment?

13. How to use registers for data storage?

14. How to design shift-register counters? Ring counters?

6.9 References

[1] Armstrong, D. B., "A Programmed Algorithm for Assigning Internal Codes to Sequential Machines," *IRE Transactions on Electronic Computers*, EC-11, No. 4 (August 1962), 466–472.

[2] Beraru, J., "A One-step Process for Obtaining Flip-flop Input Logic Equations," *Computer Design*, 6, No. 10 (October 1967), 58–63.

[3] Beraru, J., "Map Entry Method for Sequential Counter Logic," *Electro-technology* (December 1968), 51–52.

[4] Booth, Taylor L., *Sequential Machines and Automata Theory*. New York: John Wiley & Sons, Inc., 1967.

[5] Chao, Chester, "Synchronous Flip-flop Counters," *Instruments & Control Systems* (December 1968), 78–80.

[6] Chu, Yoahan, *Digital Computer Design Fundamentals*. New York: McGraw-Hill Book Company, Inc., 1962.

[7] Givone, Donald D., *Introduction to Switching Circuit Theory*. New York: McGraw-Hill Book Company, 1970.

[8] Golumb, S. (ed.), *Digital Communication with Space Applications*. Englewood Cliffs, N.J.: Prentice-Hall, Inc., 1964.

[9] Hartmanis, J., "On the State Assignment Problem for Sequential Machines I," *IRE Transactions on Electronic Computers*, EC-10, No. 2 (June 1961), 157–165.

[10] Hill, Frederick J., and Gerald R. Peterson, *Introduction to Switching Theory and Logical Design.* New York: John Wiley & Sons, Inc., 1968.

[11] Humphrey, W. S., *Switching Circuits with Computer Applications.* New York: McGraw-Hill Book Company, 1958.

[12] Kintner, P. M., *Electronic Digital Techniques.* New York: McGraw-Hill Book Company, 1968.

[13] Kohavi, Zvi, *Switching and Finite Automata Theory.* New York: McGraw-Hill Book Company, 1970.

[14] Marcus, Mitchell P., *Switching Circuits for Engineers.* London: Prentice-Hall International, 1962.

[15] Marcus, Mitchell P., "A One-map Method for Obtaining Flip-flop Input Expressions," *Computer Design*, 7, No. 3 (March 1968), 54–58.

[16] McDonagh, Kevin, "Design Better Sequential Circuits," *Electronic Design* (December 19, 1968), 80–85.

[17] Mealy, G. H., "A Method for Synthesizing Sequential Circuits," *Bell Sys. Tech. Journal*, 34 (September 1955), 1045–1080.

[18] Miller, Raymond E., *Switching Theory, Volume II: Sequential Circuits and Machines.* New York: John Wiley & Sons, Inc., 1965.

[19] Moore, E. F., "Gedanken Experiments on Sequential Machines," in *Automata Studies.* Princeton, N.J.: Princeton University Press, 1956, pp. 129–153.

[20] Prather, Ronald E., *Introduction to Switching Theory: A Mathematical Approach.* Boston, Mass.: Allyn and Bacon, Inc., 1967.

[21] Stearns, R. E., and J. Hartmanis, "On the State Assignment Problem for Sequential Machines," *IRE Transactions on Electronic Computers*, EC-10, No. 4 (December 1961), 593–603.

[22] Tien, Paul S., "Sequential Counter Design Techniques," *Computer Design*, 10, No. 2 (February 1971), 49–56.

[23] Torng, H. C., *Introduction to the Logical Design of Switching Systems.* Reading, Mass.: Addison-Wesley Publishing Company, Inc., 1964.

6.10 Problems

6-1. For each of the three systems that are tabulated in Prob. 5-2, determine the transitional behavior of each flip-flop and plot the transition maps that describe the system.

6-2. Plot the transition maps that describe each of the following counters. Assume that next states are below present states. The last state in each column counts back to the state at the top of the column.

a. Present State			b. Present State				c. Present State				
A	B	C	W	X	Y	Z	A	B	C	D	E
1	0	1	0	0	0	0	0-	0	0	0	0
1	1	1	0	1	1	1	0	0	0	1	0
0	0	0	1	1	1	0	0	0	1	0	0
0	0	1	1	1	0	0	0	0	1	1	0
1	1	0	1	0	0	0	1	1	1	0	1
0	1	1	1	0	0	1	1	0	0	0	1
			1	0	1	1	1	1	0	1	1
			0	0	0	1	0	1	1	0	1
			0	1	0	1	0	1	0	1	0
			0	1	1	0	1	0	1	0	1
			1	0	1	0	0	0	0	0	1
							0	1	1	1	1
							1	0	0	0	0
							0	0	0	1	1
							0	0	1	1	1
							1	1	1	0	0
							1	0	1	1	1
							1	0	1	1	0
							0	1	1	1	0
							0	1	0	0	0
							1	0	0	1	1
							1	1	1	1	1

6-3. Plot the transition maps for a five-bit binary down-counter.

6-4. Plot the transition maps for each of the following count sequences. First consider a counter that goes *down* the sequence and resets to the top. Repeat for a counter that counts *up* the sequence and resets to the bottom.

a. A	B	C	D	b. W	X	Y	Z	c. A	B	C
0	0	0	0	0	0	0	0	1	0	1
0	0	0	1	1	0	0	0	1	0	0
0	0	1	1	1	0	0	1	0	1	1
0	1	1	1	1	1	0	1	0	1	0
1	1	1	1	1	1	1	1	1	1	1
1	1	1	0	0	1	1	1	0	0	0
1	1	0	0	0	1	1	0	0	0	1
1	0	0	0	0	0	1	0			
1	0	0	1	0	1	0	0			
0	1	1	0	1	1	1	0			
				1	0	1	0			
				1	1	0	0			

6-5.   Each of the sets of transition maps in Fig. P6-5a–c describes the behavior of a counter. Reconstruct the counting sequences. Give a state table and state diagram for each counter showing all specified present-state/next-state relationships.

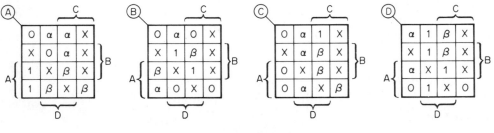

Figure P6-5a

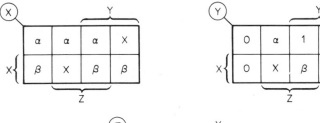

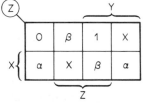

Figure P6-5b

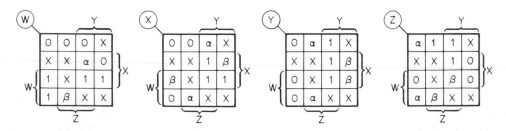

Figure P6-5c

6-6. Find the errors in each of the transition maps in Fig. P6-6.

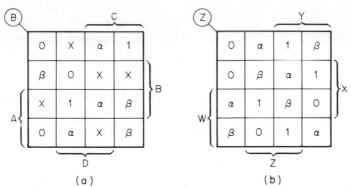

(a)   (b)

Figure P6-6a, b

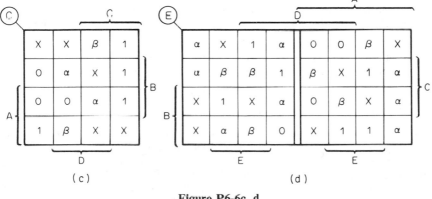

(c)   (d)

Figure P6-6c, d

6-7. The following J-K control equations are not necessarily in their most reduced form. Identify and eliminate the unnecessary terms. Remember that J_A is ANDed with $\bar{A}$ internal to the flip-flop, etc.

a. $J_X = (XYZ + WX + \bar{Z}) \cdot CL$
b. $K_A = (B\bar{C} + A\bar{B}D + \bar{A}C) \cdot CL$
c. $J_D = (AB\bar{D} + \bar{A}BD + D) \cdot CL$
d. $K_W = [(W \oplus X) + (Y \odot \bar{Z}) + (\bar{Y} \odot \bar{W})] \cdot CL$

*6-8. For each of the counters that are specified by the state tables of Prob. 5-2 perform the following design steps:

a. Develop the minimized SOP control equations for use with S-R flip-flops. Consider only fully synchronous counting.
b. Determine the behavior of the counter from each of its unspecified

present states. Draw a complete state diagram. If there are any redundant conditions that allow both the S and the R input for a flip-flop to be enabled or if the counter does not exhibit bush behavior, modify the control equations to eliminate these undesirable characteristics.

c. Develop the fully synchronous, minimized, POS, S-R control equations.

d. Use asynchronous logic to simplify the counter, if possible.

e. Determine the fully synchronous, minimized, SOP, J-K control equations. Repeat for POS control equations. Select the cheapest set.

f. Determine the behavior of the J-K counter from each of its redundant states. Draw a complete state diagram. If necessary, modify the J-K control equations to assure bush behavior.

g. Draw a complete RTL logic diagram for the J-K counter.

h. Draw a complete TTL logic diagram for the J-K counter.

i. Use asynchronous logic to simplify the counter, if possible. Consider both single-clock and separated-clock J-K flip-flops.

j. Draw an RTL logic diagram for the asynchronous, single-clock, J-K counter. Repeat for TTL flip-flops.

k. Develop the fully synchronous SOP and POS control equations for use with D and T flip-flops.

*6-9. Repeat Prob. 6-8 for each of the counting sequences given in Prob. 6-2. Consider both down-counting and up-counting.

*6-10. Repeat Prob. 6-8 for each of the counting sequences given in Prob. 6-4. Consider both down-counting and up-counting.

*6-11. The control equations for several synchronous counters are listed below. Determine the state-to-state behavior for each counter. Plot that behavior on transition maps. Also give a state diagram and state table for each counter.

a. $J_A = (B + \bar{C}\bar{D}) \cdot CL$ $K_A = 1 \cdot CL$
    $J_B = (A \oplus \bar{D}) \cdot CL$ $K_B = ACD \cdot CL$
    $J_C = (\bar{B} + \bar{D}) \cdot CL$ $K_C = (A + B) \cdot CL$
    $J_D = 1 \cdot CL$ $K_D = \bar{C} \cdot CL$

b. $J_W = (Y \oplus Z) \cdot CL$ $K_W = Y \cdot CL$
    $J_X = (WY + Z) \cdot CL$ $K_X = \bar{Y} \cdot CL$
    $J_Y = (XZ + \bar{W}) \cdot CL$ $K_Y = 1 \cdot CL$
    $J_Z = \bar{Y} \cdot CL$ $K_Z = 1 \cdot CL$

c. $J_A = B \cdot CL$ $K_A = B \cdot CL$
    $J_B = C \cdot CL$ $K_B = C \cdot CL$
    $J_C = A \cdot CL$ $K_C = A \cdot CL$

d. $J_A = C \cdot CL$ $K_A = B \cdot CL$
   $J_B = A \cdot CL$ $K_B = \bar{A} \cdot CL$
   $J_C = B \cdot CL$ $K_C = \bar{B} \cdot CL$
   $J_D = 1 \cdot CL$ $K_D = 1 \cdot CL$

Note that d is a fully synchronous, ten-state counter that requires no control gating.

*6-12. The following state tables describe multimode counters. Use the analysis procedures given in Prob. 6-8 to fully investigate the design of these counters.

a.

Present State	Next State	
	$M = 0$	$M = 1$
0000	0000	1111
0001	0001	1110
0011	0010	1101
0100	0011	1100
0110	0100	1011
0111	0101	1010
1001	0110	1001
1010	0111	1000
1100	1000	0111
1101	1001	0110
1111	1010	0101

b.

Present State	Next State			
	$XY = 00$	$XY = 01$	$XY = 10$	$XY = 11$
000	111	010	111	100
001	110	100	000	110
010	101	110	111	111
011	100	000	000	011
100	011	001	010	001
101	010	011	101	000
110	001	101	010	101
111	000	111	101	010

*6-13. A six-bit counter is to operate in the following manner:
   a. For all present states from m_0 to m_{31}, the counter is to double its contents when a clock pulse is received.
   b. For all present states from m_{32} to m_{55}, the counter is to decrement by 2 whenever a clock pulse is received.

c. Present states m_{56} to m_{63} go back to m_0 when the next clock pulse is received.

Use tabular or computer-aided minimization to solve for the synchronous J-K control equations for this counter. Consider both SOP and POS forms.

*6-14. Make a state assignment for each of the sequential systems described by the following tables and the state diagram of Fig. P6-14. Show the assign-

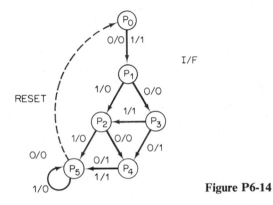

Figure P6-14

ment on a map. Then design a J-K version of each system. Show a TTL logic diagram. Investigate the behavior of the system from its redundant states, if any are present.

a.

Present State	Next State	
	$I = 0$	$I = 1$
A	A	B
B	A	C
C	B	C
D	D	E
E	D	C
F	E	F

b.

Present State	Next State		Output, F	
	$X = 0$	$X = 1$	$X = 0$	$X = 1$
a	a	b	0	1
b	c	d	1	0
c	b	d	0	0
d	c	a	1	1

c.

Present	Next State	
State	$M = 0$	$M = 1$
a	d	e
b	a	b
c	e	e
d	c	f
e	f	c
f	b	a

6-15. For comparison, redesign each of the generalized sequential systems of Prob. 6-14, making a state assignment that violates the Armstrong-Humphrey "rules." Compare the cost of the control logic obtained with and without following good state assignment practices.

6-16. Reduce each of the following RESET systems to their minimum number of states. Clearly indicate all states that are combined.

a.

Present State	Next State		Output, F	
	$I = 0$	$I = 1$	$I = 0$	$I = 1$
a	b	c	0	1
b	e	d	0	1
c	d	f	0	0
d	g	h	0	0
e	j	g	0	0
f	h	i	0	0
g	a	a	1	0
h	a	a	1	0
i	a	a	1	0
j	a	a	1	1

b.

Present State	Next State		Output, X	
	$Y = 0$	$Y = 1$	$Y = 0$	$Y = 1$
A	D	G	1	1
B	A	A	0	0
C	A	A	0	1
D	B	C	1	0
E	A	A	0	0
F	A	A	0	1
G	E	F	1	0

c.

Present State	Next State		Output, H	
	$X = 0$	$X = 1$	$X = 0$	$X = 1$
A	E	F	1	1
B	A	A	0	1
C	A	A	0	0
D	A	A	0	1
E	B	D	1	0
F	C	G	1	0
G	A	A	0	0

6-17. Reduce the following sequential systems to their minimum number of states.

a.

Present State	Next State		Output, F	
	$I = 0$	$I = 1$	$I = 0$	$I = 1$
a	b	d	0	0
b	f	g	0	1
c	e	h	0	0
d	a	e	1	1
e	h	j	0	1
f	a	e	1	1
g	e	c	0	1
h	c	e	1	1
j	b	a	0	1

b.

Present State	Next State		Output, H	
	$I = 0$	$I = 1$	$I = 0$	$I = 1$
A	B	D	0	0
B	E	C	0	1
C	E	D	1	1
D	E	C	0	1
E	C	C	1	0
F	C	G	1	0
G	F	B	1	1

*6-18. Repeat the systems design process of Ex. 6-19 for a RESET system that examines successive groups of four input values and identifies the erroneous 8421 BCD code words, $m_{10}, ..., m_{15}$. Design a J-K version of the reduced system.

6-19. A four-state sequential system is described in the following table:

Present State	Next State		Output, F	
	$I = 0$	$I = 1$	$I = 0$	$I = 1$
a	a	d	0	0
b	b	c	0	1
c	c	b	1	0
d	d	a	1	1

Three possible state assignments for this system are shown in Fig. P6-19a. Prove, by designing a state-counter controller for each case, that these assignments yield the same cost for the controller. In fact, they are identical insofar as state assignment considerations are concerned. Note that states a and d are diagonal opposites in each assignment.

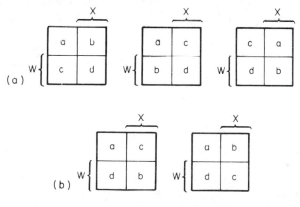

Figure P6-19a, b

There are two other state assignments that *are* distinct from those given above. They are shown in Fig. P6-19b and have (a,b) and (a,c) as diagonal opposites. Design state-counter controllers for each of these assignments. Compare costs among the three different assignments and see how the cheapest assignment agrees with the Armstrong-Humphrey suggestions.

*6-20. A three-bit counter has the following control logic. Determine its complete state diagram.

$$J_A = B \cdot CL \qquad\qquad K_A = C \cdot CL$$
$$D_B = (\bar{A} + \bar{C}) \cdot CL$$
$$T_C = AB \cdot CL$$

*6-21. Repeat Prob. 6-20 for the following control logic.

$$J_W = \bar{W}\bar{X} \cdot CL \qquad\qquad K_W = (X + Y) \cdot CL$$
$$T_X = (X\bar{Y} + \bar{W}) \cdot CL$$
$$D_Y = (W \oplus Y) \cdot CL$$

*6-22. A generalized four-bit shift register is shown in Fig. 6-45. For each of the following control expressions, determine the complete state diagram for the counter.

a. $J_A = (B \oplus C) \cdot CL$ $K_A = 1 \cdot CL$
b. $D_A = (A + C) \cdot CL$
c. $J_A = C \cdot CL$ $K_A = \bar{D} \cdot CL$
d. $J_A = K_A = D \cdot CL$

*6-23. Determine the complete state diagram for a four-bit switch-tail ring counter having the feedback logic given in Eq. (6-26).

*6-24. By thorough examination of the counting sequence diagram of Fig. 6-47, determine all possible ten-state shift register counting sequences. Then, develop the input control logic (both type D and J-K) for each counter. Select the cheapest one and show a TTL logic diagram for that counter. Generate its complete state diagram. If necessary, modify its input logic to assure bush behavior.

*6-25. A stepping motor is an electromechanical device that is often used in conjunction with digital systems. It rotates one increment forward or backward in response to a change in the control counter that activates the motor. A common type of stepping motor has four control windings. Its increment of rotation is $1.8°$, making a complete rotation consist of 200 steps. Even though the motor has four windings, only two variables are involved in the control counter, since the windings operate in pairs. The pair of windings may be designated as A and B. One winding of the A pair follows the TRUE/FALSE value of A; the other follows $\bar{A}$. The control sequences for positive and negative rotation are tabulated below. The operations repeat every four increments.

Present State		Next State			
		Forward		Backward	
A	B	A_+	B_+	A_+	B_+
0	0	0	1	1	0
0	1	1	1	0	0
1	0	0	0	1	1
1	1	1	0	0	1

a. Design a control counter for use with this stepping motor. Let the mode-control signal be F, where $F = 1$ signifies forward rotation and $F = 0$ signifies reverse rotation.
b. Show an RTL implementation using J-K flip-flops.
c. Show a complete state diagram for the control counter. Is it a bush?
d. Can the control counter be operated asynchronously? Explain.

6-26. Prove that the transition states that occur during the operation of an asynchronous binary up-counter cannot have a higher minterm number than that of the stable state that they follow. This property allows an asynchronous counter to be used in counting repetitions so as to stop when a maximum number is reached. Asynchronism will never cause the counter to appear to reach a given value earlier than it should.

More counter design problems are included in the problems sections of Chapters 7, 8, 9, and 10.

7 Digital Design at the Systems Level

7.0 Introduction

The preceding chapters have introduced the basic "tools of the trade" for designing both combinational and sequential digital devices that perform at the *functional* level. It is now appropriate, therefore, to consider methods whereby these functional building blocks can be joined together to perform some desired operation. This step, the *system design*, is the most important part of a design effort, for no matter how cleverly each subsystem is designed, the overall success or failure of the device that is designed depends on the manner in which the various functional parts of the system act together in order to accomplish the desired function.

System design begins with an analysis of the overall performance requirements of the device that is to be designed. The general task must be broken into a set of subtasks that are to be performed sequentially and/or simultaneously in order to implement the larger function. This process is much like the writing of a computer program. The programmer breaks the overall computation into a sequence of instructions that, when executed by the computer, perform the desired calculations. The subtasks that the system designer identifies correspond to the separate instructions that make up the computer program.

Once the subtasks have been specified, the functional subsystems that perform them must be designed. This design effort generally involves both combinational and sequential design techniques. Typical subsystems include adders, code-to-code converters, counters, shift registers, etc. MSI devices can often be used to perform such function-level tasks.

370

Finally, the design is completed by developing a control system that supervises the operation of the functional subsystems so as to obtain the desired system operation. This supervisory device, commonly called a *controller*, is a functional subsystem itself. Its function is to interpret the various input signals that the digital system receives and, in response to these signals, to cause each of the subsystems to "do their own thing" in the proper sequence. The net result is a digital system that performs as specified.

A generalized controller is shown in Fig. 7-0. It receives input signals and timing signals and generates a sequence of output signals that control the

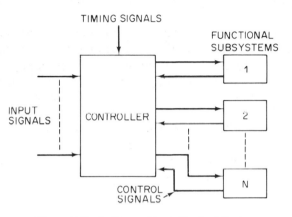

Figure 7-0. A Generalized Control System

activity of the functional subsystems. The input signals are normally used to vary the control sequence. Some of the functional subsystems may also return control information to the controller, as shown in Fig. 7-0. These signals carry information concerning the execution of each subtask, e.g., "task complete" or "subsystem busy." The controller will generally contain both memory and combinational logic, making it a sequential system in itself.

As an example, consider the control of a traffic light at a north-south/ east-west intersection. The overall task is to allow traffic to flow in both directions. The various subtasks that can be identified are

   a. Turn on N-S Green.

   b. Turn on E-W Green.

   c. Turn on N-S Red.

   d. Turn on E-W Red.

   e. Turn on N-S Amber.

   f. Turn on E-W Amber.

Some of these subtasks must be performed in parallel. The operational sequence in which these are to be performed is

A. *a* and *d*.

B. *e* and *d*.

C. *b* and *c*.

D. *f* and *c*.

The relative timing is not given, but it is normal to assume that the system will spend more time in the performance of operations *A* and *C* than in *B* or *D*.

This simple example illustrates the importance of good systems design, for, even if the signal lights and the switches that turn them on and off are designed properly, performing the subtasks in the wrong sequence could be disastrous. Consider the effects of performing *a* and *b* at the same time.

Since the operation of the system can be considered as the sequential performance of a set of subtasks, the control of the system is essentially a sequential design problem and can be handled by using the techniques presented in Chapter 6. This chapter considers the application of these sequential design techniques to the control of digital systems, and the techniques that are used for the completion of the systems design.

7.1   Controller Specification

As stated in Sec. 7.0, the portion of a digital system that supervises the operation of its functional subsystems is commonly called the *controller*. The controller must provide the sequence of logical signals that activate each functional part of the system. This operational sequence can be broken into *control intervals* during which the execution of a given subtask or group of subtasks is initiated and completed. In the traffic-light example given earlier there were four separate control intervals corresponding to the four pairs of subtasks that must be performed during the traffic control cycle. The traffic control system moves from each control interval to the next in a repetitive manner.

The concept of a general sequential system and of system states is introduced in Chapter 5. In designing a controller, each different control interval is assigned a different state. Thus, the control sequence may be considered as a present-state/next-state operation. Memory, usually in the form of flip-flops, is used to store the present state. During each control interval the controller initiates the proper functional activity and interprets its own present state and the current control inputs so that, at the end of the timing period, it can move to the proper next state. The controller must have some means of recognizing when the current control interval is over. Control intervals may last for one or more clock periods or may begin or end asynchronously, i.e., with no coordination to the system clock.

The states present in the control sequence can be named, e.g., states *A–D* in the traffic-light example, and a table formed that gives the next state for each present state. The state table for the traffic-light controller is given in Table 7-0.

Table 7-0:   The Traffic-Light Controller

Present State	Next State	Operations Performed
A	*B*	*a, d*
B	*C*	*e, d*
C	*D*	*b, c*
D	*A*	*f, c*

In most systems the control sequence is not as restricted as that of the traffic-light example. Rather, the input signals that are received by the controller are used to modify the control sequence. These control signals can be externally generated or can be generated internally by the subsystems that make up the digital device. This type of control system is called a *conditional controller*, in that its control sequence will vary depending on the logical conditions present at its control inputs. A conditional controller is a type of multimode counter.

The fundamental difference between a nonconditional controller and a conditional controller is that the latter has more than one next state available from at least one of its possible present states. The particular next state that is reached following the current control interval will depend on the current condition of the control inputs. The state diagrams and state tables that describe a conditional controller must show the variety of next states that may follow each present state. The input conditions that are associated with each state-to-state transfer must also be indicated, along with the various output signals that the controller produces.

The tabular or diagrammatic techniques that are introduced in Chapters 5 and 6 permit the state-to-state behavior of a controller to be described in a clear, compact form. Drawing a state diagram and/or filling out a state table are good practices to develop, for just as preparing a flow chart helps in writing a computer program, these two visual aids help the digital designer in separating his system into suitable subsystems and in establishing a correct control sequence.

The following examples serve to illustrate the separation of a system into functional subsystems and the initial organization of the controller.

EXAMPLE 7-0:   Consider a controller that repeatedly initiates the operation of some system for a specified number of cycles. The number of cycles is to be set by a group of switches. A block diagram for the system is shown in Fig. E7-0a. The operation begins when a *RESET* signal is provided by the

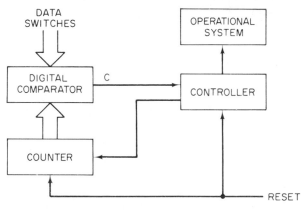

Figure E7-0a

user. Thereafter, the repetitive operation is initiated by the controller. When each repetition of the operation is completed, the controller increments a counter that is used to count the number of repetitions. The contents of the counter are compared to the number set into the data switches and a logical control signal, C, is generated. C is TRUE when the numbers compare. The system is halted when $C = 1$ and then waits until a new *RESET* signal is received.

One possible approach to the design of this system uses four control states, P_0, P_1, P_2, and P_3. During P_0 the repetitive operation is initiated. The controller remains in P_0 until the operation is completed. It then moves to P_1, wherein the counter is incremented. The controller then moves to P_2, which is a *decision state*. During P_2 the output signal from the comparison circuitry is used to determine whether or not the desired number of operations have been performed. If not, the controller returns to P_0. Otherwise, the controller goes on to P_3, the final state. The *RESET* signal forces a return to P_0. It also resets the counter.

This control sequence can be tabulated as shown below:

Present State	Next State		Operation
	$C = 0$	$C = 1$	
P_0	P_1	P_1	Initiate cycle
P_1	P_2	P_2	Increment counter
P_2	P_0	P_3	Decide
P_3	P_3	P_3	Halt

Alternatively, the control sequence may be diagrammed as shown in Fig. E7-0b. The dotted line shows the state change that is associated with the *RESET* signal. *RESET* is used to force the controller into state P_0, usually by means of the direct S or R control inputs to the flip-flops that make up the memory section of the controller.

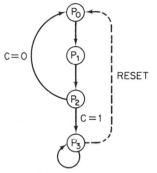

Figure E7-0b

The operations described at the right of the state table given above can be related to the output signals that are generated by the controller and are transmitted to the subsystems that are to perform the desired activity. If I_1 is used as the signal that initiates the operation that is to be repeated and I_2 is used as the signal that increments the counter, the state table can be rewritten as

Present State	Next State		Outputs			
			I_1		I_2	
	$C = 0$	$C = 1$	$C = 0$	$C = 1$	$C = 0$	$C = 1$
P_0	P_1	P_1	1	1	0	0
P_1	P_2	P_2	0	0	1	1
P_2	P_0	P_3	0	0	0	0
P_3	P_3	P_3	0	0	0	0

showing that each cycle of the repetitive operation is initiated during P_0 and that the counter is incremented during P_1. Note that

$$I_1 = P_0(\bar{C} + C) = P_0, \qquad I_2 = P_1(\bar{C} + C) = P_1$$

from inspection of the state table. Also, it may be possible to initiate the repetitive operation *and* to increment the counter at the same time. If so, states P_0 and P_1 can be combined, with I_1 acting as the control signal for both actions.

EXAMPLE 7-1:   A simple arithmetic system is to compare two five-bit binary numbers. The numbers are initially placed in two registers, A and B. The larger of the two numbers is to be moved into the A register. Separate the overall operation into subtasks and show an initial control sequence.

Solution: The subtasks can be separated as follows:

1. Load the two numbers into A and B.

2. Compare them.

3. Move the larger into A.

4. HALT.

The controller can have four states, say P_0–P_3. The state table for the controller can be

Present State	Next State		Operation
	$C = 0$	$C = 1$	
P_0	P_1	P_1	Load A and B
P_1	P_3	P_2	Compare
P_2	P_3	P_3	Move B to A
P_3	P_3	P_3	Halt

where $C = 0$ is a control signal indicating that the number in the A register is equal to or greater than the number in the B register. This condition eliminates the necessity for the $B \longrightarrow A$ transfer, allowing the controller to move from P_1 directly to P_3. If $C = 1$, however, the number in the B register is the larger, and the controller goes to P_2 to transfer the contents of B into the A register. The controller then moves on to P_3.

A state diagram for the controller is shown in Fig. E7-1.

EXAMPLE 7-2:   A digital system that plays a simple version of the dice game "craps" is to be designed. The protocol of the game is as follows:

1. A player throws two dice, obtaining a number ranging from 2 to 12.

2. If the number is 7 or 11, he wins.

3. If the number is anything else, he continues throwing the dice. The first number is remembered (his "point").

4. If the player throws 7 or 11 on his second or any subsequent roll, he loses.

5. If the player throws his original value a second time, prior to throwing 7 or 11, he wins.

The system can be broken into several functional subsystems:

1. The digital device that simulates the throw of the dice, giving a statistically correct value from 2 to 12.

2. The combinational logic that detects the values of 7 and 11. Its output may be called F. This indicates a win if F is TRUE on the first roll and a loss if F is TRUE thereafter.

3. Storage for the first number, assuming that 7 or 11 is not thrown. This storage register can be called the X register.

4. Comparison logic for indicating that a secondary roll equals the originally thrown value. This control signal may be called C. When TRUE, C indicates a win.

The controller can begin with six states, P_0–P_5. The state diagram is shown in Fig. E7-2. The operations that are performed while in each state are

State	Operation
P_0	Get first roll; store in X
P_1	Decide if first number is 7 or 11
P_2	Win (a HALT state)
P_3	Get secondary rolls
P_4	Compare secondary roll to 7, 11, and the number in X
P_5	Lose (a HALT state)

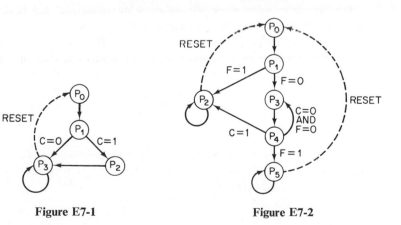

Figure E7-1   Figure E7-2

The design of Exs. 7-0, 7-1, and 7-2 is continued in later portions of this chapter.

7.2   Controller Design

Once a control sequence has been developed and tabulated or diagrammed, the sequential system that implements the control operation can be designed. Usually, a given control sequence should first be studied to see whether or not it contains one or more control states that can be combined with other control states in order to reduce the overall number of states required by the controller. For example, there may be two or more operations that can be performed simultaneously rather than sequentially. If these operations can be initiated by the same control signal, then the controller can accomplish them all during a single control interval. This combination of operations and the resultant *state reduction* may permit a simplification of the controller and allow a savings in cost. The state reduction techniques that are given in Sec. 6.4.1 can be used during this analysis.

The application of state reduction techniques to controller design is a special case, however, and the usefulness of the technique is not as demonstrable as it is in application to some other types of sequential systems. Often, a designer includes extra states in the control sequence in order to make the operation of the total digital system more straightforward, easier to monitor, or easier to troubleshoot when failures occur. These extra states may be "reducible" in a theoretical sense, but they are desirable from an operational standpoint.

As a result, the application of state reduction methods to control sequences is not always a good idea. State reduction should be considered as a *possible* design tool. In some situations state reduction can be of use in simplifying controller cost. But controller cost is often a minor part of the overall cost of a digital system, and saving cost at the expense of functional clarity and ease of maintenance is not always justifiable, especially when the low cost of modern digital components is considered. In fact, the major goal in controller design should always be the development of a control sequence that implements the desired system operation in a logical, reliable, and straightforward manner, rather than attempting to save cost at every juncture.

After reduction to the minimum number of states that seems appropriate, the controller is ready for implementation. Three different types of controller designs are considered herein:

1. The ring-counter controller.

2. The state-counter controller.

3. The microprogrammed controller.

The particular approach that is the best choice depends, to a large extent, on the number of states that the controller must take on.

For a controller having N states, the minimum number of flip-flops that the controller can have is

$$M \geq \log_2 N \qquad\qquad (7\text{-}0)$$

The M flip-flops can assume 2^M different combinations, with each combination representing a different control state. If a state assignment corresponding to the control sequence is made and a multimode counter having M flip-flops is designed in accordance with the general sequential design techniques given in Sec. 6.4, the controller can be built with the minimum number of flip-flops. This is the *state-counter controller* approach; it is discussed further and illustrated by example in Sec. 7.2.2.

For controllers having a small number of states (about ten or less), a good method of controller design is the use of the *ring-counter controller*. This approach uses N flip-flops, one per control state. The ring-counter approach would appear to increase system cost since N is always greater than M, but the ring-counter controller offers a savings in design effort, an increase in operational simplicity, and a potential decrease in the combinational logic required to implement the complete controller, all of which combine to offset the increase in flip-flop requirements. The use of a ring-counter controller for prototype or one-of-a-kind designs is also well justified, since little design effort is required, and the control sequence of a ring-counter controller is easily modified. Section 7.2.1 discusses the design of ring-counter controllers.

A third type, the *microprogrammed controller*, is a relatively new approach to the control of digital systems. It is an outgrowth of control techniques that have been used in digital computers for some time. It offers great flexibility at intermediate cost and is a good choice for a system having a large number of control states, especially if the system is to be produced in quantity. It is discussed briefly in Sec. 7.2.3.

7.2.1. The Ring-Counter Controller. The major advantage of the use of a ring-counter controller is the simplicity with which it can be designed. For many systems this type of controller can be designed directly from the state diagram or state table that describes the control sequence. The action of the control inputs, even if there are many of them, can easily be included.

In a ring-counter controller, just as in the counter from which it takes its name, one flip-flop is assigned to each control state. The flip-flops are usually given the name that corresponds to that state. When the controller is in a given control state, say P_1, only the P_1 flip-flop is SET; all the other flip-flops in the controller are RESET. Moving the controller from one state to the next involves two actions: (1) turning *off* the flip-flop that represents the present state and (2) turning *on* the flip-flop that corresponds to the proper next state. The control logic is designed to cause these two actions to be enabled *during each present state*. The J-K or S-R flip-flop types are best suited to ring-counter controller designs. Several examples showing the design of ring-counter controllers follow.

EXAMPLE 7-3:   Consider the controller described by the following state table:

Present State	Next State		Output, F	
	$I = 0$	$I = 1$	$I = 0$	$I = 1$
a	b	c	0	0
b	c	d	0	1
c	d	a	1	0
d	a	d	1	1

Design a ring-counter controller to act as specified above.

Solution: The controller will require four flip-flops, labeled *a*, *b*, *c*, and *d*. The design process can begin with the flip-flop that corresponds to state *a*. State *a* is reached from either state *c* (for $I = 1$) or state *d* (for $I = 0$). Thus, whenever the *c* flip-flop is SET and $I = 1$, the *a* flip-flop should receive a SET command, causing the controller to enter state *a*. Similarly, the *a* flip-flop should be set when $d\bar{I}$ is present. Clearly, then, for an S-R or J-K flip-flop,

$$S_a = J_a = (cI + d\bar{I}) \cdot CL$$

where *CL* is the clocking signal to the controller. Since state *a* never follows itself, the *a* flip-flop will *always* be reset by the next clock pulse. Hence

$$R_a = a \cdot CL$$

or

$$K_a = 1 \cdot CL$$

This logic takes care of both the entry into and the exit from state *a*. Similarly, considering each of the other control states in turn gives

$$S_b = a\bar{I} \cdot CL \qquad\qquad J_b = a\bar{I} \cdot CL$$
$$R_b = b \cdot CL \qquad\qquad K_b = 1 \cdot CL$$
$$S_c = (aI + b\bar{I}) \cdot CL \qquad J_c = (aI + b\bar{I}) \cdot CL$$
$$R_c = c \cdot CL \qquad\qquad K_c = 1 \cdot CL$$

The *d* flip-flop introduces another aspect of ring-counter controller design, since whenever the controller is in state *d* and $I = 1$, the controller should *remain* in state *d*. Thus, the SET logic for the *d* flip-flop is

$$S_d = J_d = (bI + c\bar{I}) \cdot CL$$

A dI term need not be included, since the system will remain in state d so long as the d flip-flop is not reset. To prevent its being reset under the $I = 1$ condition, the R_d or K_d logic for the d flip-flop is specified as

$$R_d = d\bar{I} \cdot CL \qquad K_d = \bar{I} \cdot CL$$

Thus, when in state d and $I = 1$, the system will remain in state d. The output signal, F, can be written as

$$F = bI + c\bar{I} + d\bar{I} + dI$$

which simplifies to

$$F = \underline{bI} + \underline{c\bar{I}} + d$$

The logic diagram for this controller is shown in Fig. E7-3. RTL logic is assumed. Note how the *RESET* signal is applied to the direct-coupled inputs to the flip-flops in order to force the controller into state a.

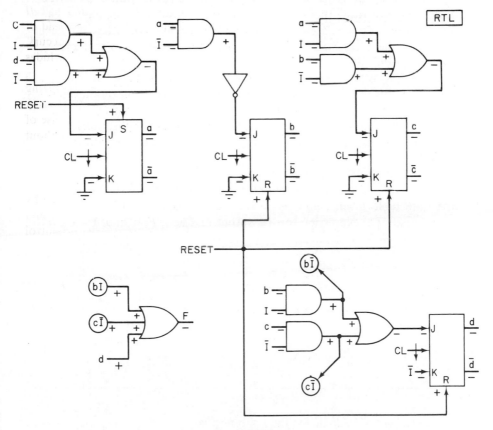

Figure E7-3

Examination of the control logic for setting the *a* flip-flop and resetting the *d* flip-flop of the ring-counter controller of Ex. 7-3 shows a problem that can arise in a ring-counter controller (or any other type of sequential system, for that matter) whose level inputs are not properly synchronized to the system clock. Consider what might happen if the controller is in state *d* and the *I* control signal changes from *I* = 1 to *I* = 0 just before an active clocking transition. The logic for resetting the *d* flip-flop includes $\bar{I}$, and if the change to *I* = 0 occurs far enough ahead of the *CL* transition for the S_d or K_d input to be properly set up, the *d* flip-flop will be reset, moving the controller out of state *d*.

At the same time, the $d\bar{I}$ input to S_a or J_a should provide for the setting of the *a* flip-flop, thereby moving the controller into state *a*, as specified by the original state table. There are two levels of logic (and two levels of delay) between the $\bar{I}$ input and the S_a or J_a terminal to the *a* flip-flop, however. Thus, the $\bar{I}$ signal may cause the *d* flip-flop to be reset, while the delayed $d\bar{I}$ signal may not reach the setting input to the *a* flip-flop far enough ahead of the *CL* transition to cause the *a* flip-flop to be set. This condition, if it occurs, can leave the controller with *no* state enabled, a condition that precludes further operation.

The possibility of this type of trouble can be eliminated by synchronizing all level inputs that are received by the controller. Synchronization can easily be accomplished by using D flip-flops, as described in Sec. 5.6.2. The use of level synchronization is recommended for all controller applications. Without it, asynchronous changes in controlling inputs can lead to erratic controller operation.

EXAMPLE 7-4:   Design a ring-counter controller for the system described by the state table given below. Give the control logic for J-K flip-flops and a combinational expression for the output function, *F*. *X* and *Y* are control inputs.

Present State	Next State				Output, *F*			
	$XY = 00$	$XY = 01$	$XY = 10$	$XY = 11$	00	01	10	11
a	*a*	*b*	*c*	*d*	1	0	0	0
b	*a*	*a*	*c*	*d*	0	1	0	0
c	*a*	*b*	*b*	*d*	0	0	1	0
d	*a*	*b*	*c*	*d*	1	1	1	1

Solution: The logic for the *a* flip-flop is

$$J_a = [b(\bar{X}\bar{Y} + \bar{X}Y) + c(\bar{X}\bar{Y}) + d(\bar{X}\bar{Y})]\cdot CL$$
$$= (b\bar{X} + c\bar{X}\bar{Y} + d\bar{X}\bar{Y})\cdot CL$$

and

$$K_a = (\bar{X}Y + X\bar{Y} + XY)\cdot CL$$
$$= (X + Y)\cdot CL$$

Similarly, the reduced control logic for the other three flip-flops is

$$J_b = (a\bar{X}Y + c(X \oplus Y) + d\bar{X}Y)\cdot CL$$
$$K_b = 1\cdot CL$$
$$J_c = (aX\bar{Y} + bX\bar{Y} + dX\bar{Y})\cdot CL$$
$$K_c = 1\cdot CL$$
$$J_d = (XY(a + b + c))\cdot CL$$
$$K_d = (\bar{X} + \bar{Y})\cdot CL$$

The output can be written as

$$F = a\bar{X}\bar{Y} + b\bar{X}Y + cX\bar{Y} + d$$

EXAMPLE 7-5: Design a ring-counter controller for the repetitive controller of Ex. 7-0.

Solution: The state diagram for the controller is repeated in Fig. E7-5. The ring-counter control equations are

$$J_0 = P_2\bar{C}\cdot CL \qquad K_0 = 1\cdot CL$$
$$J_1 = P_0\cdot CL \qquad K_1 = 1\cdot CL$$
$$J_2 = P_1\cdot CL \qquad K_2 = 1\cdot CL$$
$$J_3 = P_2C\cdot CL \qquad K_3 = 0 \qquad \text{(stays SET)}$$

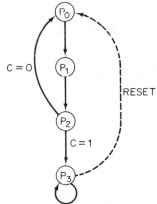

Figure E7-5

The *RESET* signal sets P_0 and resets P_1, P_2, and P_3 via the direct inputs to these flip-flops. The clock period must be long enough so that the repetitive operation is completed prior to leaving P_0.

EXAMPLE 7-6: Design a ring-counter controller for the comparison system of Ex. 7-1.

Solution:

$$J_0 = 0 \qquad\qquad K_0 = 1 \cdot CL$$
$$J_1 = P_0 \cdot CL \qquad\qquad K_1 = 1 \cdot CL$$
$$J_2 = P_1 C \cdot CL \qquad\qquad K_2 = 1 \cdot CL$$
$$J_3 = (P_1 \bar{C} + P_2) \cdot CL \qquad K_3 = 0$$

RESET gives a direct SET to P_0 and direct RESETs to P_1, P_2, and P_3.

EXAMPLE 7-7: Design a ring-counter controller for the digital craps system of Ex. 7-2.

Solution:

$$J_0 = 0 \qquad\qquad K_0 = 1 \cdot CL$$
$$J_1 = P_0 \cdot CL \qquad\qquad K_1 = 1 \cdot CL$$
$$J_2 = (P_1 F + P_4 C) \cdot CL \qquad K_2 = 0$$
$$J_3 = (P_1 \bar{F} + P_4 \bar{C}\bar{F}) \cdot CL \qquad K_3 = 1 \cdot CL$$
$$J_4 = P_3 \cdot CL \qquad\qquad K_4 = 1 \cdot CL$$
$$J_5 = P_4 F \cdot CL \qquad\qquad K_5 = 0$$

RESET turns P_0 on and turns all other flip-flops off. The logic diagram for this controller is shown in Fig. E7-7 (RTL). All the flip-flops are shown as being clocked by the same signal. In the actual design the user of the system should be provided with a switch that enables him to "throw" the dice whenever he wishes. The output signal from this switch, rather than the clocking

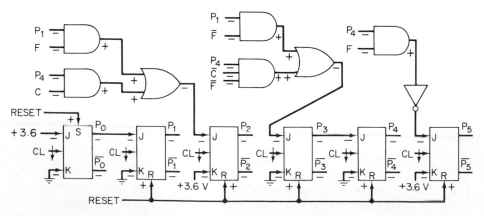

Figure E7-7

signal, should be used to move the system from states P_0 and P_3 (the "throw" states). This modification is shown in Sec. 7.3, wherein the system design of the craps simulator is completed.

———

The ring-counter controller offers advantages other than its design simplicity. Since each state is represented by a flip-flop, both the TRUE and complemented forms for the state variables are available; i.e., a variable, say P_0, is available in both positive-TRUE and negative-TRUE form without necessitating any combinational logic or inversion. Also, only one flip-flop need be considered in determining the present state of the controller, thereby eliminating the need for combinational state-decoding logic.

These advantages make the ring-counter controller a very good choice when working with most small- to medium-sized digital systems. For large systems, however, the need for one flip-flop per state becomes expensive. For these systems a state-counter controller offers a reduction in the number of flip-flops used by the controller, although it also requires a corresponding increase in design effort and in the combinational logic used by the controller. In fact, the increase in combinational complexity resulting from the use of fewer storage devices may offset the savings in flip-flop costs.

7.2.2. State-Counter Controller Design. An alternative to the ring-counter controller and its need for one flip-flop per state is the state-counter controller. This type of control counter uses the present states of *all* the flip-flops in its memory to specify the complete present state, thereby allowing 2^M, rather than M, states for M flip-flops. Each control state is assigned a particular flip-flop combination. The state assignment techniques given in Sec. 6.4.2 can be used to identify a "good" state assignment for use during the controller design process.

EXAMPLE 7-8: Design a state-counter controller for the repetitive system of Ex. 7-0.

Solution: The state table and reversed state table for this control system are

Present State	Next State(s)	Present State	Previous State(s)
P_0	P_1	P_0	P_2
P_1	P_2	P_1	P_0
P_2	P_0, P_3	P_2	P_1
P_3	P_3	P_3	P_2, P_3

Note that state changes that are accomplished by using the direct control inputs to the state-counter flip-flops (such as a *RESET* signal) are not included in either of these tables.

Rule 1 suggests a (P_2, P_3) adjacency. Rule 2 points to (P_0, P_3). A suitable assignment is shown in Fig. E7-8a, where W and X are the two flip-flops that make up the state-counter controller. The assignment is

$$P_0 = \bar{W}\bar{X}, \qquad P_1 = W\bar{X}, \qquad P_2 = WX, \qquad P_3 = \bar{W}X$$

Inserting this assignment into the original state table gives

Total Present State			Next State			Behavior	
C	W	X	W_+	X_+		W	X
0	0	0 (P_0)	1	0 (P_1)		α	0
0	0	1 (P_3)	0	1 (P_3)		0	1
0	1	0 (P_1)	1	1 (P_2)		1	α
0	1	1 (P_2)	0	0 (P_0)		β	β
1	0	0 (P_0)	1	0 (P_1)		α	0
1	0	1 (P_3)	0	1 (P_3)		0	1
1	1	0 (P_1)	1	1 (P_2)		1	α
1	1	1 (P_2)	0	1 (P_3)		β	1

The W and X transition maps are shown in Fig. E7-8b. From these maps, the control logic is found:

$$J_W = \bar{X} \cdot CL \qquad K_W = X \cdot CL$$
$$J_X = W \cdot CL \qquad K_X = \bar{C}W \cdot CL$$

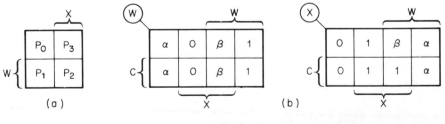

Figure E7-8a, b

The output signals from the controller were $I_1 = P_0$ and $I_2 = P_1$. The logic for these signals is

$$I_1 = \bar{W}\bar{X} \qquad I_2 = W\bar{X}$$

with I_1 acting to start the operation that is to be repeated and I_2 incrementing the counter that counts the number of repetitions.

EXAMPLE 7-9: Design a state-counter controller for the comparison system of Ex. 7-1.

Solution: The state table and reverse state table are

Present State	Next State(s)		Present State	Previous State(s)
P_0	P_1		P_0	—
P_1	P_2, P_3		P_1	P_0
P_2	P_3		P_2	P_1
P_3	P_3		P_3	$P_1, P_2, P_3 \begin{cases} P_1, P_2 \\ P_2, P_3 \\ P_1, P_3 \end{cases}$

The rule 1 groupings are (P_1,P_2), (P_2,P_3), and (P_1,P_3), two of which will be satisfied by *any* state assignment that uses only two state variables. Rule 2 points to the desirability of a (P_2,P_3) adjacency. Thus, any assignment with P_2 and P_3 next to each other is a "good" one. For example, consider Fig. E7-9a. The state table that results from this assignment is

Total Present State			Next State		Behavior	
C	W	X	W_+	X_+	W	X
0	0	0	0	1	0	α
0	0	1	1	1	α	1
0	1	0	1	1	1	α
0	1	1	1	1	1	1
1	0	0	0	1	0	α
1	0	1	1	0	α	β
1	1	0	1	1	1	α
1	1	1	1	1	1	1

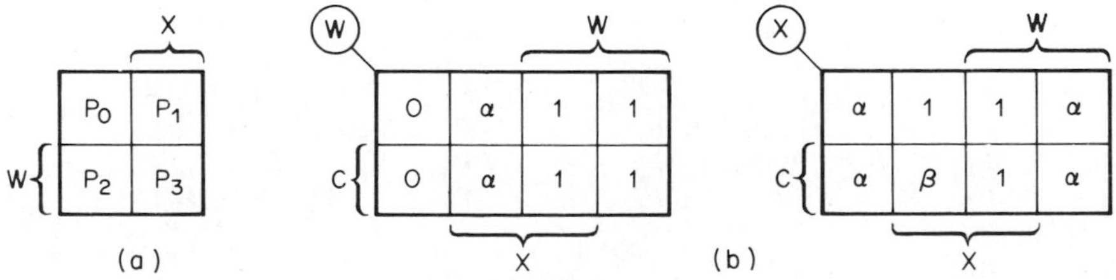

Figure E7-9a, b

The transition maps are shown in Fig. E7-9b, leading to the following control equations:

$$J_W = X \cdot CL \qquad K_W = 0$$
$$J_X = 1 \cdot CL \qquad K_X = C\bar{W} \cdot CL$$

Note that once set, the W flip-flop is never reset. This is a result of the transfer into P_3, a HALT state.

The logic diagram for a TTL implementation of the controller is shown in Fig. E7-9c. The *RESET* signal returns the system to state P_0 by direct resetting of both W and X.

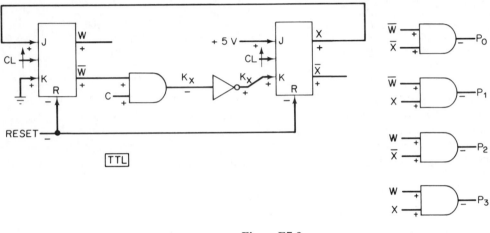

Figure E7-9c

EXAMPLE 7-10: Design a state-counter controller for the digital craps system of Ex. 7-2.

Solution: The rule 1 and 2 tables are

Present State	Previous State(s)		Present State	Next State(s)	
P_0	—		P_0	P_1	
P_1	P_0	$\{P_1, P_2$	P_1	P_2, P_3	
P_2	P_1, P_2, P_4	$\{P_1, P_4$	P_2	P_2	
P_3	P_1, P_4	$\{P_2, P_4$	P_3	P_4	$\{P_2, P_3$
P_4	P_3		P_4	P_2, P_3, P_5	$\{P_2, P_5$
P_5	P_4, P_5		P_5	P_5	$\{P_3, P_5$

The rule 1 groupings are (P_1,P_2), (P_2,P_4), (P_1,P_4) twice, and (P_4,P_5). The rule 2 groupings are (P_2,P_3) twice, (P_2,P_5), and (P_3,P_5).

Since there are six states, the controller requires three flip-flops, say W, X, and Y. A state assignment that satisfies as many of the adjacencies suggested by rules 1 and 2 as possible is shown in Fig. E7-10a. The controller design problem involves five variables: W, X, Y, C, and F. C and F are the

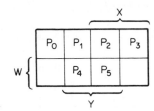

Figure E7-10a

control inputs that influence the next-state behavior. The complete state table for the controller is

| Present | Next State | | | |
State	$CF = 00$	$CF = 01$	$CF = 10$	$CF = 11$
P_0	P_1	P_1	P_1	P_1
P_1	P_3	P_2	P_3	P_2
P_2	P_2	P_2	P_2	P_2
P_3	P_4	P_4	P_4	P_4
P_4	P_3	P_5	P_2	X
P_5	P_5	P_5	P_5	P_5

The fourth entry in P_4 is a can't-happen. This arises because of the operation of the previous control states. The system will not reach P_4 if the first roll results in a 7 or 11, since that is a WIN and state P_2 is reached. Thus, P_4 is reached only if the first number is other than 7 or 11. That value is stored in a register for comparison with further rolls. The $C = 1$ condition indicates a comparison between the original value and a subsequent roll, which is the secondary WIN condition. $F = 1$ indicates that a secondary roll gave a value of 7 or 11. Since these conditions are mutually exclusive, the P_4CF combination is a can't-happen, and next-state behavior can be unspecified.*

The state assignment can now be inserted into the table, giving the following counting sequence:

---

*For a complete discussion of the analysis of sequential machines with unspecified next states, see Kohavi [3].

Total Present State			Next State			Behavior		
C F W X Y			W_+	X_+	Y_+	W	X	Y
0 0 0 0 0 (P_0)			0	0	1	0	0	α
0 0 0 0 1 (P_1)			0	1	0	0	α	β
0 0 0 1 0 (P_3)			1	0	1	α	β	α
0 0 0 1 1 (P_2)			0	1	1	0	1	1
0 0 1 0 0			X	X	X	X	X	X
0 0 1 0 1 (P_4)			0	1	0	β	α	β
0 0 1 1 0			X	X	X	X	X	X
0 0 1 1 1 (P_5)			1	1	1	1	1	1
0 1 0 0 0			0	0	1	0	0	α
0 1 0 0 1			0	1	1	0	α	1
0 1 0 1 0			1	0	1	α	β	α
0 1 0 1 1			0	1	1	0	1	1
0 1 1 0 0			X	X	X	X	X	X
0 1 1 0 1			1	1	1	1	α	1
0 1 1 1 0			X	X	X	X	X	X
0 1 1 1 1			1	1	1	1	1	1
1 0 0 0 0			0	0	1	0	0	α
1 0 0 0 1			0	1	0	0	α	β
1 0 0 1 0			1	0	1	α	β	α
1 0 0 1 1			0	1	1	0	1	1
1 0 1 0 0			X	X	X	X	X	X
1 0 1 0 1			0	1	1	β	α	1
1 0 1 1 0			X	X	X	X	X	X
1 0 1 1 1			1	1	1	1	1	1
1 1 0 0 0			0	0	1	0	0	α
1 1 0 0 1			0	1	1	0	α	1
1 1 0 1 0			1	0	1	α	β	α
1 1 0 1 1			0	1	1	0	1	1
1 1 1 0 0			X	X	X	X	X	X
1 1 1 0 1			X	X	X	X	X	X
1 1 1 1 0			X	X	X	X	X	X
1 1 1 1 1			1	1	1	1	1	1

The transition maps for the three flip-flops are shown in Fig. E7-10b. The control equations are

$$J_W = X\bar{Y} \cdot CL \qquad K_W = \bar{F}\bar{X} \cdot CL$$
$$J_X = Y \cdot CL \qquad K_X = \bar{Y} \cdot CL$$
$$J_Y = 1 \cdot CL \qquad K_Y = (\bar{C}\bar{F}\bar{X} + \bar{F}\bar{W}\bar{X}) \cdot CL$$

The state identification logic can be simplified by using the unassigned states. The original state assignment was as shown in Fig. E7-10c. By using the redundancies, the state decoding logic can be reduced to

$$P_0 = \bar{X}\bar{Y}, \qquad P_1 = \bar{W}\bar{X}Y, \qquad P_2 = \bar{W}XY,$$
$$P_3 = X\bar{Y}, \qquad P_4 = W\bar{X}, \qquad P_5 = WX$$

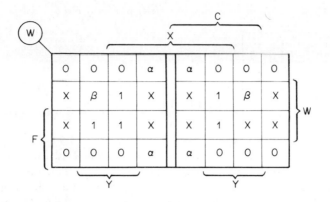

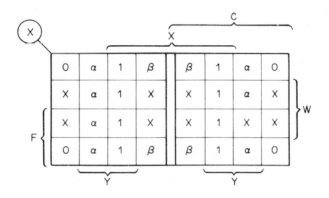

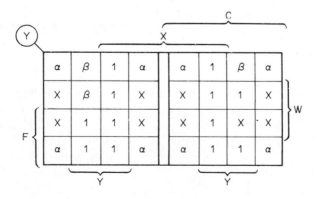

Figure E7-10b

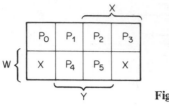

Figure E7-10c

Some control sequences may allow the apparently advantageous use of asynchronous control logic in operating a state-counter controller. While this may cut the cost of the controller, the transition states that arise during asynchronous state changes may cause the controller to malfunction. For example, if a transition state that occurs between two stable control states happens to reset the entire digital system, then the system will certainly not operate properly. As a result, asynchronous control logic should rarely be used to implement a state counter.

7.2.3. The Microprogrammed Controller (Refs. [1], [2], [5]).

This third approach to controller design replaces the conventional sequential controller made from flip-flops and gates by a sequence of control instructions, called *microinstructions*, that are stored in a random-access memory. This control technique, first proposed by Wilkes in 1951, has found considerable application in the control of large digital systems such as computers. It has only recently found application in the control of smaller digital systems.

The conventional controller uses combinational logic to interpret its present state and the current values of its control inputs in order to determine what output signals should be generated and what should be done next. The microprogrammed controller, on the other hand, follows a sequence of instructions that are "listed" in its memory. Each instruction tells the controller what to do now and what to do next. In fact, a microprogrammed controller is very much like a small, stored-program digital computer. Its sequence of microinstructions are like a machine-language program that is executed by the controller.

Microprogramming breaks each of the original control intervals into one or more subintervals, with each subinterval corresponding to the performance of one microinstruction. The subintervals consist of two distinct phases. One is the *fetch* phase, during which the next microinstruction is extracted from the memory portion of the microprogrammed controller; the other is the *execution* phase, during which the operations that are specified by the newly fetched microinstruction are actually performed. The fetch/execute sequence is repeated for each microinstruction.

During a given control interval, the microinstructions that correspond to that control state are executed, one after the other. The "present state" of the microprogrammed controller is represented by the address of the microinstruction that is currently being executed. Often, several addresses (i.e., several microinstructions) correspond to the same original control state.

The transfer from the present state to the next state involves the changing of the address and the retrieval from memory of the next microinstruction. The commands are normally executed in sequence, with the microinstructions for each successive control state placed one after the other in the memory. Conditional state transfers are accomplished via conditional branches in the microinstruction sequence.

The organization of a typical microprogrammed controller is shown in Fig. 7-1. The memory unit stores the various microinstructions. In many

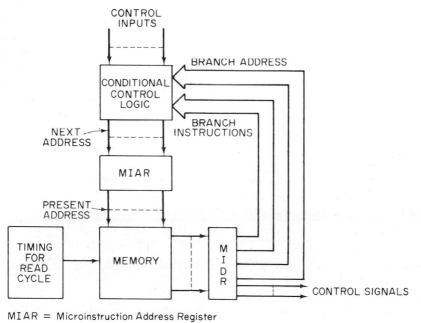

MIAR = Microinstruction Address Register
MIDR = Microinstruction Data Register

Figure 7-1. A Microprogrammed Controller

cases the operation of the controller never changes, thereby allowing the use of *read-only* memory for storage of the control information. In other situations a normal read/write memory may be used so that the microinstruction sequence can be changed in order to change the controller's operational characteristics. An example of this would be a traffic-light controller that would be reprogrammed during peak traffic periods so as to expedite flow into or out of a congested business district. The reprogramming could be accomplished over a dial-up telephone line.

The present state of the microprogrammed controller is represented by the address that is held in the microinstruction address register (MIAR). This address points to the memory location from which the current microinstruction is to be taken. During the fetch portion of the microinstruction subinterval, a read cycle places the new command in the microinstruction data register (MIDR).

Each microinstruction generally contains three types of information. The *output signals* that are required during each control interval are contained in the command. Typical subsystem control functions are reset signals, shift signals, etc. Usually, each output signal is represented by a bit in the microinstruction. When required, these bits are made TRUE to indicate which of these output signals is to take place during the current control interval. *Branching conditions* are also included in the microinstructions. These bits tell the conditional address selection logic what conditions to use in making the branching decision. Typical branching conditions include the logical state of certain input signals or the output from comparison circuitry. If a branch is to be allowed, the microinstruction will also contain the *branch address*. A conditional branch will normally cause the next address to be chosen from either the branch address that is contained in the microinstruction or the current address plus 1. The branching decisions are made by the conditional control logic shown in Fig. 7-1. This decision logic interprets the control inputs and the branching conditions in order to pick the next address from these two possibilities. Multiple decisions can be made as a sequence of two-way decisions or by including more than one branching address in the microinstruction.

The number of microinstructions will generally exceed the number of control states, since some control operations (particularly those involving branches) may require several microinstructions. The system *RESET* signal is generally used to force the MIAR to the address of the first microinstruction in the control sequence, normally address 0. The first instruction is then fetched from the memory, and the operation of the controller is under-way.

The major advantage of the use of a microprogrammed controller is the flexibility that it affords. The control sequence can be changed simply by changing the information that is stored in the controller's memory unit. The design of the controller is quite straightforward, with design procedures closely paralleling the programming of a small computer in its own machine language. In fact, some computer-aided microprogramming languages (much like microlevel assembly languages) have been developed. Through modification of the microprogram, new control states can be added, old states deleted, and new output signals or branching signals included.

The need for state reduction is somewhat minimized, so long as the memory of the microprogrammed controller has sufficient storage. Also, the selection of a suitable state assignment is of no concern, since the use of a random-address memory unit allows transfer from any address (present state) to any other address (next state) within the same amount of time.

The need for a memory unit is the major drawback of the use of a microprogrammed controller. The cost of the memory unit and its associated addressing logic and read/write electronics is all pure overhead, since the memory unit only tells the controller what to do next; it plays no part in the actual execution of the control sequence. Until recently, the cost of both read/write and read-only memory was sufficiently high to make the use of microprogrammed control for small digital systems uneconomical. Only when spread over the entirety of a large computer or some other complex digital system could the cost of a microprogrammed controller be justified. Recently, however, the widespread availability of low-cost memories of both read/write and read-only form has brought the memory costs down so that microprogrammed control is economically feasible, even in medium-sized digital systems. Even so, the hardware costs involved in microprogramming are still somewhat higher than those of conventional "hard-wired" controllers, and the use of microprogramming is recommended only for controlling digital systems having a fairly complex control sequence, especially if the system is to be built in quantity.

Microprogrammed controllers are not without other disadvantages. The speed of operation of a microprogrammed controller is generally lower than its conventionally constructed counterpart, since the time spent during the fetch phases of each microinstruction subinterval is added to the actual execution time. Also, the use of read-only memory produces a microprogrammed controller that is quite inflexible, making that type of memory quite difficult to use during the early developmental stages of a design project. A better approach is to use a read/write memory during development and to replace it with a read-only memory only when the microinstruction sequence has been well "debugged."

Designing a microprogrammed controller really involves two tasks. One is the procurement and construction of the basic controller hardware, which is shown in Fig. 7-1. The other is the specification of the microprogram itself. The programming task should usually be considered first, since an estimate on the size of the memory unit will be needed. The major concerns are how many microinstructions will be needed, and how many bits will be required for each microinstruction. These two values specify the word size and number of words that are required of the memory unit.

The "writing" of the microprogram is much like any other computer programming problem. The complete state table for the controller must be specified as usual. The "programmer" begins at address 0 and specifies control functions, output signals, branch conditions, and branch addresses, working his way through the control sequence. The microinstructions are then loaded into the memory, and proper controller operation is verified.

Several very simplified examples of the application of microprogramming to the control of small digital systems are given in the remainder of this subsection. These examples are really somewhat below the level at which microprogrammed control should actually be considered; they are included for illustration only. As memory costs decrease and controller complexity increases, however, the use of microprogrammed controllers should be given more and more consideration.

Readers who are particularly interested in microprogramming are referred to Husson [2].

EXAMPLE 7-11: Design a microprogrammed controller for the repetitive control system of Ex. 7-0.

Solution: This controller has one control signal, C, and two output signals, I_1 and I_2. One possible microinstruction sequence is shown below. States P_0, P_1, and P_3 are each represented by one microinstruction. State P_2, a branching state, is represented by two microinstructions.

		Contents of Memory		
State	Address of Memory	Control Signals I_1 I_2	Branch Control C U	Branch Address
P_0	000	1 0	0 0	000
P_1	001	0 1	0 0	000
P_2	{010	0 0	1 0	100
	{011	0 0	0 1	000
P_3	100	0 0	0 1	100

The first two bits in each microinstruction act directly as the I_1 and I_2 control outputs. I_1 is made TRUE during P_0 (address 000) and I_2 is made TRUE during P_1. The next two bits of each microinstruction control the branching decision. No branch takes place in P_0 or P_1, but two possible branches are considered during P_2. The C column specifies a transfer to the branch address if the C signal is TRUE. The U column specifies an unconditional branch. Thus, if C is TRUE during the first microinstruction of P_2 (address 010), then the controller will branch to P_3 (address 100). Otherwise the controller will sequence to the next address (011), from which an unconditional branch back to P_0 (address 000) is taken. The conditional address selection logic considers the C input signal and the C and U outputs from the MIDR in order to select either the next sequential address (the current address plus 1) or the branch address.

State P_3 (address 100) is made to act as a HALT state by specifying an unconditional branch back to itself. The system *RESET* will force the MIAR back into address 000.

A flow diagram of this control sequence is shown in Fig. E7-11. The memory portion of the microprogrammed controller requires five addresses,

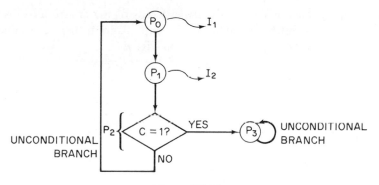

Figure E7-11

each storing seven bits of information. If the control sequence will never be changed, read-only memory can be used. The complete contents of the memory unit are tabulated below:

Address	Contents
000	1000000
001	0100000
010	0010100
011	0001000
100	0001100

EXAMPLE 7-12: Design a microprogrammed controller for the digital craps system of Ex. 7-2.

Solution: One possible design for this controller uses *two* branch conditions. A possible microinstruction sequence is

| | | *Contents of Memory* | | | |
State	Address of Memory	Control Signals	C F U	Branch Address
P_0	000		0 0 0	000
P_1	001		0 1 0	110
P_3	010		0 0 0	000
	011	To be	1 0 0	110
P_4	100	specified	0 1 0	111
	101		0 0 1	010
P_2	110		0 0 1	110
P_5	111		0 0 1	111

No control signals are shown, since the details of the controller/subsystem interface have not yet been supplied. The state diagram that corresponds to this microinstruction sequence is shown in Fig. E7-12.

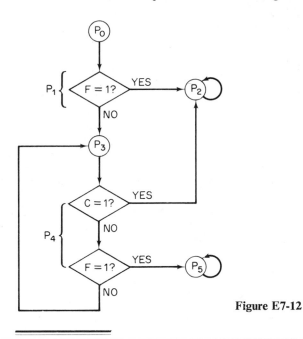

Figure E7-12

7.3 Completing the System: Examples

Once the controller and the functional subsystems that make up the complete system have been designed, the details of the interfaces between these units must be specified and the controlling interconnections described. This is the "nitty-gritty" of the design process. Voltage symbolism and loading requirements must be met; timing problems must be resolved. This final design step presents the designer with a great opportunity to use his ingenuity. Experience is a useful resource, since it is expected that the more design experience that one has, the better his ability to tie together a complete system design will be.

To illustrate the complete design process, several examples of system designs follow, with each design presented in fairly great detail. The best illustration, however, is design practice. Often, one's best teacher is the system that looks great on paper and yet does not operate properly when actually constructed.

EXAMPLE 7-13: Design a complete repetitive control system, as introduced in Ex. 7-0.

Solution: To complete the design, the details of the interface between the controller and the external device that is to be repetitively operated must be known. Example 7-5 considered the design of a fully synchronous ring-counter controller for this system. A more typical situation would involve some asynchronism, in that the external device may not always take the same amount of time to complete its cyclic operation. In that case, the controller would have to wait in state P_0 for the longest possible cycle time (an *open-loop* system) or would have to receive an *operation complete* signal from the system (a *closed-loop* system).

In this example the latter case is considered. The external device is assumed to have the following interface requirements:

START: an input signal that initiates the cyclic operation by providing a falling transition.

COMPLETE: a switch closure (SPDT) that indicates when the operational cycle is complete; the closure is reopened when the system receives the next *START* signal.

The ring-counter controller of Ex. 7-5 must be modified in order to keep it in state P_0 until the *COMPLETE* signal is received. Since the *COMPLETE* signal is an asynchronous switch closure, some synchronizing circuitry is necessitated. A useful device for handling signals of this type is the *pulse-catcher*. A pulse-catcher receives an unsynchronized switch closure and emits a single, properly sychronized pulse. Once closed, the switch must be opened and reclosed in order to generate another output pulse. The output pulses are of one clock period in duration and are synchronized to the clocking transitions. A typical design for a pulse-catcher uses two J-K flip-flops, as shown in Fig. E7-13a, in RTL form.

The switch closure gives a direct set to the S_1 flip-flop. The next clock transition then sets the S_2 flip-flop, since S_1 enables J_2. The $S_1 S_2$ combination

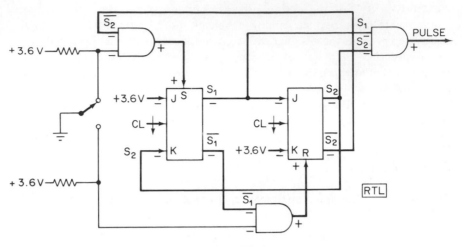

Figure E7-13a

makes the output from the *PULSE* gate become TRUE. The second clock pulse following the switch closure resets S_1, since S_2 enables K_1. The pulse-catcher then remains in the $S_1 S_2 = 01$ state until the switch is reopened, at which time the other pole of the switch is ANDed with $\overline{S_1}$ to give a direct reset to S_2. This circuit will emit one pulse per open→close→open switch cycle, no matter how long or how short the switch stays closed.

The arm of the SPDT switch is connected to 0 V and resistively pulled up to +3.6 V so that both outputs from the switch will be FALSE (rather than both TRUE) during the opening or closing process when the arm of the switch is between poles. The negative-TRUE polarity required for input to an RTL AND gate forces this arrangement.

The pulse-catcher will properly interface the *COMPLETE* signal from the external system with the ring-counter controller. The *START* signal to the external signal must be provided during state P_0. The α_{P_0} transition that is produced upon entry into state P_0 can be used for this purpose. The falling transition that is required as the *START* signal requires connection to P_0 with negative-TRUE voltage symbolism. Thus, a falling *START* signal will be produced each time the P_0 flip-flop is set. The complete interface connections are shown in Fig. E7-13b.

The P_0 and P_1 flip-flops of the ring-counter controller can now be connected. Their control logic is as diagrammed in Fig. E7-13c.

When *PULSE*, the pulse-catcher output, is TRUE, it enables J_{P_1} and K_{P_0}, thereby forcing a change from control state P_0 into state P_1 upon the next clock pulse. The output from the pulse-catcher becomes FALSE after that one clock period, so that J_{P_1} is no longer enabled. Either α_{P_1} or β_{P_1} can be used to increment the repetition counter. After one clock period P_1 is reset and P_2 is set.

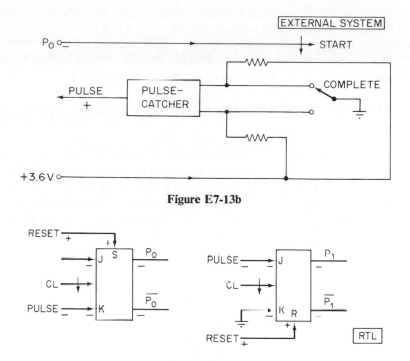

Figure E7-13b

Figure E7-13c

The α_{P_1} or β_{P_1} signals are derived from the outputs of the P_1 flip-flop of the ring-counter controller. These signals are related to the P_1 outputs themselves by combining the voltage change that is required to cause the repetition counter to increment and the logical change that is represented by the α or β symbol. This can easily be accomplished by using the before/after table described in Sec. 6.2.5. Earlier, the use of asynchronous control logic in controller design was cautioned against. This applies to the operation of the control counter itself, but it is often quite advantageous to use asynchronous transitions which are taken *from* the controller (which is fully synchronous itself) to activate functional subsystems that are external to the controller.

The number of repetitions that are to be performed is provided to the controller by a set of switches. This number could be provided in binary, but a human operator would probably prefer decimal numbering. If the number of repetitions is limited to 99, two decades of BCD code can be used. Thumbwheel switches [4] are useful for providing the encoded number, since they convert their decimal setting into BCD code with no external gating for decoding. Thumbwheel switches act as four-pole, ten-position switches and can be obtained in a variety of BCD codes. The 8421 BCD code is assumed in this example.

Since the number of repetitions is represented by two decades of 8421 BCD code, the repetition counter should operate in this same code in order to simplify the comparison logic. Two 8421 BCD up-counters are required. This type of counter is designed in Chapter 6. MSI decade counters (four flip-flops and their 8421 BCD counting control logic in a single package) could also be used. The units counter can be incremented by β_{P_1}. The tens counter can then be incremented by β_8, the resetting transition of the high-order bit of the units counter. Both counters are reset to 0 by the *RESET* signal.

The output of each bit of the counters is compared with the corresponding bit from the number set into the thumbwheel switches. Eight COINCI-DENCE gates are used for this comparison. The outputs from these gates are ANDed together so that when all eight bits of the repetition counter compare with the number set into the switches, the complete signal, *C*, is made TRUE. The counter/comparison logic is shown in Fig. E7-13d.

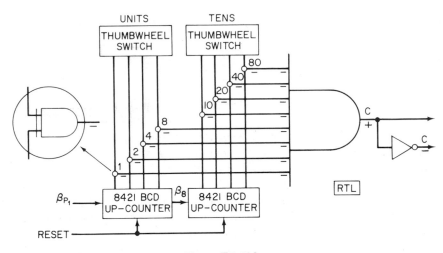

Figure E7-13d

Comparison between the repetition counter and the specified number of repetitions is made during P_2. Since a cycle of operation takes place during P_0 and the counter is incremented during P_1 (starting with 0), the number present in the counter during P_2 is the number of repetitions that have been completed. If $C = 1$ when the controller is in P_2, then the system has finished its task and should move to P_3. Otherwise it should return to P_0 to initiate another cycle of the external operation. The complete controller logic is shown in Fig. E7-13e. The *RESET* signal should be provided by a closure to $+3.6$ V. A clocking oscillator must also be provided.

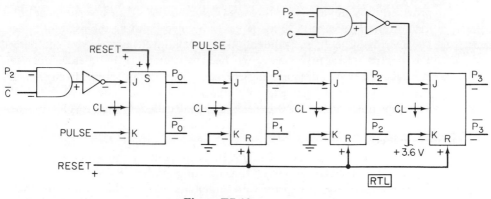

Figure E7-13e

EXAMPLE 7-14: Design a complete digital comparison system that acts according to the control sequence that is given in Ex. 7-1. Assume that the numbers A and B are both five bits long. Consider the design as a fully parallel system and as a parallel/serial system.

Solution: Fully parallel system: The state-counter controller designed in Ex. 7-9 can be used. Its TTL implementation is shown in that example. The A and B registers are loaded during the first control state, P_0. The two five-bit numbers are first selected with toggle switches and then gated into the A and B registers by the P_0 signal. The direct inputs to the flip-flops that make up these two registers can be used for this purpose (see Fig. E7-14a).

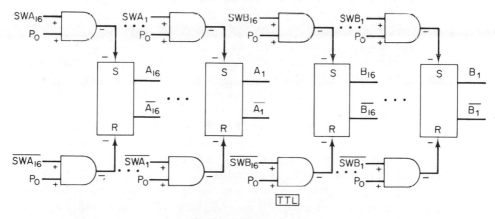

Figure E7-14a

SWA_{16} is the signal name for the output from the switch that provides the high-order bit of the A word, A_{16}. Its complement is $\overline{SWA}_{16}$, which can be provided by using a double-throw switch or by using an inverter. The total load on the P_0 source is 20 loads, probably necessitating the use of a buffer.

During state P_1 the contents of the A and B registers are compared. The comparison can be accomplished by using the parallel comparison arrangement that is designed in Sec. 4.5.2. The output signals from the parallel comparator of that example are G (for $A > B$) and L (for $A < B$). The controller of Ex. 7-10 expects a signal called C that is TRUE when $A < B$. Thus $C = L$, and the logic for the G output of the comparator can be omitted.

If C is FALSE at the end of P_1, the state-counter controller will transfer into P_3. If C is TRUE, however, the controller goes to P_2 in order to transfer the contents of the B register (the larger number, in that case) into the A register. This transfer can be accomplished by using the clocked inputs to the A register flip-flops. Type D flip-flops are the most convenient for this purpose. The flip-flops can be clocked by β_{P_2}, the transition at the end of the P_2 state. Since the TTL type D flip-flop requires a rising transition, the clock inputs of the A register should be connected to P_2 with negative-TRUE symbolism. The complete A register is shown in Fig. E7-14b. The B register can be constructed with unclocked D-C SR flip-flops (latches).

Second Solution: Parallel/serial system: As an alternative to the fully parallel system described above, the comparison system can be modified to make the transfer from B to A during P_2 a *serial transfer* rather than a parallel transfer. This is not necessarily better, but it *is* different and it serves to further illustrate control techniques.

To accomplish the serial transfer, the A and B registers must be connected together to form a ten-bit shift register (see Fig. E7-14c). The direct-coupled logic for storing the data during P_0 is still required.

The *SHIFT* pulses that move the contents of the B register into the A register should be generated only during P_2. There must be exactly five of them, after which the controller should move to P_3 and HALT. An auxiliary binary up-counter can be used to count the *SHIFT* pulses, signaling the controller when it reaches m_4 following the fourth *SHIFT* pulse. The logic diagram for this counter is shown in Fig. E7-14d. The m_4 signal will be TRUE during the clock period prior to the fifth clock pulse. Thus, the move from P_2 to P_3 will be enabled while m_4 is TRUE, and the state change will actually take place following the fifth shift pulse.

The $P_2 \rightarrow P_3$ state transition involves the setting of the X flip-flop of the state-counter controller of Ex. 7-9, since $P_2 = W\bar{X}$ and $P_3 = WX$. If the setting of the X flip-flop is inhibited until m_4 becomes TRUE, then the controller will remain in P_2 for five clock periods, thereby permitting five *SHIFT* pulses to be generated.

Unlike the ring-counter controller, which permits easy modification of

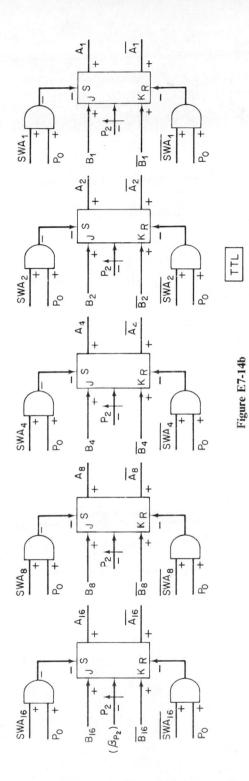

Figure E7-14b

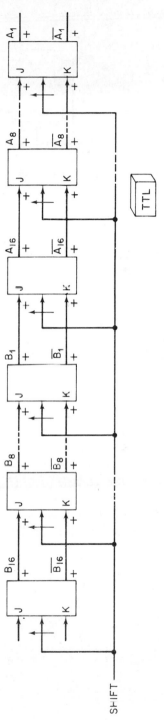

Figure E7-14c

405

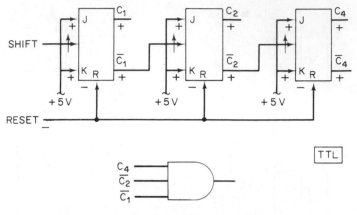

Figure E7-14d

the control sequence, this change in operation requires a complete redesign of the state-counter controller. The new control diagram is shown in Fig. E7-14e. The new state table is

Present State	Next State			
	$Cm_4 = 00$	01	10	11
P_0	P_1	P_1	P_1	P_1
P_1	P_3	P_3	P_2	P_2
P_2	P_2	P_3	P_2	P_3
P_3	P_3	P_3	P_3	P_3

Rule 1 suggests (P_1,P_2), (P_1,P_3), and (P_2,P_3) as adjacencies, with (P_1,P_2) appearing twice. Rule 2 suggests (P_2,P_3) twice. Thus, the assignment given in Fig. E7-14f is a better choice for the modified controller than the assign-

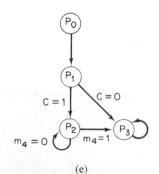

(e)

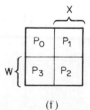

(f)

Figure E7-14e, f

ment that was given in Ex. 7-9. With this assignment the complete state table becomes

Total Present State				Next State		Behavior	
C	m_4	W	X	W_+	X_+	W	X
0	0	0	0 (P_0)	0	1	0	α
0	0	0	1 (P_1)	1	0	α	β
0	0	1	0 (P_3)	1	0	1	0
0	0	1	1 (P_2)	1	1	1	1
0	1	0	0	0	1	0	α
0	1	0	1	1	0	α	β
0	1	1	0	1	0	1	0
0	1	1	1	1	0	1	β
1	0	0	0	0	1	0	α
1	0	0	1	1	1	α	1
1	0	1	0	1	0	1	0
1	0	1	1	1	1	1	1
1	1	0	0	0	1	0	α
1	1	0	1	1	1	α	1
1	1	1	0	1	0	1	0
1	1	1	1	1	0	1	β

Transition mapping gives the following control equations for the new state-counter controller:

$$J_W = X \cdot CL \qquad K_W = 0$$
$$J_X = \bar{W} \cdot CL \qquad K_X = (\bar{C}\bar{W} + m_4 W) \cdot CL$$

The K_X logic includes m_4, thereby keeping the controller in P_2 for five clock periods.

Logic must be provided to AND the system clock and the P_2 signal in order to generate the *SHIFT* pulses. The synchronous nature of P_2 (it becomes TRUE and goes FALSE in coincidence with a clock pulse) permits the use of a simple gating circuit for this purpose. This logic for generation of the *SHIFT* pulses is shown in Fig. E7-14g. The timing diagram shown in Fig.

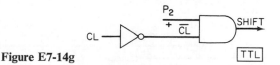

Figure E7-14g

E7-14h for this gating circuit shows that a clock pulse is generated for each period that P_2 is TRUE. The inverter in the gating circuitry assures that the *SHIFT* pulses coming out of the AND gate have the same polarity as the original clock pulses, since the TTL AND gate gives a change in voltage symbolism. The m_4 signal becomes TRUE following the fourth *SHIFT* pulse and causes the X flip-flop to be reset on the fifth shift pulse. This moves the controller to $P_3 = W\bar{X}$, the final state. The complete controller design is shown in Fig. E7-14i.

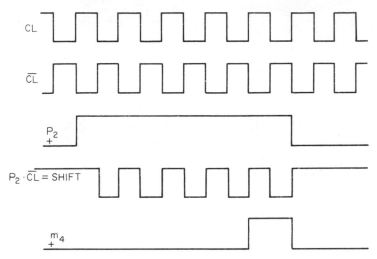

Figure E7-14h

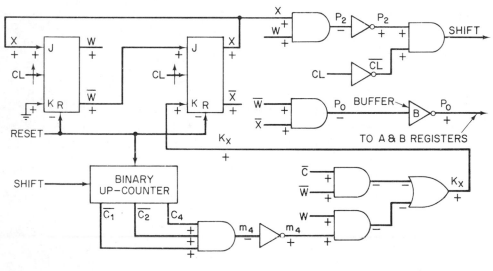

Figure E7-14i

EXAMPLE 7-15:   Complete the design of the digital craps playing system of Ex. 7-2.

Solution: The ring-counter controller that was designed for this system in Ex. 7-7 can be used. During states P_0 and P_3 a simulated dice throw must be implemented. Rather than having the system clock automatically move the controller from P_0 or P_3, the human player can be allowed to inject his own randomness into the "throwing" of the dice if he is given a *THROW* push-button switch (SPDT) that allows him to select the time for the state change. The switch must be connected to a pulse-catcher in order to generate one properly synchronized *THROW* pulse for each depression of the push button. The ring-counter controller of Ex. 7-7 can be modified to include the *THROW* signal, as shown in Fig. E7-15a. The *THROW* pulse enables both K_{P_0} and K_{P_3}, ending whichever of those states was present. The *THROW* pulse will also enable the setting or P_1 following P_0 or the setting of P_4 following P_3.

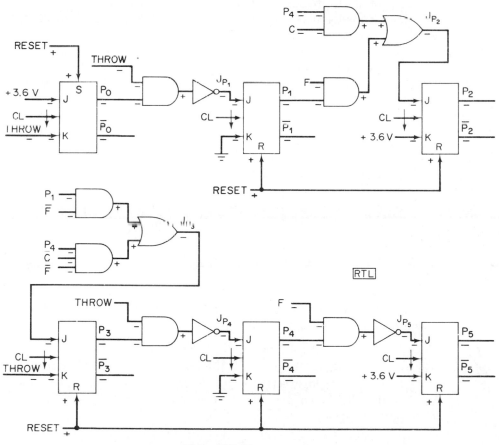

Figure E7-15a

The dice simulator must provide a random number from 2 to 12. A simple m_2–m_{12} binary counter will not do, however, since the probabilities of occurrence of the various numbers are not equal. To get a statistically correct number, two counters operating from m_1 to m_6 must be used and their values summed, just as two dice are thrown and totaled.

The design of a synchronous counter that counts from m_1 to m_6 is easily accomplished by following the techniques given in Chapter 6. Two three-bit counters are required. The first counter is incremented by the system clock. The second counter can be clocked by the β_4 transition from the high-order bit of the first counter. Thus, one counter goes from m_1 to m_6 repeatedly, while the other counter is incremented once for each six clock pulses. The binary outputs of these two counters can be added (using two full adders and one half-adder) to obtain a statistically correct value from 2 to 12. A four-bit MSI adder could also be used for this purpose. The four-bit total can be labeled D_8, D_4, D_2, and D_1. If the clock runs at a high rate (say 1 MHz), the randomness with which the user's *THROW* signal arrives will act to introduce *chance* into the number selection.

The first number that is "thrown" must be gated into the X register (four flip-flops) on leaving P_1. The β_{P_1} signal can act to clock these flip-flops (see Fig. E7-15b). The value of the "dice" will be stored in X upon exit from state P_1.

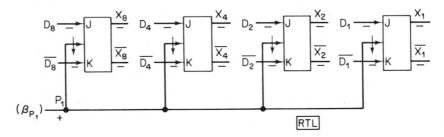

Figure E7-15b

The 7 or 11 logic must examine the D_8–D_1 signals during P_1 in order to detect the WIN condition. This signal has been named F. The K-map for F is shown in Fig. E7-15c, showing that

$$F = D_4 D_2 D_1 + D_8 D_2 D_1$$

This same logical signal indicates a LOSE condition if it is TRUE during P_4.

The comparison logic acts during P_4 to determine whether the new number from the "dice" is equal to the number stored in the X register. The logic for this, the C signal, is

$$C = (D_1 \odot X_1) \cdot (D_2 \odot X_2) \cdot (D_4 \odot X_4) \cdot (D_8 \odot X_8)$$

which, in RTL terms, is as shown in Fig. E7-15d. When the dice generator with its two binary counters and its binary adder, the F logic, the C logic, and the controller are connected together, the digital crap shooter is ready to roll (no pun intended).

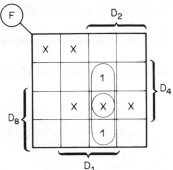

Figure E7-15c

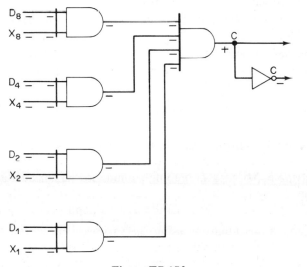

Figure E7-15d

7.4 Summary of the System Design Method

Chapters 1–7 have presented what is intended to be a consistent, reasonably organized approach to the design of digital systems. In a sense, the total instructional material presented thus far describes a design philosophy rather than a cookbook method. This philosophy, in a few words, involves designing toward proper system operation by using a reasonable blend of design effort and system cost. The emphasis has been upon explaining design techniques that work well in practice and that, taken in concert, make the design of a digital system a relatively straightforward task.

The design process has been broken into a sequence of design steps, each of which relies on the design tools described herein *and* the ingenuity of the designer. The following outline summarizes the suggested design process:

1. Break the overall task that the system must perform into functional-level subtasks. Typical subtasks include arithmetic operations, shifting, and counting.

2. Design the functional subsystems that must perform each subtask. Combinational and sequential design techniques play a major role at this point. MSI devices should be utilized where possible. The input, output, and timing requirements of each subsystem should be defined.

3. Develop the state table or state diagram that shows the sequence in which the subtasks must be performed in order to accomplish the system function.

4. Reduce the number of control states (if desired) by using state reduction techniques.

5. Design the controller. In small systems the ring-counter controller is a good choice. In larger systems a state-counter controller may be preferable. Microprogramming should also be considered, especially in larger systems that are to be produced in quantity. Remember that a good controller generally assures good system operation.

6. Complete the design by using the various control signals generated by the controller to activate the functional subsystems in the proper sequence. Consider the effects that supposedly can't-happen conditions can have on system performance.

7. Document the design. Include logic diagrams, timing diagrams, and any other information that might be of use to the user. This last step is often the great shortcoming of the practicing designer.

This textbook is certainly not a complete reference for digital design procedures. Rather, only a skeletal framework of related design techniques is presented. With these techniques alone, however, one can design complete digital systems. It is expected that the designer will broaden his skills, "fleshing out" the simplified design approach that is given herein. As better techniques for state reduction, state assignment, and combinational and sequential implementation become available, they can easily be incorporated into the generalized design sequence that is listed above.

The primary goal of this text is the establishment of a foundation of design techniques upon which a designer can base his expanding skills. It is hoped, however, that a foundation based upon the techniques and suggestions that are given herein will continue to help the reader to design good digital systems in a direct and logical method.

7.5 Summary

Chapter 7 concludes the major instructional portion of this textbook. The complete systems design process is presented and illustrated. Several approaches to the design of the controlling portion of a digital system are presented. The controller design process is closely related to the present-state/next-state design techniques presented in Chapter 6. The assembly of the controller and the related functional subsystems into the total digital system is also explained and illustrated.

7.6 Did you Learn?

1. How to separate a system's overall operation into functional subtasks?

2. How to use state tables and state diagrams to specify a control sequence?

3. How to design the three types of controllers? When and why to use each type?

4. How to use external control signals and interval counters to vary the length of control intervals?

5. How to build and use a pulse-catcher?

6. The general philosophy of system design that is summarized in Sec. 7.4?

7.7 References

[1] Berndt, Helmut, "Functional Microprogramming as a Logic Design Aid," *IEEE Transactions on Computers*, C-19, No. 10 (October 1970), 902–907.

[2] Husson, Samir S., *Microprogramming Principles and Practices.* Englewood Cliffs, N.J.: Prentice-Hall, Inc., 1970.

[3] Kohavi, Zvi, *Switching and Finite Automata Theory.* New York: McGraw-Hill Book Company, 1970.

[4] Krieg, Harold, "The Thumbwheel Revolution," *Electronic Products* (October 1970), 94–96.

[5] Vandling, G. C., and D. E. Waldecker, "The Microprogram Technique for Digital Logic Design," *Computer Design*, 8, No. 8 (August 1969), 44–51.

7.8 Problems

7-1.   The following state diagrams (Figs. P7-1a and d) and state tables describe several sequential systems. Design a ring-counter version of each system. Show an RTL logic diagram. Include output generating logic where applicable.

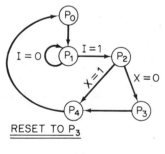

(a)

Present	Next State			
State	$XY = 00$	01	10	11
a	a	b	c	d
b	b	b	c	d
c	a	b	d	a
d	c	c	c	a

(b)

Figure P7-1

Present State	Next State		Output, F	
	$I = 0$	$I = 1$	$I = 0$	$I = 1$
A	B	C	0	0
B	C	D	1	0
C	D	E	0	1
D	A	A	1	1
E	E	F	1	0
F	F	D	0	1

(c)

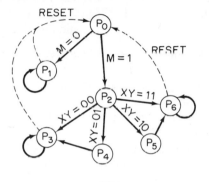

(d)

Present State	Next State*			Output, F*		
	$X = 1$	$Y = 1$	$Z = 1$	$X = 1$	$Y = 1$	$Z = 1$
a	a	b	c	0	0	1
b	b	b	d	0	1	0
c	e	a	b	1	0	0
d	e	e	e	0	1	1
e	c	d	d	1	1	1

(e)

*Assume that only one of the X, Y, and Z control signals can be TRUE at any time.

Figure P7-1 (Cont.)

*7-2. Design a ring-counter controller for each of the reduced systems of Prob. 6-16.

*7-3. Design a state-counter controller for each of the reduced systems of Prob. 6-17.

*7-4. The stepping motor is introduced in Prob. 6-25. Design a digital system that receives a two-digit number and a forward/backward control signal and will automatically move a stepping motor through the number of steps specified by the number.

*7-5.   Design a five-state clocked system that will operate in the following fashion:

    a. Begin in P_0, remain for one clock period.
    b. Move to P_1 for two clock periods.
    c. Move to P_2 for three clock periods.
    d. Move to P_3 for four clock periods.
    e. Move to P_4 for five clock periods.
    f. Return to P_0.

Design a ring-counter controller that acts as described above. An auxiliary binary counter should be used to keep the system in each state for its specified number of clock periods. Show an RTL logic diagram for the system.

7-6.   Binary multiplication can be implemented as a sequence of shifts and additions involving the multiplier and the multiplicand. The controller for such a process must properly sequence the ADD/SHIFT cycles. Assume that two ten-bit binary numbers are to be multiplied. The control sequence requires that an ADD cycle take place if the most-significant bit of the multiplier is TRUE. If the bit is FALSE, no ADD cycle is executed. After the ADD/not-ADD decision, a SHIFT cycle is made. The controller must perform one conditional ADD operation following the ninth SHIFT cycle and then HALT. Let a control input I represent the value of the incoming bit of the multiplier. Design the controller for this system; use both ring-counter and state-counter techniques.

7-7.   A digital data acquisition system consists of an analog-to-digital converter (see Chapter 11) and a digital tape recorder. The converter generates a 12-bit binary word. Its conversion cycle is initiated by a positive transition. The conversion takes no more than 1 msec. The conclusion of the conversion is signaled by a positive *END* pulse of 1-μsec duration. The data are available at the time of the *END* pulse, and remain unchanged until the next conversion cycle is started. The tape unit records six-bit characters in an incremental write mode. The write cycle is initiated by a positive transition. It takes no more than 2 msec to write a character. No feedback signal is available from the tape unit. The controller for this system must cause the analog-to-digital converter to make a conversion once each 10 msec and the data that result to be recorded in two successive tape characters. After 1000 samples (2000 characters) have been recorded, the controller must cause an end-of-record character (111111) to be written on the tape and an inter-record gap to be written. The gap is initiated by a positive transition on a separate control input to the tape unit. A *HALT* input should cause the controller to terminate the current record with an EOF character (111110) and then stop. The controller is restarted by a *START* signal.

8 Binary Arithmetic

8.0 Introduction

The preceding chapters have been an attempt to explain a consistent, applications-oriented (rather than theoretical) approach to the design of digital systems. This chapter and the chapters that follow are somewhat different in that they consider typical applications of digital systems design technology rather than the design techniques themselves.

The first applications area is the use of digital systems to perform arithmetic operations. This topic is considered because of the frequency with which computational applications arise in digital systems practice. The digital computer is the most common digital system that is in use at the current time, and many smaller digital systems perform arithmetic tasks as part of their specialized operations. This chapter is not addressed to computer design techniques, however. It considers only the simple arithmetic operations of addition and subtraction in binary notation.

The basic concepts of binary addition and subtraction are introduced in Chapter 1. Also, in Chapters 2–7, various examples and problems have considered the implementation of these processes. Chapters 8 and 9 expand on these introductory concepts. Timing considerations, the representation and manipulation of negative numbers, arithmetic operations with decimal numbers, and other related topics are discussed. Emphasis is placed on the logical characteristics of the arithmetic operations and on the hardware systems that are used to implement them.

8.1 Binary Addition Revisited

The basic building block for implementing binary addition is the full adder. Full adders can be used in combination to perform parallel binary addition, or a single full adder can be used sequentially to perform serial addition. Given enough hardware and/or time, binary numbers of any length can be added by using only the full-addition operation.

8.1.1. Parallel Addition.

Parallel addition in a digital system generally involves the use of registers to store the original addends and the sum. A typical register-adder arrangement is shown in Fig. 8-0. The numbers that

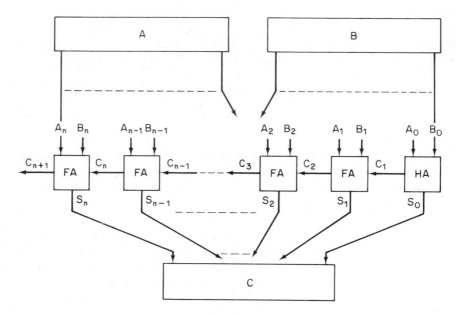

Figure 8-0. The Three-register Binary Adder

are to be added must first be loaded into the *A* and *B* registers. Once the sum and carry signals that are generated by the full adders have settled, the sum is transferred into the *C* register. An algebraic notation for this operation is

$$(A) + (B) \Rightarrow C \tag{8-0}$$

where the parentheses are interpreted as "the contents of the register" and the double arrow indicates that the final sum is transferred into the *C* register. The plus sign should not be confused with the OR function. This suggests the use of the $\lor$ symbol for the OR operation whenever the addition operation is also involved.

The low-order bits of *A* and *B* can be added by a half-adder, since there will not be a carry into the low-order stage. The half-adder and the full adders can be built from separate gating components, as shown in Sec. 4.3.6, or they can be MSI devices.

8.1.2. The Binary Accumulator. The three-register addition system of Fig. 8-0 can be modified in order to reduce its register requirements if the sum is returned to one of the addend registers. This eliminates the need for the C register. Symbolically, the addition operation becomes

$$(A) + (B) \Rightarrow A \tag{8-1}$$

The number that was originally stored in the A register is lost, since it is replaced by the sum.

This arrangement is called an *accumulator*. The summation (or accumulation) of several numbers can be accomplished by placing the numbers in the B register, one after the other, and executing a series of addition operations. The accumulating sum acts as the A operand.

In many cases the B register is eliminated as well, with the B operand being provided from some external source (a set of switches, for example). This arrangement is shown in Fig. 8-1. The A register is not loaded directly.

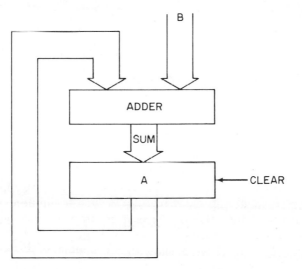

Figure 8-1. The Binary Accumulator

Rather, it is cleared to 0 by a control signal, and the first accumulation cycle adds the beginning number into A. The operational sequence for summing a series of binary numbers is

Operation	Result
$0 \Rightarrow A$	Clear A
$(A) + N_1 \Rightarrow A$	$A = N_1$
$(A) + N_2 \Rightarrow A$	$A = N_1 + N_2$
$(A) + N_3 \Rightarrow A$	$A = N_1 + N_2 + N_3$
$(A) + N_4 \Rightarrow A$	$A = N_1 + N_2 + N_3 + N_4$

and so on. This sequence is closely related to the operational sequence that is used by many computers to implement binary multiplication as a series of additions [1].

The accumulator of Fig. 8-1 can be built by using a series of full adders. The hardware arrangement for one of the stages that make up this type of accumulator is shown in Fig. 8-2. This representative hardware diagram is

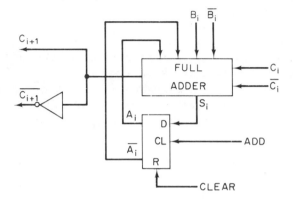

Figure 8-2. A Bit-slice through a Binary Accumulator

called a *bit slice*, since it shows the complete hardware arrangement for a single bit of the accumulator.

The B_i and C_i signals are received in double-rail form. C_i is the carry from the lower-ordered stage. The sum is returned to the A_i flip-flop, one of the flip-flops that make up the A register. A type D flip-flop is shown. The ADD signal causes the value of the sum, S_i, to be transferred into A_i. A $CLEAR$ signal is also provided.

Equations (4-16) and (4-18) show that a NAND/NOR implementation for a full adder costs 25 c.u. plus one inverter for generating $\overline{C_{i+1}}$. Thus, each stage of the accumulator costs 26 c.u. of gating and one type D flip-flop (or one J-K flip-flop plus one inverter). But since one of the operands that are summed is already in the A_i register, the logic for controlling the A_i flip-flops can be simplified.

Figure 8-3a shows a 1-bit slice of this generalized J-K accumulator arrangement. The accumulation operation can be considered as a present-state/next-state design problem involving the flip-flops of the A register. This behavior is tabulated in Table 8-0. The A_{i+} and C_{i+1} columns correspond to the sum and carry outputs from a normal full adder.

The transition map for A_i and the K-map for C_{i+1} are shown in Figs. 8-3b and c. The J_i and K_i control equations are

Table 8-0: Flip-flop Behavior During Binary Accumulation

Present State $A_i\ B_i\ C_i$			Next State A_{i+}	Carry Output C_{i+1}	Behavior A_i
0	0	0	0	0	0
0	0	1	1	0	α
0	1	0	1	0	α
0	1	1	0	1	0
1	0	0	1	0	1
1	0	1	0	1	β
1	1	0	0	1	β
1	1	1	1	1	1

$$J_i = K_i - D_i\bar{C}_i \mid \bar{B}_iC_i \tag{8-2}$$

or

$$J_i = K_i = B_i \oplus C_i \tag{8-3}$$

The logic for C_{i+1} is

$$C_{i+1} = A_iB_i + A_iC_i + B_iC_i \tag{8-4}$$

which is the carry logic required by a normal full adder. The use of Eqs. (8-2) and (8-4) reduces the per-bit gating cost of the accumulator to 15 c.u. plus an inverter for $\overline{C_{i+1}}$, a total of 16 c.u. The use of an EXCLUSIVE-OR

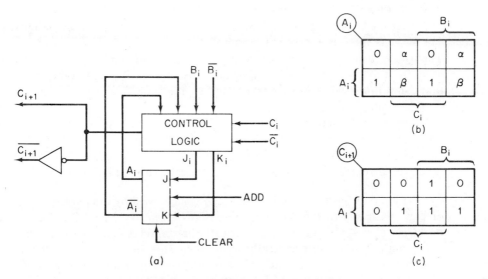

Figure 8-3. The J-K Binary Accumulator

gate to implement J_i and K_i as in Eq. (8-3) permits a further cost reduction. In both cases, the modified J-K accumulator is cheaper than the full-adder accumulator design given in Fig. 8-2.

A third accumulator design uses two successive control states to perform the addition operation. This arrangement, the *gated-carry accumulator* or *two-step accumulator*, reduces the per-stage gating cost to 12 c.u. Because of its partially sequential nature, this accumulator operates more slowly, however, and its controller must be modified to provide two successive control pulses during the addition operation. This, in part, offsets the savings in gating costs.

The logic diagram for a single stage of a gated-carry accumulator is shown in Fig. 8-4. The A_i flip-flop is used with J_i and K_i tied together, making

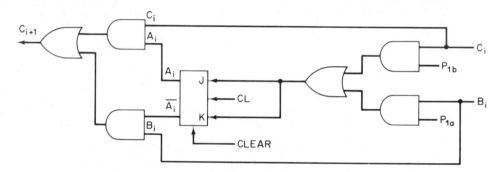

Figure 8-4. A Single Stage of the Gated-carry Accumulator

it act as a type T flip-flop. The two control intervals are called P_{1a} and P_{1b}, replacing the ADD signal used by the two previous accumulator designs.

If the B_i input is TRUE at the end of P_{1a}, then the A_i flip-flop is toggled. The new value of A_i is then used to determine the value of C_{i+1}, the output carry from the ith stage. The carry signals settle during P_{1b}, and at the end of that control interval, A_i is toggled again if C_i, the carry from the next lower-ordered stage, is TRUE.

The two-step action is detailed in Table 8-1. The first state change is reflected in the A_{i+} column. The C_{i+1} signal is then generated. The secondary state change is shown as A_{i++}. Note that the C_{i+1} and the eventual value of A_i correctly represent the sum and carry for each input combination.

8.1.3. Overflow.   The fixed-length registers used by the parallel adder of Fig. 8-0 or any of the accumulator designs given in Sec. 8.1.2 can cause problems during the addition process. If the numbers that are combined during an addition are sufficiently large, their sum may exceed the length of either of the input operands. This condition is indicated whenever the addition process produces a TRUE carry signal out of the high-order position

Table 8-1: The Gated-Carry Accumulator

Present State	First Change*	Carry Output†	Second Change‡
A_i B_i C_i	A_{i+}	C_{i+1}	A_{i++}
0 0 0	0	0	0
0 0 1	0	0	1
0 1 0	1	0	1
0 1 1	1	1	0
1 0 0	1	0	1
1 0 1	1	1	0
1 1 0	0	1	0
1 1 1	0	1	1

*After P_{1a}, caused by $P_{1a} \cdot B_i$

†$C_{i+1} = C_i A_{i+} + \overline{A_{i+}} B_i$

‡After P_{1b}, caused by $P_{1b} \cdot C_i$

of the adder or accumulator. For example, if a six-bit adder combines $A = 101101_2$ and $B = 111010_2$, then the sum is

$$\overset{\text{high-order carry}}{\overbrace{}}$$
$$1\ 1\ 1\ 0\ 0\ 0 \longleftarrow \text{carries}$$
$$101101$$
$$+111010$$
$$\overline{100111} \longleftarrow \text{Sum}$$

The sum cannot be properly expressed within only six bits. This condition is commonly called *overflow*.

A common example of overflow occurs when an automobile is driven over 100,000 miles. The odometer "rolls-over" at 99,999.9, with the result that the mileage accumulation begins from zero again. The same thing happens when an adder or an accumulator reaches its upper limit. For an n-bit device this limit is $2^n - 1$, the all-1s binary number. Addition of one more unit to this value causes a high-order carry (2^n in value) and resets the sum to zero. The value at which the device resets to zero is called the *modulus*.

EXAMPLE 8-0: What is the modulus of a car's odometer?

Answer: 100,000 (Miles)

EXAMPLE 8-1: What is the modulus of an eight-bit accumulator? What is the largest number that it can contain?

Answer: The modulus is $2^8 - 256$. The largest number is 255.

EXAMPLE 8-2: A ten-bit accumulator contains a value of 1011101011_2. If the value 0100110101_2 is added to it, what is the result?

Solution:

$$
\begin{array}{r}
1011101011 \\
+\ 0100110101 \\
\hline
10000100000
\end{array}
$$

Overflow takes place. Since the accumulator can store only the ten low-order bits, the *apparent* sum is 0000100000_2.

Overflow can be accommodated in a parallel adder if the C register is extended to be one bit longer than the A or B registers. This solution cannot be used in an accumulator, however, since the same register stores an operand *and* the sum. In any case, a longer sum register causes problems in digital systems that store or transfer data.

The primary concern about overflow is that it must be detected. Otherwise a grossly incorrect sum might be passed on for use during further arithmetic processing. Many systems halt whenever overflow is detected, signaling the condition to the operator for remedial action. Possible actions include scaling the data to prevent overflow or the use of a floating-point representation for numerical quantities [1].

The detection of the overflow condition when adding normal binary numbers is simple, since the generation of a high-order carry, C_{n+1}, always means that arithmetic overflow has occurred. The detection of overflow when signed numbers are added is more difficult. This problem is considered in Sec. 8.4.3.

8.1.4. Timing Considerations. The parallel connection of a group of full adders in order to perform binary addition has one major drawback: carries must be propagated from low-order stages to higher stages. Thus, the delays that are present in each full adder may accumulate, making the time required for a binary adder or accumulator to settle become quite large relative to the delay through a single gate.

Carry propagation may not always be present. Consider $A = 101010_2$ and $B = 010101_2$. Their sum is $C = 111111_2$ with no carries. The worst-case combination of operands is $A = 11\cdots11_2$ and $B = 00\cdots01_2$. Here, the two low-order 1s produce a carry signal that must be propagated all the way to the high-order bit before the proper sum is obtained.

This propagational delay can be studied by considering the logical operation of the generalized binary adder that was given in Fig. 8-0. The addition process begins when the A and B operands are loaded into their respective registers. All the full adders receive their A and B inputs at that time. Consider the Boolean expression for the sum output of the ith stage:

$$S_i = (A_i \bar{B}_i + \bar{A}_i B_i)\bar{C}_i + (\bar{A}_i \bar{B}_i + A_i B_i)C_i \tag{8-5}$$

Equation (8-5) shows that the proper value of the ith sum cannot be determined until C_i, the carry into the ith full adder, is known. The expression for C_i can be written as

$$C_i = A_{i-1}B_{i-1} + C_{i-1}(A_{i-1} + B_{i-1}) \tag{8-6}$$

Equation (8-6) shows that if *both* A_{i-1} and B_{i-1} are TRUE, then C_i will not depend on C_{i-1}, and the carry into the ith stage will settle within two gate delays after the A and B operands are applied to the adder. If A_{i-1} and B_{i-1} are both FALSE, then there can never be a carry into the ith stage, and carry propagation delay is of no concern. If, however, either A_{i-1} or B_{i-1} is TRUE, the C_i signal will depend on the logical value of C_{i-1}, which, in turn, will depend on the values of A_{i-2}, B_{i-2}, and possibly C_{i-2}. Each successive level of dependence adds two more gate delays to the carry-settling time. This delay is inherent in the two-level AND/OR sequence of Eq. (8-6). Thus, in the worst case, the carry propagation time through n full adders will be $2 \times n \times t_g$.

The carry into the nth stage of an n-bit binary adder may have to propagate through as many as $n - 1$ stages. Thus C_n will settle in no more than $2(n - 1)t_g$. The sum output, S_n, will settle $2t_g$ later, as will the high-ordered carry, C_{n+1}. Depending on the actual operands, the sums and carries may settle in less than $2nt_g$, but the allowance of that amount of time assures completion of the summation process in all cases.

As n becomes large, the addition process becomes slower and slower. Two approaches can be used to speed it up. One is to use faster gates, so that the basic gate delay is decreased. Emitter-coupled logic is often used for this purpose, since it offers small propagational delays. A second approach uses additional combinational logic to bypass the cumulative propagation delays that are present in the normal carry path. This technique is commonly called *carry lookahead*; it is the subject of the next subsection.

8.1.5. Carry Lookahead.

To accelerate the propagation of carries from low-ordered full-adder stages to higher stages, extra combinational logic may be included in the binary addition system. The worst-case carry originates in the low-order half adder. It is present if both A_0 and B_0 are TRUE, since

$$C_1 = A_0 B_0 \tag{8-7}$$

The carry out of the next stage is defined as

$$C_2 = A_1 B_1 + C_1(A_1 + B_1) \tag{8-8}$$

which follows from Eq. (8-6). Inserting Eq. (8-7) into Eq. (8-8) gives

$$C_2 = A_1 B_1 + A_0 B_0 A_1 + A_0 B_0 B_1 \tag{8-9}$$

This expression for C_2 will always settle within only two gate delays, since it depends only on A_1, B_1, A_0, and B_0, all of which are applied to the adder at the same time. Continuing the carry lookahead logic gives

$$C_3 = A_2 B_2 + A_1 B_1 A_2 + A_1 B_1 B_2 + A_0 B_0 A_1 A_2$$
$$+ A_0 B_0 A_1 B_2 + A_0 B_0 B_1 A_2 + A_0 B_0 B_1 B_2 \tag{8-10}$$

Implementation of C_3 in the form given in Eq. (8-10) will allow C_3 to settle in only $2t_g$, just as Eqs. (8-9) and (8-7) did for C_1 and C_2.

The lookahead process can be continued indefinitely, although practicalities such as gating cost and fan-in/fan-out limitations rapidly become unmanageable. Theoretically, however, the extension of carry lookahead logic can permit all carries to settle in $2t_g$. The sum bits will settle $2t_g$ later, giving a worst-case settling time of $4t_g$.*

A slightly different approach to the implementation of the carry lookahead logic uses a reexpression of Eq. (8-6), the carry equation:

$$C_{i+1} = G_i + P_i C_i \tag{8-11}$$

where G_i is called the *carry generate* term for the ith stage and P_i is the *carry propagate* term. The logical expressions for G_i and P_i are

$$G_i = A_i B_i \tag{8-12}$$

and

$$P_i = A_i + B_i \tag{8-13}$$

When TRUE, G_i indicates that a carry is generated in the ith full adder, making C_{i+1} become TRUE within $2t_g$ after the values of A_i and B_i are known. If neither G_i nor P_i is TRUE, C_{i+1} is always FALSE. This condition occurs whenever $A_i B_i = 00$, the *carry stop* condition. If only P_i is TRUE, C_{i+1} will not settle until $2t_g$ after C_i is known.

The G_i and P_i terms can be used to implement the carry lookahead logic. The low-ordered carry given in Eq. (8-7) can be reexpressed as

$$C_1 = G_0 = A_0 B_0 \tag{8-14}$$

The carry into the third stage is

$$C_2 = G_1 + C_1 P_1 \tag{8-15}$$

or, combining Eqs. (8-14) and (8-15),

$$C_2 = G_1 + G_0 P_1 \tag{8-16}$$

---

*Note, as well, that an n-bit binary adder can be arranged so as to settle in only $2t_g$ if the entire adder is considered as a single multiple-output combinational problem. It will have $2n$ input variables, however, making its solution, even with computer assistance, quite difficult for typical values of n (say 8 or more).

which is logically equivalent to Eq. (8-8). If C_2 is implemented as in Eq. (8-16), it will settle within $2t_g$ after G_0, G_1, and P_1 settle, a total delay of $4t_g$ after the A_i and B_i values are known.

Continuing this type of lookahead expansion gives

$$C_3 = G_2 + G_1P_2 + G_0P_1P_2 \qquad (8\text{-}17)$$

Equation (8-17) shows that a carry is made from the third full adder if a carry is generated within the third stage *or* a carry is generated in the second stage and propagated through the third stage *or* a carry is generated in the low-ordered stage and propagated through both of the next two stages. Since all the terms on the right-hand side of Eq. (8-17) settle within $2t_g$ after the operands are applied, C_3 will settle within $4t_g$. Continuing to C_4 gives

$$C_4 = G_3 + G_2P_3 + G_1P_2P_3 + G_0P_1P_2P_3 \qquad (8\text{-}18)$$

In general terms, the carry into the ith stage of a binary adder can be written as

$$
\begin{aligned}
C_i = G_{i-1} + G_{i-2}P_{i-1} &\mid G_{i-3}P_{i-2}P_{i-1} + \cdots \\
&+ G_2P_3P_4 \cdots P_{i-1} + G_1P_2P_3 \cdots P_{i-1} + G_0P_1P_2 \cdots P_{i-1} \qquad (8\text{-}19)
\end{aligned}
$$

With this carry lookahead technique, all carries settle in $4t_g$, allowing the total sum to settle in $6t_g$. The implementation of Eq. (8-19) requires a single i-input OR gate and $i - 1$ AND gates whose number of inputs ranges from 2 to i. The maximum loading factor applies to the P_{i-1} carry propagate signal. It requires $i - 1$ unit loads for the implementation of C_i alone (not to mention loading for C_{i+1}, etc.). Thus, as carry lookahead is implemented for larger adders, fan-in and loading constraints serve as limiting factors.

As a result, carry lookahead is generally implemented in groups of bits with higher-level lookahead between groups (Refs. [1] and [2]). As an example of the lookahead arrangement, a six-bit adder with full carry lookahead is shown in Fig. 8-5. The total delay for all carry signals to settle is only $4t_g$. The sum settles in no more than $6t_g$.

The popularity of carry lookahead techniques has led to the MSI implementation of carry lookahead networks for groups of several bits. In fact, complete multibit binary adders that incorporate lookahead techniques for faster operation are available as MSI components [3].

8.1.6. Serial Addition.

The parallel addition or accumulation operations that have been considered thus far in Sec. 8.1 can be replaced by serial operations, as was shown in general form in Sec. 5.2. The serial addition of two n-bit operands will require n clock periods or control cycles, but it will reduce the hardware requirements from $n - 1$ full adders and one half-adder to a *single* full adder. The control logic required by the system will be increased, however, and both operands must be available in serial form. This serialization is usually provided by shift registers.

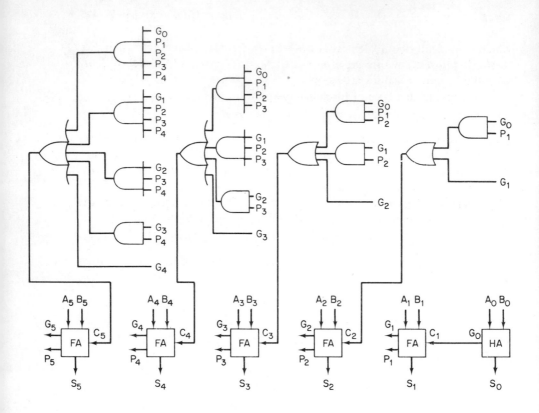

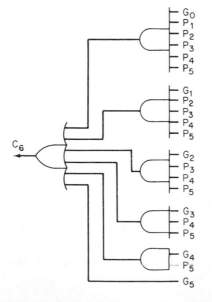

Figure 8-5. Six-bit Binary Adder with Full Carry Lookahead

A generalized serial binary accumulator is shown in Fig. 8-6. The *CLEAR* signal resets the accumulator register, A, the input number is loaded into the B register, and the serial addition begins. The A and B registers are shifted toward their least-significant end, so that the addition begins with A_0 and B_0 being sent to the full adder.

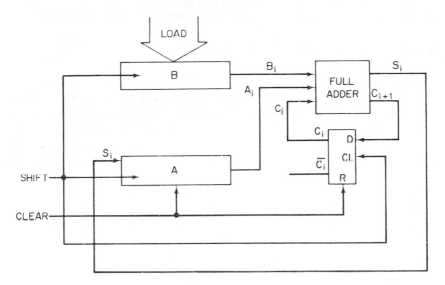

Figure 8-6. A Serial Binary Accumulator

The C_i flip-flop is reset by the *CLEAR* signal so that no carry input is initially present. During the first clock period the full adder combines A_0 and B_0, generating S_0 and C_1. S_0 is returned to the A register (or to a separate C register if an adder, rather than an accumulator, is desired). The carry output, C_{i+1}, is sent to the *CARRY* flip-flop. It acts as a one-bit shift register. Thus, when the first *SHIFT* pulse is received, the A and B registers move to present A_1 and B_1 to the full adder, and the carry flip-flop stores C_1. The full adder then has one clock period to generate S_1 and C_2 from these input operands.

This process continues for n periods, at which time the complete sum is present in the A register. The C_i flip-flop, if it is SET after the nth *SHIFT* pulse, indicates arithmetic overflow (a high-order carry).

The time required for the serial addition process is n times that required for each sequential addition operation. The A_i, B_i, and C_i signals may change following each *SHIFT* pulse. Their output changes will lag the *SHIFT* pulse by t_p, the propagational delay of a flip-flop. The S_i and C_{i+1} outputs from the full adder will lag its inputs by $2t_g$. Once these output signals have settled, the setup time for the control inputs to the A register and the C_i flip-flop must be allowed. Thus, the time that must be allowed between successive *SHIFT* pulses is $t_p + 2t_g + t_s$. The complete serial addition takes $n(t_p + 2t_g + t_s)$ sec.

It should be noted that the t_p and t_s delays must also be allowed for in the parallel adder or accumulator, but they are normally quite small relative to $2nt_g$, the settling time for fully parallel addition.

8.2 Binary Subtraction

In each of the binary adder and accumulator examples that were considered in Sec. 8.1, the addition operation can be replaced by subtraction with no loss of generality. Accumulation becomes repeated subtraction, a common requirement for implementing binary division [1]. Carry lookahead becomes *borrow lookahead*, with *borrow generate*, *borrow propagate*, and *borrow stop* conditions being of concern. Serial subtraction makes repeated use of a single full subtractor and so on.

The major difference between subtractive operations and additive operations is the possibility of a negative result, i.e., the generation of $A - B$ when B is greater than A. This relates to overflow in an adder, in that it generates a result that cannot be properly expressed within the bit lengths of the original operands.

Consider $A = 1001_2$ and $B = 1101_2$. Subtracting B from A gives

$$
\begin{array}{r}
1\,1\,1 \leftarrow \text{borrows} \\
A = 1001 = 9_{10} \\
-B = 1101 = -13_{10} \\
\hline
\text{Difference} = 1100 = 12_{10}\ (?)
\end{array}
$$

This condition was introduced in Sec. 1.5.2. The high-order borrow points to the negative result that has been generated. The answer appears in *complemented* form, a representation that is discussed in detail later in this section. The answer is certainly not in natural binary form, since $9_{10} - 13_{10}$ cannot give 12_{10}.

One approach to a problem such as this is to compare A and B prior to the subtraction. The smaller number is then subtracted from the larger, with the difference being noted as positive if $A > B$. If $B < A$, however, the magnitude of the difference is correct, but it must be given a negative sign. This requires the inclusion of an additional bit, commonly called the *sign bit*, in each operand. The sign bit is generally FALSE for positive numbers and TRUE for negative numbers. With a sign bit included, the original numbers may be written as $A = 0,1001_2$ and $B = 0,1101_2$. The comma separates the sign bit from the normal magnitude bits.

EXAMPLE 8-3: Calculate $C = A - B$, where $A = 0,1001_2$ and $B = 0,1101_2$.

Solution: Begin by subtracting B from A (magnitudes only):

$$\begin{array}{r}
1\;1\;1 \longleftarrow \text{borrows} \\
A = \quad 1001 \\
-B = -1101 \\
\hline
1100
\end{array}$$

The high-order borrow indicates that B is greater than A. Recalculate, forming $C - B - A$:

$$\begin{array}{r}
B = \quad 1101 \\
-A = -1001 \\
\hline
C = \quad 0100
\end{array}$$

Assign a negative sign to the difference, making

$$C = 1,0100$$

The decimal equivalent to this subtraction is $+9 - (+13) = -4$.

The inclusion of a sign bit increases the numerical range that can be represented. A normal four-bit binary number ranges from 0_{10} to 15_{10}. When a sign bit is included, the numerical range of the *five-bit* number is -15_{10} to $+15_{10}$, a total of 31 different values. A five-bit binary number would normally have 32 values, but the sign bit allows two different forms for zero: $0,00000$ ($+0$) and $1,0000$ (-0). The double representation for 0 reduces the range from 32 total values to only 31.

This type of representation for signed quantities is called *sign and magnitude* representation, or simply *signed binary*. It corresponds to the way in which quantities are normally expressed for human use and, as a result, is a good way to represent data for input to and output from a digital system. The next section considers several aspects of arithmetic operations with signed binary numbers.

8.3 Signed Binary Arithmetic

The signed binary notation can be extended to all the operands that are processed in a digital system. This complicates the addition or subtraction process, since the operands may not have the same sign. Consider the addition of two signed binary numbers, say A and B. If both A and B are positive, their magnitudes can be added as usual, giving a positive sum. Overflow will be indicated if a high-order carry appears. The same situation applies if A and B are both negative. Their magnitudes are added, with the sum being given a negative sign (a TRUE sign bit) and overflow arising as usual.

If A_s and B_s are the sign bits of the A and B operands, then the logic for adding A and B is

$$ADD = \bar{A}_s \bar{B}_s + A_s B_s \tag{8-20}$$

$$= A_s \odot B_s \tag{8-21}$$

The first term in Eq. (8-20) indicates that both operands are positive; the second term indicates that both are negative. In either case the sum is given the same sign as A or B.

If A and B have different signs, the addition operation must be replaced by subtraction. Only the magnitudes of the A and B operands are used. A common approach begins by subtracting the magnitude of B from the magnitude of A. If there is no high-order borrow, then $|B| \leq |A|$. The net difference is then assigned the same sign as the original A operand. If the subtraction produces a high-order borrow, however, $|B| > |A|$ and the subtraction must be reversed. The magnitude of A is subtracted from the magnitude of B, with the difference being given the sign of B.

Whenever A and B have different signs their sum must be smaller than either of them, so overflow cannot occur.

Note that the sign bits of the operands are not involved in the addition or subtraction process. The sign of the sum is determined by *logical* comparison of the signs of A and B and by the subsequent processing that takes place.

EXAMPLE 8-4: Consider the addition of $A = 0,10101_2$ and $B = 1,11101_2$.

Solution: Since $A_s \neq B_s$, the addition is actually a subtraction. Begin by forming $|A| - |B|$:

```
  1 1 1 ←— borrows
   10101
 −11101
 ───────
   11000
```

The high-order borrow indicates that $|B| > |A|$. As a result, $|B| - |A|$ must be calculated:

```
   11101
 −10101
 ───────
   01000
```

The final difference is given the sign of B, showing that

$$0,10101 + 1,11101 = 1,01000$$

The decimal equivalent is

$$+21 + (-29) = -8$$

The addition of signed binary numbers requires the implementation of both addition and subtraction. A conditional adder/subtractor can be used. The design of such a device is the subject of Prob. 8-8. A second approach is the implementation of subtraction by addition of complements, a technique described in Sec. 8.4.

The following example considers the control problems that arise when signed binary operands are added.

EXAMPLE 8-5: Two signed binary operands are to be added by a digital system. The operation begins by loading the two numbers into a pair of registers, say A and B. The system is to generate their sum, storing it in a third register, say C. It then halts, indicating overflow if present. Design the controller for this system.

Solution: A conditional adder/subtractor is to be used. It receives the magnitudes of A and B and, if its ADD control signal is TRUE, presents $|A| + |B|$ at its output terminals. If ADD is FALSE, it presents $|A| - |B|$. The adder/subtractor indicates overflow (for add) or high-order borrow (for subtract) by making a signal, F, become TRUE.

The control sequence is shown by the state diagram in Fig. E8-5. P_0 is the LOAD state. During P_0 the signs of A and B are compared. If they are

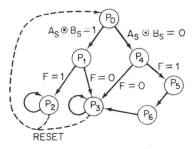

Figure E8-5   RESET

equal, $|A|$ and $|B|$ are added during P_1. The ADD control signal is made TRUE during P_1 ($ADD = P_1$). The sum is given the sign of A. If overflow occurs, the system transfers from P_1 to P_2. Otherwise it goes to P_3, the normal HALT state.

If A and B have opposite signs, the system goes from P_0 to P_4, wherein $|A| - |B|$ is formed. Since ADD is TRUE only in P_1, the conditional adder/subtractor will subtract. If no high-order borrow appears, then the sign of A is transferred into C_s, and the system moves to P_3. If a high-order borrow does appear, the system moves to P_5. P_5 causes the magnitude portions of the A and B registers to be swapped. A subtraction is made during P_6, with the sign of B being transferred into C_s. The system then halts in P_3. The system cannot overflow if A and B have opposite signs.

The subtraction of B from A can be accomplished by changing the sign bit of the B operand and performing the addition operations described above. This is equivalent to generating $A + (-B)$.

8.4 Sign and Complement Arithmetic

Example 8-5 shows how the inclusion of a sign along with the normal binary magnitude can greatly complicate binary addition. An alternative to the signed magnitude notation uses the sign bit as part of the arithmetic process, allowing the sign of the result to be computed during the addition process rather than being determined by a separate logical operation. This representation is called *sign and complement notation*, and, in the binary case, falls into two common forms.

8.4.1. Sign and 2's Complement. The representation of positive numbers in sign and 2's complement form (called 2's complement hereafter) is the same as that used in signed magnitude notation. A FALSE sign bit is added to the high-order end of the word. For a negative number the representation is quite different, however, since the sign bit changes the magnitude portion of the word.

The sign bit of a 2's complement operand has a negative binary value equal to -2^{n-1}, where n is the total number of bits present in the operand. Each operand contains $n - 1$ magnitude bits and one sign bit. The sign bit acts as a negative bias, setting the net value of the word to -2^{n-1} when the magnitude portion of the operand is 0. To represent a less-negative value the magnitude portion of the operand is chosen so that its positive value, when added to the -2^{n-1} weight of the sign bit, equals the desired negative value. Negative values can range from -2^{n-1} (sign plus 0) to -1 (sign plus the maximum positive magnitude allowed with $n - 1$ magnitude bits).

EXAMPLE 8-6: Give a five-bit, 2's complement representation for -11_{10}.

Solution: A five-bit 2's complement operand has four magnitude bits, so the sign bit has a weight of -2^4 or -16_{10}. Since $-16_{10} + 5_{10} = -11_{10}$, the 2's complement operand is 1,0101.

EXAMPLE 8-7: Give six-bit, 2's complement representations for -32_{10}, -15_{10}, -11_{10}, -5_{10}, and -1.

Answers:
$$-32 = 1,00000 = -32 + 0$$
$$-15 = 1,10001 = -32 + 17$$
$$-11 = 1,10101 = -32 + 21$$
$$-5 = 1,11011 = -32 + 27$$
$$-1 = 1,11111 = -32 + 31$$

With 2's complement notation an n-bit operand can take on values from $+(2^{n-1} - 1)$ to -2^{n-1}, a range covering 2^n values. This range is 1 greater than that allowed with signed magnitudes, since there is only one way to represent 0 in 2's complement: $0,00\cdots0$. Of interest is the lower limit of the 2's complement range, $-2^{n-1} = 1,00\cdots0$. This value has no positive equivalent, since the largest positive number is $+(2^{n-1} - 1) = 0,11\cdots1$.

The value of the sign bit is equal to the modulus of the magnitude area of the 2's complement operand. Thus, if the sign bit is treated as part of the arithmetic portion of the operands, carries into the sign column will act as modular overflow and will not affect the net value of the sum. Normal arithmetic overflow can still occur, however, and its detection when 2's complement notation is being used is discussed later in this section.

Arithmetic operations with 2's complement operands are best introduced by example.

EXAMPLE 8-8:   Add -9_{10} and $+2_{10}$ by using five-bit, 2's complement operands.

Solution:
$$-9 = -16 + 7 = 1,0111$$
$$+2 = 0,0010$$

Their sum is

$$\begin{array}{r} 1,0111 \\ +0,0010 \\ \hline 1,1001 \end{array}$$

which is -7, since $-16 + 9 — -7$. This may be interpreted as

$$-9 + 2 = (-16 + 7) + 2 = -16 + 9 = -7$$

EXAMPLE 8-9:   Add -9 to -4.

Solution:
$$-9 = -16 + 7 = 1,0111$$
$$-4 = -16 + 12 = 1,1100$$

Their sum is

$$\begin{array}{l} 1,0111 \\ 1,1100 \\ \hline 1,0011 = -16 + 3 = -13 \end{array}$$

Example 8-9 shows that a high-order carry out of the sign column may be generated during the addition process. In the 2's complement case this carry is ignored. It *does not* indicate arithmetic overflow.

The results of Ex. 8-9 can be interpreted as

$$
\begin{aligned}
-9 + (-4) &= (-16 + 7) + (-16 + 12) \\
&= -32 + 19 \\
&= -16 + 3 \\
&= -13
\end{aligned}
\tag{8-22}
$$

which is the correct answer.

The major advantage to the 2's complement representation is the simplicity with which arithmetic operations take place. Numbers may be added without concern for sign, since the sign bits are treated as part of the normal arithmetic operand. The sign of the sum is determined during the addition process. Only addition is required, even when operands have different signs. What of overflow, however?

If two positive numbers are added and their sum overflows the magnitude portion of the 2's complement operand, a carry will be made into the sign bit. This will make the sum appear to be negative. Similarly, whenever two negative numbers are added, their sum bits produce a high-order carry and, unless a carry from the magnitude area restores the sign bit to 1, the sum will appear to be positive. These are the only two overflow conditions, since the addition of numbers having opposite signs cannot produce an overflow.

Overflow detection can take two approaches. One technique compares the signs of A and B and their sum, C. If A and B are positive and the sum is negative, then overflow has occurred. Similarly, if A and B are negative and the sum has a positive sign, then negative overflow (a negative value less than -2^{n-1}) has occurred. This overflow detection logic is

$$
OVERFLOW = \overline{A_s}\,\overline{B_s}\,C_s + A_s B_s \overline{C_s}
\tag{8-23}
$$

The second detection technique compares the carry into the sign bit with the carry out of the sign bit. If these are not equal, then overflow has occurred. This logic is

$$
\begin{aligned}
OVERFLOW &= C_{n+1}\overline{C_n} + \overline{C_{n+1}}C_n \\
&= C_{n+1} \oplus C_n
\end{aligned}
\tag{8-24}
$$

EXAMPLE 8-10:   The addition of $+9$ and $+12$ produces overflow if five-bit, 2's complement operands are used. Demonstrate the detection of the overflow by using Eqs. (8-23) and (8-24).

Solution:

$$
\begin{aligned}
+9 &= 0,1001 \\
+12 &= 0,1100
\end{aligned}
$$

Their sum is

$$
\begin{array}{r}
C_{n+1}\searrow\swarrow C_n \\
01\ \ 100 \leftarrow \text{carries} \\
{}^{A_s}\!\searrow 0,1001 \\
{}_{B_s}\!\searrow +0,1100 \\
\hline
C_s \rightarrow 1,0101
\end{array}
$$

Since $\overline{A}_s \cdot \overline{B}_s \cdot C_s$ is TRUE, Eq. (8-23) will detect the overflow. Similarly, since $C_{n+1} \neq C_n$, Eq. (8-24) will be TRUE.

EXAMPLE 8-11: Show the detection of overflow when -9 and -12 are added.

Solution:

$$
\begin{array}{l}
10\ \ 100 \leftarrow \text{carries} \\
1,0111 = -9 \\
1,0100 = -12 \\
\hline
0,1011
\end{array}
$$

Clearly, both $A_s \cdot B_s \cdot \overline{C}_s$ and $C_{n+1} \cdot \overline{C}_n$ are TRUE.

The conversion from signed magnitude to 2's complement notation is a common requirement. In many cases subtraction is accomplished by converting the subtrahend into its 2's complement negation and adding, thereby eliminating the need for a binary subtractor.

Two's complementation can be implemented in several ways. One approach is to subtract the magnitude from the all-zero's number and change the logical value of the sign.

EXAMPLE 8-12: Convert $+9_{10}$ into its 2's complement equivalent.

Solution:

$$+9 = 0,1001$$

Subtract 1001 from 0000, ignoring the high-order borrow, and give the result a minus sign:

$$
\begin{array}{l}
1111 \leftarrow \text{borrows} \\
0000 \\
-1001 \\
\hline
1,0111 = -16 + 7 = -9 \\
\nwarrow\ \text{forced minus sign}
\end{array}
$$

EXAMPLE 8-13: Convert -12_{10} to its six-bit 2's complement equivalent.

Solution:

$$-12 = -32 + 20 = 1,10100$$

$$\begin{array}{r} \underset{\longleftarrow \text{ borrows}}{1\,1\,1\,0\,0} \\ 00000 \\ -10100 \\ \hline 0,01100 = +12 \\ \underset{\longleftarrow \text{ forced plus sign}}{} \end{array}$$

The technique illustrated in Exs. 8-12 and 8-13 has the drawback that it requires the subtraction operation. The elimination of the need for subtraction is one of the reasons for choosing 2's complement notation, but if the complementation process itself requires a subtraction, then there is no gain.

An algorithmic approach to the conversion process eliminates the need for a subtraction. The 2's complementation algorithm specifies that the binary number that is to be complemented is examined beginning with its least significant bit. All low-order 0s *and* the first low-order 1 bit are kept unchanged. All other bits (including the sign bit) are complemented. This produces the negation of the original number.

EXAMPLE 8-14: Complement $-9_{10} = 1,0111$.

Solution: The low-order bit is a 1. It is kept unchanged. All other bits are complemented, yielding 0,1001, which is +9.

EXAMPLE 8-15: Complement $+12_{10} = 0,01100$.

Answer: $1,10100 = -32 + 20 = -12$.

This conversion algorithm can be implemented in either parallel or serial form. Figure 8-7a shows a parallel implementation for this conversion technique. The low-order bit is never changed. If the low-order bit is a 1, then the next higher bit is inverted. An EXCLUSIVE-OR gate can be used to perform this conditional inversion. If $A_0 = 0$, then A_1 passes through the gate unchanged, since $A_1 \oplus 0 = A_1$. If, however, $A_0 = 1$, then A_1 is complemented, since $A_1 \oplus 1 = \overline{A_1}$. A_2 is complemented if either A_1 or A_0 is TRUE, and so on. The OR gates shown in Fig. 8-7a have increasingly large fan-in requirements. Figure 8-7b shows a modification of the complementation circuit that keeps the fan-in at two per OR gate. The latter technique is slower, however, because of the serial propagation path from the low-order bit to the sign bit. Another parallel complementation technique is examined in Prob. 8-19.

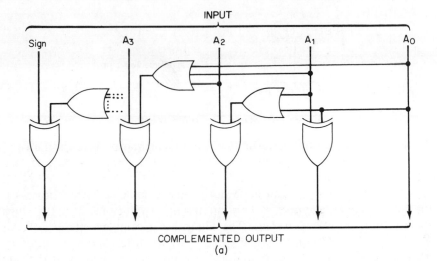

COMPLEMENTED OUTPUT
(a)

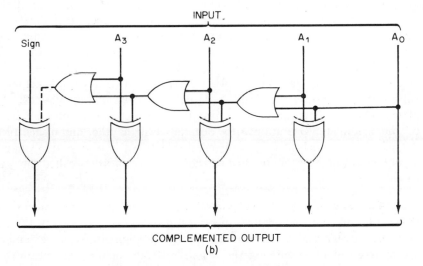

COMPLEMENTED OUTPUT
(b)

Figure 8-7. Combinational 2's Implementation

Figure 8-8 shows the state diagram of a controller for a serial 2's complementation system. While the controller is in P_0, any low-order 0s of the input operand are shifted out with no change. The first TRUE bit moves the controller to P_1, with a TRUE value being passed on to the output. The bits of the input operand are inverted between input and output when the controller is in the P_1 state.

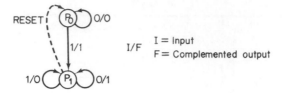

Figure 8-8. Serial 2's Complementation

A third method for performing 2's complementation uses the 1's complementation process as an intermediate step. This technique is discussed later in this section.

A form of overflow can arise during the complementation process if the $1,00 \cdots 0$ operand appears. This number has a value of -2^{n-1}, for which there is no positive equivalent. The algorithmic complementation of this number yields $1,00 \cdots 0$, the same value. This condition can be detected by comparing the signs of the original number and its complement. If they are the same, then the complementation process has produced an overflow, and some remedial action, such as setting the result to $0,11 \cdots 1$, the largest positive number, must be taken.

Two's complement binary notation is an example of a general mathematical technique called *radix complementation*. The decimal equivalent is 10's complement, since the radix for the decimal system is 10. Chapter 9 discusses the use of the 10's complement representation. The radix complement notation simplifies arithmetic operations involving both positive and negative quantities. Its major drawback is the need for a complementation subsystem. The use of the *diminished-radix complement* simplifies the complementation process, although it produces an increase in the complexity of the arithmetic operations. In decimal systems the diminished-radix complement is called the 9's complement. Its use is also discussed in Chapter 9. In binary notation the diminished-radix representation is called the *one's complement*.

8.4.2. Sign and 1's Complement. As in the 2's complement case, the sign bit of a 1's complement operand is given a negative weight. Positive numbers are represented in normal binary with a 0 sign bit. Negative numbers are represented by combining the negative weight of the sign bit and the positive value given in the magnitude portion of the operand. The sign bit of an n-bit, 1's complement operand has a weight of $-(2^{n-1} - 1)$, one unit greater than the negative value of the 2's complement sign.

EXAMPLE 8-16: Express -9_{10} as a five-bit, 1's complement number.

Solution: The sign bit has a weight of $-(2^4 - 1)$ or -15. Since $-9 = -15 + 6$, the answer is
$$-9 = 1,0110$$

EXAMPLE 8-17: Express -21_{10} as a six-bit, 1's complement number.

Solution: The sign bit of a six-bit, 1's complement operand has a weight of -31. Since $-21 = -31 + 10$, the answer is

$$-21 = 1,01010$$

EXAMPLE 8-18: What are the six-bit, 1's complement numbers for 31, -15, -7, and -1?

Answers:

$$-31 = 1,00000$$
$$-15 = -31 + 16 = 1,10000$$
$$-7 = -31 + 24 = 1,11000$$
$$-1 = -31 + 30 - 1,11110$$

The most negative 1's complement number is $1,00\cdots0$, having a value of $-(2^{n-1} - 1)$. This has the same magnitude as that of the most positive number, $0,11\cdots1$. The range of an n-bit, 1's complement number is only $2^n - 1$ values, therefore, similar to the range of the signed binary notation. The "missing" value is lost because of double representation for 0 that is inherent with 1's complementation notation. One expression for 0 is $0,00\cdots0$. The second is $1,11\cdots1$, the all-1's operand. This has a value of $[-(2^{n-1} - 1) + (2^{n-1} - 1)]$, which is 0. The two expressions are often referred to as $+0$ and -0, even though their arithmetic values are the same.

One's complement addition is similar to 2's complement addition in that the sign bits are treated as part of the arithmetic operand. Carries from the magnitude field are added into the sign column. If no high-order carry appears at the output of the sum column, then the answer is correct (unless overflow has occurred). A carry out of the sign column can occur only when one of the operands is negative. If such a carry does appear, the use of 1's complement notation necessitates an additive correction of the sum.

With 2's complement notation, carries out of the sign column indicate modular overflow and no correction is necessary. The diminished nature of the 1's complement sign bit causes high-order overflow to occur with values that are one less than the modulus. Thus, the sum must be increased by one unit whenever a high-order carry occurs.

The high-order carry is returned as $+1$ to the low-order column, a process commonly called *end-around carry* (not to be confused with a football play of the same name). The end-around carry operation corrects the sum to its proper value.

EXAMPLE 8-19: Add -9 and $+4$ by using five-bit, 1's complement notation.

Solution:

$$-9 = -15 + 6 = 1,0110$$
$$+4 = 0,0100$$

The sum is

$$1,0110$$
$$0,0100$$
$$\overline{1,1010} = -15 + 10 = -5}$$

EXAMPLE 8-20:   Add -7 and $+10$ in 1's complement.

Solution:

$$-7 = -15 + 8 = 1,1000$$
$$+10 = 0,1010$$

The sum is

$$11\ \ 000 \longleftarrow \text{carries}$$
$$1,1000$$
$$+0,1010$$
$$\overline{0,0010}$$

The end-around carry must be made to increase the sum by 1:

$$0,0010$$
$$+0,0001 \ \longleftarrow \text{end-around carry}$$
$$\overline{0,0011} = +3$$

EXAMPLE 8-21:   Show a 1's complement addition of -6 and -7.

Answer:

$$11\ \ 000 \longleftarrow \text{carries}$$
$$1,1001$$
$$1,1000$$
$$\overline{1,0001}$$
$$\longrightarrow +1$$
$$\overline{1,0010} = -15 + 2 = -13$$

The addition of the end-around carry can be implemented by replacing the low-order half-adder present in a parallel binary adder by a full adder and then connecting the carry out of the sum column back around to the carry input of the low-order full adder, as shown in Fig. 8-9. This connection may slightly increase the settling time, however, since the extra addition cycle can cause secondary carry propagations. This increase is the subject of Prob. 8-15.

The need for the end-around carry makes 1's complement notation

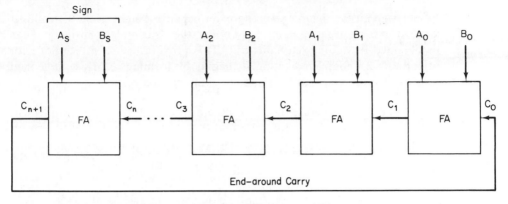

Figure 8-9. Parallel 1's Complement Adder

more difficult to use arithmetically than 2's complement. As a result, most digital systems that perform arithmetic operations upon signed binary numbers use 2's complement notation. Why, then, is 1's complement worthy of mention?

The major advantage of 1's complement notation is the simplicity with which the complement of a number (equivalent to multiplication by -1) can be found. The chief use of this property is as a stepping-stone to finding the 2's complement, rather than as an arithmetic operation in itself.

The 1's complement of a binary number can be obtained by subtracting the number from the all-1s operand. This process can *never* produce a borrow, since each column of the minuend will contain a 1. In fact, since $1 - 0 = 1$ and $1 - 1 = 0$, the result of the subtraction will be the bit-for-bit inversion of the original binary number. As a result, the 1's complement process can be implemented without a subtraction; all the bits of the original operand are simply inverted, sign bit included. This is evident from observation of the results of Exs. 8-19–8-21.

EXAMPLE 8-22:   Several 1's complement numbers and their complements are given below:

Number	Complement
$+10 = 0,1010$	$1,0101 = -15 + 5 = -10$
$+7 = 0,0111$	$1,1000 = -15 + 8 = -7$
$-12 = -15 + 3 = 1,0011$	$0,1100 = +12$
$-27 = -31 + 4 = 1,00100$	$0,11011 = +27$

The magnitude portions of the 1's complement and 2's complement representations of the same negative binary number will differ by one unit. The 2's complement number will have the larger magnitude, since it must offset a sign bit that has a one-unit greater negative weight. If the 1's complement form for a negative number is known, the 2's complement operand can be obtained by adding one unit to it. This provides a useful technique for performing 2's complementation. It begins by forming the 1's complement (bit-by-bit inversion) and then adds one unit to that result.

EXAMPLE 8-23: Generate the 2's complement of $+9$ by the 1's complement $+1$ process. Repeat for $+18$, -15, and 0.

Solution:

$$+9 = 0,1001$$
$$\text{1's comp. of } +9 = 1,0110$$
$$\text{1's comp.} + 1 = 1,0111$$

In 2's complement, the final answer reads $-16 + 7 = -9$. The other operations are

$$+18 = 0,10010$$
$$\text{1's comp.} = 1,01101$$
$$\text{1's} + 1 = 1,01110 = -32 + 14 = -18$$

$$-15 = -16 + 1 = 1,0001$$
$$\text{1's} = 0,1110$$
$$\text{1's} + 1 = 0,1111 = +15$$

$$0 = 0,0000$$
$$\text{1's} = 1,1111$$
$$\text{1's} + 1 = 0,0000 \quad \text{(high-order carry ignored)}$$

The 1's complement operation can never produce an overflow, since the magnitude of the most negative and most positive values are the same. If 2's complementation is implemented as 1's complement $+ 1$, however, an overflow can occur just as it can when other 2's complementation techniques are used. If the 2's complement number $1,00 \cdots 0$ is 1's complemented, the result is $0,11 \cdots 1$. The addition of one unit to this number gives $1,00 \cdots 0$, which is the same as the original number. This is the complementary overflow condition; it should be detected as described earlier.

The implementation of 2's complementation as the 1's complement $+ 1$ is quite simple. Figure 8-10 shows a parallel binary addition-subtraction arrangement that performs subtraction by addition of the complement. If the *SUBTRACT* control signal is FALSE, then A and B are added in normal fashion. The EXCLUSIVE-OR gate on the left of Fig. 8-10 compares C_{n+1} to C_n. If they are dissimilar, then *OVERFLOW* is indicated.

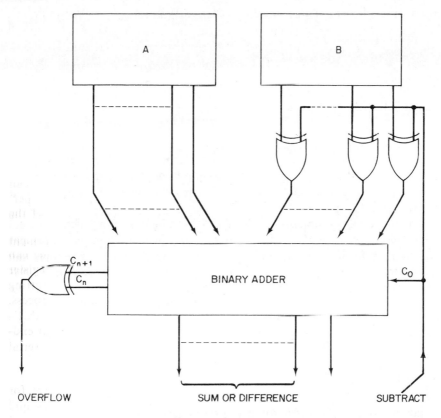

Figure 8-10. A 2's Complement Adder/Subtractor

If *SUBTRACT* is TRUE, then the EXCLUSIVE-OR gates connected to the outputs of the *B* register invert all the bits of the *B* operand, thereby forming the 1's complement of *B*. The *SUBTRACT* signal also supplies a carry to the low-order full adder. This acts as the "plus one" required by the 2's complementation process. As a result, the binary adder combines *A*, the 1's complement of *B*, and a low-order 1. This is equivalent to subtracting *B* from *A* and giving the result in 2's complement form.

EXAMPLE 8-24:   Describe the calculation of $A - B$ by the system shown in Fig. 8-9, where $A = +7$ and $B = +9$.

Solution:

$$A = 0,0111$$
$$B = 0,1001$$

Since subtraction is specified, the *B* operand is inverted and added to *A*. A low-order carry is forced. The 1's complement of *B* is 1,0110. Adding *A*, the 1's complement of *B*, and a low-order carry gives

$$
\begin{array}{r}
0,0111 \\
1,0110 \\
+ \quad 1 \\
\hline
1,1110 = -16 + 14 = -2
\end{array}
$$

which is the correct result.

———————

The 1's complementation of a number that is stored in a register can also be accomplished by toggling all the flip-flops in the register or by performing an *n*-bit rotational shift with inversion between the output of the register and its input.

If the serial addition system of Fig. 8-6 is to be used as a 2's complement adder/subtractor, the 1's complement $+ 1$ complementation technique can still be used for implementing the subtraction. The output of the *B* register is inverted to give 1's complementation. The low-order 1 is added by setting the *CARRY* flip-flop prior to the beginning of the serial addition process.

It should be noted that performing serial arithmetic operations with 1's complement operands is especially troublesome, since the need for an end-around carry will generally require the repetition of the complete serial addition process.

8.4.3. Conversions Between Forms.   Each of the three ways for encoding signed numerical quantities exhibits its unique advantages and disadvantages. Signed magnitude is preferable for data input and output since humans are most accustomed to that form for expressing quantities. The use of signed magnitude notation makes the arithmetic operations of addition and subtraction quite involved, however, since magnitude comparison and *both* addition and subtraction are required. In other words, signed magnitude is preferred outside of a digital system but not inside.

Two's complement notation is the best choice for arithmetic operations, but the 2's complementation process is somewhat difficult to implement. One's complement notation allows easy complementation but requires more involved arithmetic operations.

The optimum choice for representing quantities uses all three notations. Signed magnitude is used for data input. It is converted to 2's complement following input, and the 2's complement representation is used for all arithmetic processes. When it is required, subtraction is implemented by complementation and addition, eliminating the need for a binary subtractor. Two's complementation is implemented by forming the 1's complement (which is easier) and adding 1 via the low-order full adder. Data are converted back to signed magnitude form prior to output.

An illustration of this general arrangement is given in Fig. 8-11. The

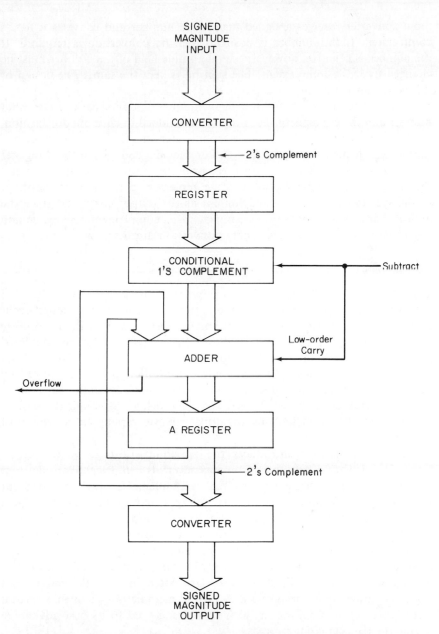

Figure 8-11. Typical Data Flow in Arithmetic Digital System

input converter receives a signed magnitude number and converts it to 2's complement. If the number is positive, then no conversion is required. If the number is negative, however, the magnitude field of the number must be changed to its 2's complement. The sign bit is kept the same. The design of this converter is the subject of Prob. 8-18.

The rest of the system follows the techniques described earlier. The input number and the contents of the A register are added in accumulator fashion. Subtraction is implemented by 2's complementing the input number via the 1's complement + 1 route. The 2's complement contents of the A register are converted into signed binary form prior to output. The logical circuit that performs the signed-binary to 2's-complement conversion at the input to the system can be duplicated for use as the output converter, since the logical characteristics of the signed binary to 2's complement conversion and the 2's complement to signed binary conversion are the same.

8.5 Summary

The binary representation for numerical quantities is quite common in digital systems. Chapter 8 shows how combinational and sequential design techniques can be used to implement arithmetic operations that involve binary operands. Binary operands can be added $[(A) + (B) \Rightarrow C]$ or accumulated $[(A) + (B) \Rightarrow A]$. Binary adders and accumulators can be built in both parallel and serial form. The combinational addition process is slowed by the gating delays that are inherent in the carry generation path. Carry lookahead techniques can be used to reduce these delays, although the cost of implementing the carry lookahead gating increases rapidly as the length of the binary operands increases.

Binary subtraction introduces the possibility of negative quantities, since $A - B$ gives a negative result whenever $A < B$. Negative quantities can be represented in signed binary, 2's complement, or 1's complement notation. Each form has its own advantages and disadvantages. Signed binary is useful for data input and output but is cumbersome in addition and subtraction. Two's complement is the best choice for arithmetic purposes, but the implementation of the 2's complementation process is somewhat involved. One's complementation is easy to implement (it results in bit-for-bit inversion), but since 1's complement arithmetic involves the end-around carry operation, it is rarely used. A typical arithmetic system may use all three representations, however, with each being used to its best advantage during the data-handling process.

8.6 Did You Learn?

1. What an accumulator does? How the two-step accumulator operates?

2. How signed quantities are represented in signed binary, 2's complement, and 1's complement notation?

3. How to convert one of these forms into another?

4. How 2's complementation can be implemented as 1's complementation $+ 1$?

5. Serial and parallel techniques for binary addition? 2's complementation? 1's complementation? Conditional addition-subtraction?

6. The advantages and disadvantages of each of the three signed notations?

7. How carry lookahead can be used to speed the binary addition process?

8. How and why all three signed notations are used during arithmetic operations?

8.7 References

[1] Flores, Ivan, *The Logic of Computer Arithmetic*. Englewood Cliffs, N.J.: Prentice-Hall, Inc., 1963.

[2] MacSorley, O. L., "High-speed Arithmetic in Binary Computers," *Proceedings of the IRE*, 49, No. 1 (January 1961), 67–91.

[3] Texas Instruments, *The Integrated Circuits Catalog for Design Engineers*, No. CC401. Texas Instruments, Inc., Dallas, 1971.

8.8 Problems

General Concepts

8-1. A ten-bit binary accumulator is cleared and then used to add the following sequence of numbers:

$$B_1 = 1000010101$$
$$B_2 = 0000100111$$
$$B_3 = 0001011000$$
$$B_4 = 0100010001$$

Show the contents of the A register following each cycle. What is the largest value that can be added to the final total without causing overflow?

8-2. A gated-carry accumulator is cleared and then used to add $B_1 = 1000101101$ and $B_2 = 0011110101$. Show the contents of the A register following each portion of the two addition cycles.

8-3. A ten-bit binary adder is built with RTL gates ($t_g = 12$ nsec).
   a. What is its worst-case settling time?

b. What is its settling time to add $A = 1011110101$ and $B = 0000010111$? (Determine the longest path through which a carry signal must be propagated.)

8-4. Write the complete carry lookahead equations for a ten-bit binary adder. Use the G_i and P_i notation given in Eq. (8-19). List the total loading requirements for each of the G_i and P_i signals. Calculate the total number of cost units used for AND gates and the total number of cost units used for OR gates. What is the maximum fan-in requirement and where does it occur?

8-5. Calculate $A - B$ for each of the following pairs of binary numbers. Give the answers in signed binary form.
  a. $A = 10110101$ $B = 10010111$
  b. $A = 10110101$ $B = 11010111$
  c. $A = 100101101$ $B = 111010011$
  d. $A = 01011111$ $B = 01100001$

8-6. Add each of the following pairs of signed binary numbers. Give the answers in signed binary form. Indicate overflow if it should occur.
  a. $A = 0,10110101$ $B = 0,11110101$
  b. $A = 0,10110101$ $B = 1,10110101$
  c. $A = 0,1011001010$ $B = 1,0110100111$
  d. $A = 1,11010111$ $B = 1,10100001$
  e. $A = 1,01011010$ $B = 0,10111000$

8-7. A digital processor uses n-bit binary operands. For values of n from 2 to 10, calculate the following quantities, assuming that signed binary notation is used. Repeat for 2's complement and 1's complement.
  a. The modulus
  b. The largest positive number
  c. The largest negative number
  d. The weight of the sign bit (if it is applicable)
  e. The coded representation for -1
  f. All possible representations for 0.

8-8. Express each of the following signed decimal quantities in signed binary, 2's complement, and 1's complement notation.
  a. $+25$ d. -16
  b. -70 e. -32
  c. -15 f. -165

8-9. Determine the signed decimal equivalent of each of the following binary operands assuming that the operands are encoded in the signed binary notation. Repeat, assuming 2's complement and 1's complement notation.
  a. $1,101101$ d. $1,00000$
  b. $1,000001$ e. $0,101101$
  c. $1,011010$ f. $1,1100000101$

8-10.   Convert each of the numbers given in Prob. 8-9 into 2's complement. Use both the subtract-from-zeros, the algorithmic, and the 1's complement + 1 methods. Convert the 2's complement numbers back into their signed decimal equivalents as a check.

8-11.   Add each of the following pairs of 2's complement numbers. Indicate overflow, should it occur. Give the decimal equivalents of each operation.
   a. $A = 1,10010101$ $B = 0,10010101$
   b. $A = 0,01110111$ $B = 1,11111111$
   c. $A = 1,00010110$ $B = 1,01101011$
   d. $A = 0,101101$ $B = 0,011011$
   e. $A = 1,11110000$ $B = 1,00011111$
   f. $A = 1,00000000$ $B = 1,10101010$

8-12.   Subtract B from A for each of the pairs of 2's complement numbers given in Prob. 8-11. Perform the subtraction by addition of the 2's complement of B. Indicate any overflows that occur. Give the decimal equivalents for each operation.

8-13.   Repeat Probs. 8-11 and 8-12 assuming that the pairs of operands are represented in 1's complement notation.

8-14.   The binary arithmetic system shown in Fig. 8-11 is used to add $+14_{10}$ and -7_{10}. Show the sequence of binary operands that is produced as the data passes through the system.

8-15.   Prove that the worst-case settling time for adding a pair of n-bit, 1's complement numbers is $2(n + 1)t_g$, as compared to $2nt_g$ for adding normal binary or 2's complement numbers. The small increase in the settling time results from the end-around carry process.

Functional Design Problems

*8-16.   Design a combinational device that will act as a conditional half-adder/subtractor. The block diagram for this device is shown in Fig. P8-16. When ADD is TRUE, the device acts as a half-adder. When ADD is FALSE,

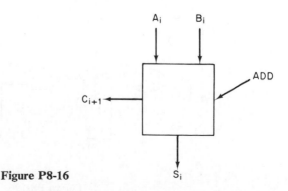

Figure P8-16

the device acts as a half-subtractor, with C_i representing the borrow output and S_i representing the difference output. Give an RTL implementation for the device.

*8-17. Design a conditional full adder/subtractor. The block diagram for this device is shown in Fig. P8-17. When ADD is TRUE, it acts as a full adder. When ADD is FALSE, it acts as a full subtractor. Give a TTL implementation for the device.

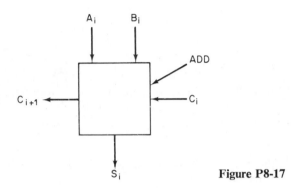

Figure P8-17

8-18. Two different combinational circuits for performing 2's complementation are given in Fig. 8-7. For n-bit operands, n ranging from 5 to 15, determine the cost of the A-O-N gating used to implement the two complementing circuits. Also compare the worst-case delay for each circuit. Assume that TTL logic ($t_g = 5$ nsec) is used.

*8-19. A space-iterative circuit using repeated OR and EXCLUSIVE-OR gates is shown in Fig. P8-19.

    a. Will this circuit perform 2's complementation?

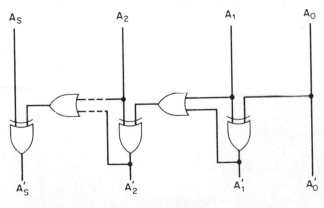

Figure P8-19

    b. Will it work for *all* 2's-complement operands? If not, for what operands does it fail? How can any failures be detected?

    c. Show the use of this circuit to complement all of the operands given in Prob. 8-10.

    d. Can the space-iterative arrangement be converted into a time-iterative equivalent? Show a complete logic circuit for this.

*8-20. Design the signed binary to 2's complement converter shown at the input to the arithmetic system of Fig. 8-11. Show an RTL logic diagram for an eight-bit converter. Verify that the same logical system will perform 2's complement to signed magnitude conversion with but one exception— the conversion of 1,0000000 from 2's complement into signed magnitude. Modify the output conversion logic so that this number, -128_{10}, will be converted into its closest signed binary equivalent, 1,1111111 or -127_{10}.

*8-21. Design a binary adder that accepts two four-bit binary numbers (double-rail) and generates their five-bit sum. Do not use the full-adder arrangement. Rather, consider the design as an eight-input, five-output combinational design problem. This adder will settle in only $2t_g$ after its inputs are applied. It acts as a hexadecimal half-adder.

Systems Design Problems

*8-22. Design a binary accumulation system that receives an eight-bit binary number and two control signals, *RESET* and *ADD*. The system is to add its binary input into a nine-bit accumulator each time the *ADD* signal is made TRUE. If overflow occurs, the system is to halt, and an *OVERFLOW* indicator is to be turned on. Once overflow occurs, the system must be *RESET* in order to resume the accumulation process. Design the complete system.

*8 23. A simple arithmetic system receives two six-bit, signed binary numbers, *A* and *B*. It is to add these numbers, generating their signed sum. The addition cycle is initiated by an *ADD* control signal. Thereafter, the system sequences through the steps required by the addition operation. An *OVERFLOW* signal is to be made TRUE at the conclusion of the addition cycle if arithmetic overflow occurs. Subtraction, when necessary, can be implemented by using a conditional adder/subtractor as per Probs. 8-16 and 8-17 or by complementing and adding as per Fig. 8-10. Give a complete design for this system. The required control sequence is described in Sec. 8.3 and illustrated in Exs. 8-4 and 8-5. Note that the system can be made to subtract by changing the sign of the *B* number.

*8-24. Design a six-bit accumulator that uses the gated-carry technique. The accumulation cycle is started by an *ADD* signal. Use one-shots to convert the *ADD* signal into the two control pulses needed by the accumulator. Disable the *ADD* signal when *OVERFLOW* occurs. Provide a *CLEAR* signal that will restart the accumulation process.

*8-25. Design a serial binary accumulation system in accordance with Fig. 8-6. Assume ten-bit operands. The accumulation cycle should be started by an *ADD* signal. The system controller should load the new operand into the *B* register, cycle through the serial addition sequence, and stop. Overflow should be detected and used to disable further addition until a *CLEAR* signal is received.

*8-26. Design an arithmetic system that receives two six-bit binary numbers, an *ADD/SUBTRACT* signal, a *START* signal, and a *CLEAR* signal. Assume that the input numbers are represented in 2's complement notation. Subtraction is to be implemented by adding the complement of the subtrahend. Overflow, either during complementation or addition, is to be detected and indicated.

*8-27. Design a serial 2's complementation system. The system is to receive a ten-bit number and serially convert it to its 2's complement. The controller should automatically stop after the proper number of clock periods. Overflow during the complementation process should be detected and indicated.

*8-28. Design a *serial* arithmetic system that will add or subtract a pair of 2's complement numbers according to the system specifications of Prob. 8-25. Use the results of Prob. 8-27 to perform 2's complementation, when required.

9 Decimal Arithmetic

9.0 Introduction

Humans, probably because of the number of toes or fingers that they normally possess, have adopted the decimal numbering system. From a logic design point of view, this is an unfortunate choice. Binary, octal, or even hexadecimal would be more appropriate.

Despite the efforts of devoted digital systems designers to correct this mistake [6], the decimal system continues to be used. As a result, it is often necessary for the digital systems designer to convert data from decimal (bad!) to binary (good!) and vice versa. An alternative is to implement all arithmetic operations in decimal form. This chapter considers these three related topics: binary-to-decimal conversion, decimal-to-binary conversion, and decimal arithmetic (addition and subtraction).

The data flow through a typical digital system that processes arithmetic data, receiving information from and returning information to a human user, is shown in Fig. 9-0. Decimal notation, in sign and magnitude form, is used for data input and output. These data are converted into signed binary upon input. The signed binary is converted to 2's complement as per Sec. 8.4.1. Output data go through a similar conversion process, being changed from 2's complement binary to signed binary and then to signed decimal.

9.1 Decimal-to-Binary Conversion

Decimal information can be represented in a variety of ways, as is discussed in Sec. 1.4. A common technique uses ten separate signals, one for each of

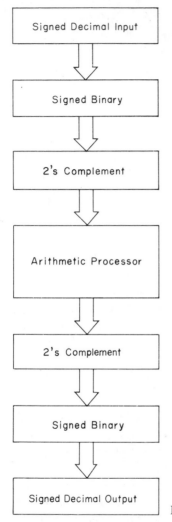

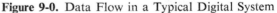

Figure 9-0. Data Flow in a Typical Digital System

the ten values that a given decimal digit can have. A more efficient approach uses four Boolean variables to represent the ten values from 0 to 9. Since four variables can actually assume $2^4 = 16$ values, there are always six redundancies in this type of encoding. The four-variable representation for a decimal digit is commonly called BCD, for *binary-coded decimal*, although Boolean-coded decimal is actually more appropriate.

The most common BCD code is 8421 BCD. This code represents each decimal value by its four-bit binary equivalent. This code is tabulated in Table 9-0. The close relationship between 8421 BCD and normal binary numbering makes the 8421 BCD code useful for arithmetic operations, as shown in Sec. 9.3. The 8421 BCD code does not use $m_{10}-m_{15}$; these code words become redundancies.

Table 9-0:   The 8421 BCD Code

Decimal Value	Code			
	A	*B*	*C*	*D*
0	0	0	0	0
1	0	0	0	1
2	0	0	1	0
3	0	0	1	1
4	0	1	0	0
5	0	1	0	1
6	0	1	1	0
7	0	1	1	1
8	1	0	0	0
9	1	0	0	1

Another common BCD code is the *excess-three code*, which is often abbreviated as *XS3 code*. This code represents each decimal value, say d, by the four-bit binary number that is three greater, m_{d+3}. Zero is encoded as $0011 = m_3$, for example. The complete XS3 code is given in Table 9-1. Like the 8421 BCD code, the XS3 code also has useful arithmetic properties. The use of the XS3 code in decimal arithmetic is also detailed in Sec. 9.3. Several other BCD codes and the various advantages and disadvantages that are associated with their use are discussed in Chapter 10.

Table 9-1:   The Excess-Three Code

Decimal Value	Code			
	A	*B*	*C*	*D*
0	0	0	1	1
1	0	1	0	0
2	0	1	0	1
3	0	1	1	0
4	0	1	1	1
5	1	0	0	0
6	1	0	0	1
7	1	0	1	0
8	1	0	1	1
9	1	1	0	0

The BCD representation uses at least four bits for each decade of the decimal number that is represented. Thus, a two-digit number requires at least 8 bits, a four-digit number requires 16 bits, etc.

EXAMPLE 9-0: Give the 8421 BCD code for 97, 103, and 436.

Answers: 10010111, 000100000011, and 010000110110.

EXAMPLE 9-1: Repeat 9-0, giving the XS3 code instead.

Answers: 11001010, 010000110110, and 011101101001.

EXAMPLE 9-2: Convert 011110000101 into decimal, assuming that 8421 code is being used. Repeat assuming XS3 code is being used.

Answers: In 8421 BCD the answer is 785. In XS3 the answer is 452.

If a non-BCD code such as one-out-of-ten is used, a conversion to BCD prior to arithmetic processing is generally advised for compactness and arithmetic simplicity.

EXAMPLE 9-3: A decimal digit is encoded by making one of ten Boolean signals, $A, B, C, \ldots, I, J$, be TRUE. A corresponds to 0, B to 1, C to 2, etc. Give the logical expressions for recoding this digit in 8421 BCD.

Solution: Let $WXYZ$ represent the 8421 BCD code. Table 9-0 shows that W is only TRUE for a decimal 8 and 9, so that

$$W = I + J$$

Similarly,

$$X = E + F + G + H$$
$$Y = C + D + G + H$$
$$Z = B + D + F + H + J$$

Once each digit of a decimal number has been encoded in some BCD code, a decimal-to-binary conversion can be made. The dibble-dabble algorithm (Sec. 1.5) is most useful. The decimal number is repeatedly halved, with the low-order remainders of 1 or 0 forming the binary equivalent.

The algorithm can be implemented in either a serial or a combinational form. The combinational form is considered first.

Figure 9-1 shows a combinational arrangement for applying the dibble-dabble algorithm to a one-digit decimal number. The logical subsystems shown in Fig. 9-1 serve as divide-by-two networks; i.e., they receive an encoded decimal value and generate the code for one-half of that value. Each logical circuit also generates a remainder signal that indicates whether or not the division by 2 left a remainder of 1. All odd decimal values produce a remainder; all even values do not. The remainders combine to form the binary equivalent, reading from bottom to top. The output from the last circuit *must* be 0, since the original value is 9 or less, and it has been halved four times.

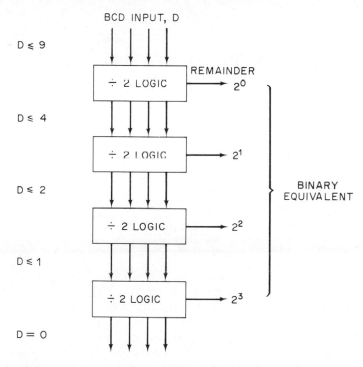

Figure 9-1. Combinational BCD-to-Binary Conversion—One Digit

The logical system given in Fig. 9-1 works fine for a one-digit decimal number. The presence of more than one digit forces only a small revision, however, as shown in Fig. 9-2. The high-order digit is treated exactly as was shown in Fig. 9-1. The lower digits must receive a signal from their higher-ordered neighbor, since each level of logic must serve to divide the *entire* decimal word (rather than each separate column) by 2.

Each decade that contains an odd value must signal that fact to its lower neighbor, since the division process will be affected. Halving a higher decade that contains an odd value adds a value of 5 to the next lower decade. Halving 18, for example, gives 09, since the odd value in the tens column adds 5 to the normal result obtained by halving the units column.

The function of each logical circuit can now be fully specified. Each box receives an encoded decimal value and an odd/even signal, say C_i, from the next higher decade. It converts the input code into the code for one-half of the decimal value, adding 5 if the next higher decade is odd. The circuit must also generate the odd/even signal, C_{i-1}, for the next lower decade. The odd/even signals that are generated in the units decade are used to form the binary number.

Table 9-2: The Combinational Halving Circuit

Input Value	Output Value		Odd/Even Output C_{i-1}
	$C_i = 0$	$C_i = 1$	
0	0	5	0
1	0	5	1
2	1	6	0
3	1	6	1
4	2	7	0
5	2	7	1
6	3	8	0
7	3	8	1
8	4	9	0
9	4	9	1

A general description of the halving circuit is given in Table 9-2. The coded representations for each decimal value must be inserted into the table. If a four-bit BCD code is used, the circuit is defined as a five-input/five-output combinational system. The odd/even input signal is designated as C_i, being TRUE if the decimal value of the next higher decade is odd. The odd/even output is named C_{i-1}. Note that the value of C_{i-1} is independent of C_i, so that ripple (similar to carry propagation) will not occur.

EXAMPLE 9-4: Design an 8421 BCD halving circuit. A general block diagram for the circuit is shown in Fig. E9-4a.

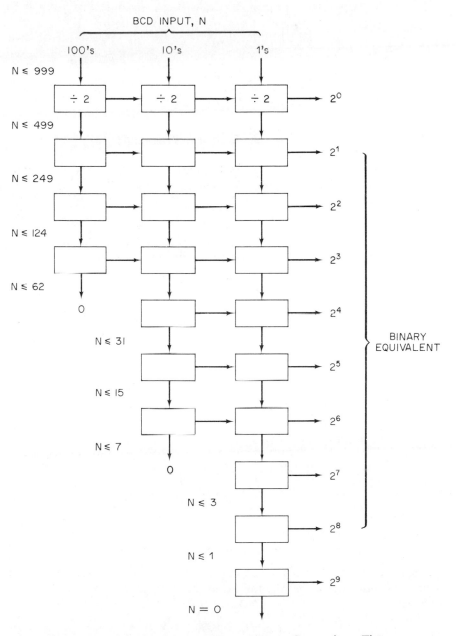

Figure 9-2. Combinational BCD-to-Binary Conversion—Three-Digit Input

461

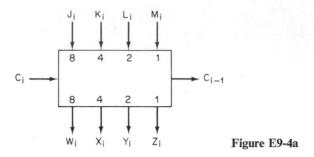

Figure E9-4a

Solution: Inserting the 8421 BCD codes into Table 9-2 and rearranging the table into normal truth table form gives (*i* subscripts omitted)

Inputs					Outputs				
J	K	L	M	C	W	X	Y	Z	C_{-1}
0	0	0	0	0	0	0	0	0	0
0	0	0	1	0	0	0	0	0	1
0	0	1	0	0	0	0	0	1	0
0	0	1	1	0	0	0	0	1	1
0	1	0	0	0	0	0	1	0	0
0	1	0	1	0	0	0	1	0	1
0	1	1	0	0	0	0	1	1	0
0	1	1	1	0	0	0	1	1	1
1	0	0	0	0	0	1	0	0	0
1	0	0	1	0	0	1	0	0	1
0	0	0	0	1	0	1	0	1	0
0	0	0	1	1	0	1	0	1	1
0	0	1	0	1	0	1	1	0	0
0	0	1	1	1	0	1	1	0	1
0	1	0	0	1	0	1	1	1	0
0	1	0	1	1	0	1	1	1	1
0	1	1	0	1	1	0	0	0	0
0	1	1	1	1	1	0	0	0	1
1	0	0	0	1	1	0	0	1	0
1	0	0	1	1	1	0	0	1	1

There are 20 specified input conditions and 12 redundancies. Minimizing the output functions gives the following results:

$$W = KLC + JC$$
$$X = \bar{J}\bar{L}C + \bar{J}\bar{K}C + J\bar{C}$$
$$Y = \bar{K}LC + K\bar{C} \mid K\bar{L}$$
$$Z = \bar{L}C + L\bar{C} = L \oplus C$$
$$C_{-1} = M$$

As an illustration, the use of this circuit to convert 39 into binary is shown in Fig. E9-4b.

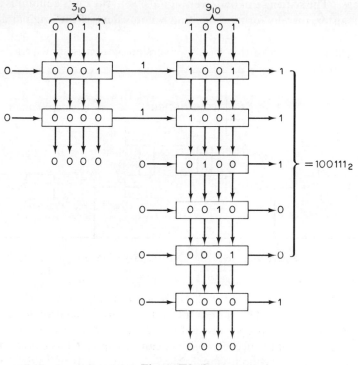

Figure E9-4b

The 8421 halving circuit that is designed in Ex. 9-4 is of sufficient use to merit its MSI manufacture and sale [5]. Similar devices for performing combinational decimal-to-binary conversion can be designed for use with any decimal code by inserting the encoded values into Table 9-2 and minimizing the output variables.

The succession of halving circuits must be continued in each column until it is assured that the value of that column has been reduced to 0. In Fig. 9-2 the largest value of the decimal input is 999. This reduces to 062 following the fourth divide-by-two operation, eliminating the need for further division into the hundreds column. Further division yields 007 as a maximum value after seven successive divisions, thereby allowing the tens column to be terminated. The units column requires ten divisions to assure a value of 0. This correlates with the need for ten bits to represent 999_{10} in binary.

The divide-by-two process can also be implemented serially (Refs. [2], [3], and [7]). Each digit can be stored in a register that, when given a clock pulse, changes state so as to contain the half of its value. Odd/even information must be propagated from decade to decade, just as in the combinational case. The sequence of odd/even signals from the units column is shifted into a register, thereby forming the binary equivalent of the original decimal value.

A general version of this serial conversion system is shown in Fig. 9-3.

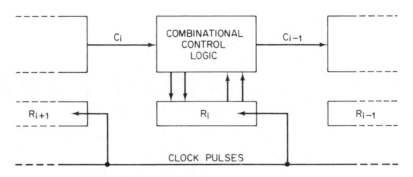

Figure 9-3. Serial BCD-to-Binary Conversion

Each register must be loaded with the code that represents the beginning value of that decade. Thereafter, the combinational control logic shown in Fig. 9-3 causes the proper divide-by-two operations to take place.

The behavior of the decade registers can be expressed as a present-state/next-state operation by modifying Table 9-2. This results in Table 9-3. Once the codes that represent each decimal value have been inserted in Table 9-3,

Table 9-3: The Serial Decimal-to-Binary Conversion Process

Present State	Next State		Odd/Even Output C_{i-1}
	$C_i = 0$	$C_i = 1$	
0	0	5	0
1	0	5	1
2	1	6	0
3	1	6	1
4	2	7	0
5	2	7	1
6	3	8	0
7	3	8	1
8	4	9	0
9	4	9	1

the present states and next states for each flip-flop in the decade register can be compared, yielding the transitional behavior that is required for implementing the division-by-two operation. Transition mapping can then be used to design the combinational control logic for the decade register.

EXAMPLE 9-5:   Design a decade counter/register for implementing the conversion from 8421 BCD to binary.

Solution: The general diagram for the counter is shown in Fig. E9-5a. W, X, Y, and Z are the flip-flops that store the present decimal value. The control logic combines the present values of W, X, Y, Z, and C, the odd/even signal from the next higher decade, in order to properly enable or disable the J and K inputs to the flip-flops. As a result, each clock pulse causes a divide-by-two operation, following Table 9-3. The control logic also generates C_+, the odd/even signal for the next lower decade. The C_+ signal from the units decade forms a serial version of the binary number, least-significant bit first.

Inserting the 8421 BCD code into Table 9-3 and rearranging the table into normal present-state/next-state form gives

Total Present State					Next State				Output
C	W	X	Y	Z	W_+	X_+	Y_+	Z_+	C_+
0	0	0	0	0	0	0	0	0	0
0	0	0	0	1	0	0	0	0	1
0	0	0	1	0	0	0	0	1	0
0	0	0	1	1	0	0	0	1	1
0	0	1	0	0	0	0	1	0	0
0	0	1	0	1	0	0	1	0	1
0	0	1	1	0	0	0	1	1	0
0	0	1	1	1	0	0	1	1	1
0	1	0	0	0	0	1	0	0	0
0	1	0	0	1	0	1	0	0	1
1	0	0	0	0	0	1	0	1	0
1	0	0	0	1	0	1	0	1	1
1	0	0	1	0	0	1	1	0	0
1	0	0	1	1	0	1	1	0	1
1	0	1	0	0	0	1	1	1	0
1	0	1	0	1	0	1	1	1	1
1	0	1	1	0	1	0	0	0	0
1	0	1	1	1	1	0	0	0	1
1	1	0	0	0	1	0	0	1	0
1	1	0	0	1	1	0	0	1	1

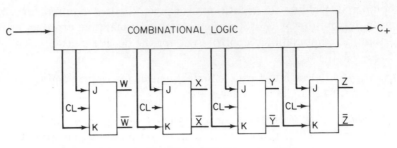

Figure E9-5a

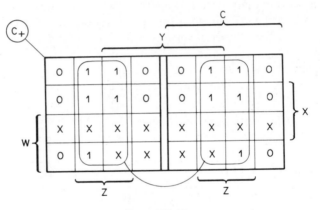

Figure E9-5b

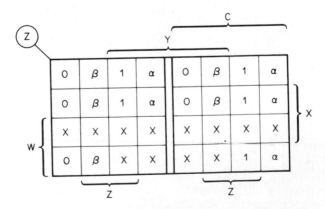

Figure E9-5c

The C_+ function is K-mapped and grouped as shown in Fig. E9-5b. The resultant expression is

$$C_+ = Z$$

Comparison of the present-state and next-state columns for each flip-flop gives their transitional behavior. This behavior is tabulated below:

Total Present State					Next State				Behavior			
C	W	X	Y	Z	W_+	X_+	Y_+	Z_+	W	X	Y	Z
0	0	0	0	0	0	0	0	0	0	0	0	0
0	0	0	0	1	0	0	0	0	0	0	0	β
0	0	0	1	0	0	0	0	1	0	0	β	α
0	0	0	1	1	0	0	0	1	0	0	β	1
0	0	1	0	0	0	0	1	0	0	β	α	0
0	0	1	0	1	0	0	1	0	0	β	α	β
0	0	1	1	0	0	0	1	1	0	β	1	α
0	0	1	1	1	0	0	1	1	0	β	1	1
0	1	0	0	0	0	1	0	0	β	α	0	0
0	1	0	0	1	0	1	0	0	β	α	0	β
1	0	0	0	0	0	1	0	1	0	α	0	α
1	0	0	0	1	0	1	0	1	0	α	0	1
1	0	0	1	0	0	1	1	0	0	α	1	0
1	0	0	1	1	0	1	1	0	0	α	1	β
1	0	1	0	0	0	1	1	1	0	1	α	α
1	0	1	0	1	0	1	1	1	0	1	α	1
1	0	1	1	0	1	0	0	0	α	β	β	0
1	0	1	1	1	1	0	0	0	α	β	β	β
1	1	0	0	0	1	0	0	1	1	0	0	α
1	1	0	0	1	1	0	0	1	1	0	0	1

The transition map for the Z flip-flop is given in Fig. E9-5c. Grouping the terms on this map gives

$$J_Z = (\bar{C}Y + C\bar{Y})\cdot CL \qquad K_Z = (\bar{C}\bar{Y} + CY)\cdot CL$$
$$= (C \oplus Y)\cdot CL \qquad\qquad = (C \odot Y)\cdot CL$$
$$\qquad\qquad\qquad = \bar{J}_Z\cdot CL$$

The other control equations are

$$J_W = CXY\cdot CL \qquad K_W = \bar{C}\cdot CL$$
$$J_X = (C \oplus W)\cdot CL \qquad K_X = (\bar{C} + Y)\cdot CL$$
$$J_Y = X\cdot CL \qquad\qquad K_Y = (C \odot X)\cdot CL$$

An example of the serial conversion process follows. The beginning decimal value is 357.

Clock Periods	Hundreds				Tens				Units				Binary Number
	W	X	Y	Z	W	X	Y	Z	W	X	Y	Z	
0	0	0	1	1	0	1	0	1	0	1	1	1	0
1	0	0	0	1	0	1	1	1	1	0	0	0	1
2	0	0	0	0	1	0	0	0	1	0	0	1	01
3	0	0	0	0	0	1	0	0	0	1	0	0	101
4	0	0	0	0	0	0	1	0	0	0	1	0	0101
5	0	0	0	0	0	0	0	1	0	0	0	1	00101
6	0	0	0	0	0	0	0	0	0	1	0	1	100101
7	0	0	0	0	0	0	0	0	0	0	1	0	1100101
8	0	0	0	0	0	0	0	0	0	0	0	1	01100101
9	0	0	0	0	0	0	0	0	0	0	0	0	101100101

The binary result is 101100101_2, which is the correct equivalent of 357_{10}.

9.2 Binary-to-Decimal Conversion

Binary-to-decimal conversion is very closely related to the decimal-to-binary conversion processes described in Sec. 9.1. The dabble-dibble algorithm is a commonly used technique, with its double and add sequence being implemented in either combinational or serial form. In some applications, particularly when small binary numbers (five or six bits or less) are to be converted, other techniques are used.

The most straightforward binary-to-decimal conversion technique is direct combinational conversion. The block diagram for this type of converter is shown in Fig. 9-4. The binary number acts as the input to the com-

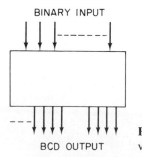

BINARY INPUT

BCD OUTPUT

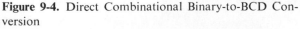

Figure 9-4. Direct Combinational Binary-to-BCD Conversion

binational circuit, and the proper decimal value is generated at its output. The design of this type of converter follows the techniques given in Chapter 4. It is usually a fully specified problem (no redundant input conditions), since a decimal value should be found for all possible binary inputs.

The design of this type of converter becomes quite difficult as the length of the binary number increases. Computer-aided minimization is advisable for binary numbers of six or more bits, with the design of a converter for use with typical computer-generated binary numbers (12–32 bits or more) being all but impossible.

Larger binary numbers require an iterative conversion process that breaks the complete conversion process into a set of logically manageable sub-operations; the dabble-dibble algorithm allows this. The hardware arrangement for implementing the algorithm is similar to that used by the dibble-dabble algorithm in Sec. 9.1.

A combinational binary-to-BCD converter for use with a six-bit binary number is shown in Fig. 9-5. The most-significant bit is applied to the upper

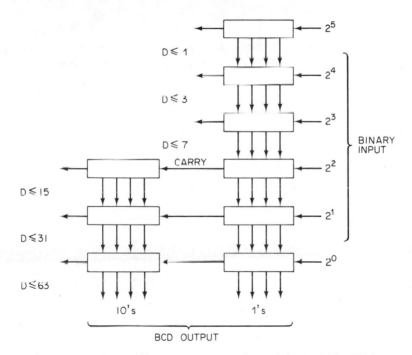

Figure 9-5. Combinational Implementation of the Dabble-dibble Conversion Algorithm

conversion circuit. Each conversion circuit serves to double its decimal input and then to add 1 to the doubled total if its "carry" input is TRUE. This doubling process is performed in a modulo-10 sense, in that whenever a value greater than 9 is generated by the double-add process, a carry of 10 is passed to the next higher decade (where it adds one unit), and the output value from the converter is reduced by 10. This process is described in general terms by Table 9-4. The bits of the binary input act as the carry signals to

Table 9-4:   Combinational Binary-to-Decimal Conversion

Decimal Input	Decimal Output		Carry Output C_+
	$C = 0$	$C = 1$	
0	0	1	0
1	2	3	0
2	4	5	0
3	6	7	0
4	8	9	0
5	0	1	1
6	2	3	1
7	4	5	1
8	6	7	1
9	8	9	1

the units decade, as shown in Fig. 9-5. The combinational conversion process can be expanded to handle binary numbers of any length. A new column of converters must be started whenever the doubled decimal number becomes too large to be contained in the current number of columns. The hardware requirements do increase quite rapidly, however, as shown in Table 9-5.

Table 9-5:   Hardware Requirements for Combinational Binary-to-Decimal Conversion

Number of Bits	Number of Conversion Circuits Required
1	1
2	2
3	3
4	5
5	7
6	9
7	12
8	15
9	18
10	22
15	44
20	74
30	156

An example of the design and use of the combinational conversion technique follows.

EXAMPLE 9-6:   Design a combinational circuit for converting from binary into 8421 BCD. Illustrate its use to convert 101110 into decimal.

Solution: The basic conversion circuit is shown in Fig. E9-6a. Its general behavior is described by Table 9-4. Inserting the 8421 BCD code into that table gives

BCD Input				BCD Output								Carry Output
				$C = 0$				$C = 1$				
W	X	Y	Z	W_+	X_+	Y_+	Z_+	W_+	X_+	Y_+	Z_+	C_+
0	0	0	0	0	0	0	0	0	0	0	1	0
0	0	0	1	0	0	1	0	0	0	1	1	0
0	0	1	0	0	1	0	0	0	1	0	1	0
0	0	1	1	0	1	1	0	0	1	1	1	0
0	1	0	0	1	0	0	0	1	0	0	1	0
0	1	0	1	0	0	0	0	0	0	0	1	1
0	1	1	0	0	0	1	0	0	0	1	1	1
0	1	1	1	0	1	0	0	0	1	0	1	1
1	0	0	0	0	1	1	0	0	1	1	1	1
1	0	0	1	1	0	0	0	1	0	0	1	1

Solving for the output signals gives

$$W_+ = X\bar{Y}\bar{Z} + WZ$$
$$X_1 = W\bar{Z} + YZ + \bar{X}Y$$
$$Y_+ = \bar{W}\bar{X}Z + XY\bar{Z} + W\bar{Z}$$
$$Z_+ = C$$
$$C_+ = W + XY + XZ$$

The complete arrangement for converting a six-bit binary number into a two-digit decimal number is shown in Fig. F9-6b. The conversion of 101110_2 into its 8421 BCD equivalent, $01000110 = 46_{10}$, is superimposed on the diagram. Note that the characteristics of the 8421 BCD code permit some simplification of the converter relative to the general arrangement given in Fig. 9-5.

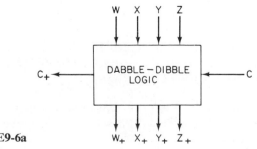

Figure E9-6a

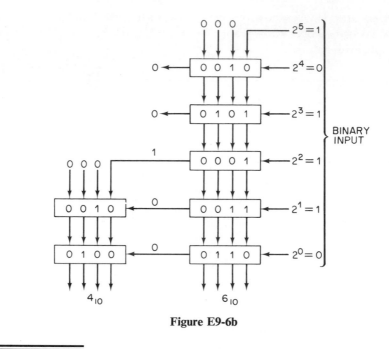

Figure E9-6b

As was the case in Ex. 9-4, the binary-to-decimal 8421 BCD conversion process is common enough to merit the MSI manufacture of the iterative conversion circuit designed in Ex. 9-6. One version of this circuit is described in Ref. [5].

A serial binary-to-decimal conversion is also possible. It closely follows the serial decimal-to-binary process described in Sec. 9.1. The present-state/next-state behavior of the conversion counters is described in Table 9-6. This table is derived from Table 9-4.

The general arrangement of the serial conversion circuitry is shown in Fig. 9-6. The binary number is placed in a shift register, MSB first. As it is shifted, each successive bit acts as the low-order carry into the units decade. The successive conversion counters double their values, adding one unit if there is a carry from the next-lower decade. Modulo-10 operation is observed.

One counter is needed for each decade. Converting an n-bit binary number requires n clock periods. The design and use of a typical sequential converter is described in Ex. 9-7.

EXAMPLE 9-7: Design a sequential converter for use in converting from binary to 8421 BCD. Illustrate its use in converting 10110101_2 into decimal.

Table 9-6: The Serial Binary-to-Decimal Conversion Process

Present Decimal Value	Next Decimal Value		Carry Output C_+
	$C = 0$	$C = 1$	
0	0	1	0
1	2	3	0
2	4	5	0
3	6	7	0
4	8	9	0
5	0	1	1
6	2	3	1
7	4	5	1
8	6	7	1
9	8	9	1

Solution: The basic counter/converter can be defined as shown in Fig. E9-7. Inserting the 8421 BCD code into Table 9-6 gives

Total Present State					Next State				Carry Output C_i	Behavior			
C	W	X	Y	Z	W_+	X_+	Y_+	Z_+		W	X	Y	Z
0	0	0	0	0	0	0	0	0	0	0	0	0	0
0	0	0	0	1	0	0	1	0	0	0	0	α	β
0	0	0	1	0	0	1	0	0	0	0	α	β	0
0	0	0	1	1	0	1	1	0	0	0	α	1	β
0	0	1	0	0	1	0	0	0	0	α	β	0	0
0	0	1	0	1	0	0	0	0	1	0	β	0	β
0	0	1	1	0	0	0	1	0	1	0	β	1	0
0	0	1	1	1	0	1	0	0	1	0	1	β	β
0	1	0	0	0	0	1	1	0	1	β	α	α	0
0	1	0	0	1	1	0	0	0	1	1	0	0	β
1	0	0	0	0	0	0	0	1	0	0	0	0	α
1	0	0	0	1	0	0	1	1	0	0	0	α	1
1	0	0	1	0	0	1	0	1	0	0	α	β	α
1	0	0	1	1	0	1	1	1	0	0	α	1	1
1	0	1	0	0	1	0	0	1	0	α	β	0	α
1	0	1	0	1	0	0	0	1	1	0	β	0	1
1	0	1	1	0	0	0	1	1	1	0	β	1	α
1	0	1	1	1	0	1	0	1	1	0	1	β	1
1	1	0	0	0	0	1	1	0	1	β	α	α	0
1	1	0	0	1	1	0	0	1	1	1	0	0	1

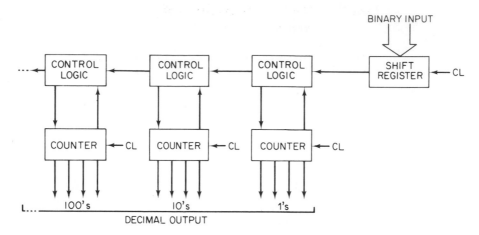

Figure 9-6. Serial Binary-to-Decimal Conversion

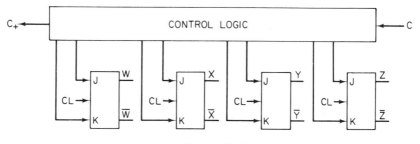

Figure E9-7

The input conditions that do not appear in the above table can be used as redundancies. Carrying out the transition mapping process gives

$$J_W = X\bar{Y}\bar{Z}\cdot CL \qquad\qquad K_W = \bar{Z}\cdot CL$$
$$J_X = (Y + W\bar{Z})\cdot CL \qquad K_X = (\bar{Y} + \bar{Z})\cdot CL$$
$$J_Y = (\bar{W}\bar{X}Z + \underline{W\bar{Z}})\cdot CL \qquad K_Y = (X \odot Z)\cdot CL$$
$$J_Z = C\cdot CL \qquad\qquad K_Z = \bar{C}\cdot CL$$

The combinational logic for the carry signal is

$$C_+ = W + XZ + XY$$

The conversion of an eight-bit binary number can generate a decimal value as large as 127. Thus, three decade counters are needed. The conversion of 10110101_2 is illustrated in the following table. It takes eight clock pulses. The result is 181_{10}, which is the correct conversion.

Clock Periods	Hundreds				Tens				Units				Binary Number
	8	4	2	1	8	4	2	1	8	4	2	1	
0	0	0	0	0	0	0	0	0	0	0	0	0	10110101
1	0	0	0	0	0	0	0	0	0	0	0	1	0110101
2	0	0	0	0	0	0	0	0	0	0	1	0	110101
3	0	0	0	0	0	0	0	0	0	1	0	1	10101
4	0	0	0	0	0	0	0	1	0	0	0	1	0101
5	0	0	0	0	0	0	1	0	0	0	1	0	101
6	0	0	0	0	0	1	0	0	0	1	0	1	01
7	0	0	0	0	1	0	0	1	0	0	0	0	1
8	0	0	0	1	1	0	0	0	0	0	0	1	

Both the serial and the combinational techniques for the decimal-to-binary conversion can be used with any decimal code. Tables 9-2 and 9-4 are the "keys" to the basic conversion process. Their use with codes other then 8421 BCD is considered in the problems at the end of this chapter. For further information on this subject see Refs. [2], [3], [4], [6], and [8].

9.3   Decimal Arithmetic (Refs. [1] and [8])

The conversion techniques that are given in Secs. 9.1 and 9.2 can be used to convert decimal data to and from binary notation. These processes allow decimal input and output along with binary arithmetic. An alternative is the continuance of the decimal data representation throughout the input, arithmetic, and output sequence. This is called *fully decimal arithmetic*. Only decimal addition and subtraction are discussed herein

Decimal addition and subtraction are introduced in early schooling. The so-called "arithmetic tables" that define these operations are learned. "Six plus five is one and a carry of one," is an example, although the terminology has changed somewhat since the "new math" concepts have been introduced. Both addition and subtraction are carried out on a column-by-column basis, just as binary operations proceed from bit to bit.

The lowest-ordered column of a pair of decimal numbers can be added by a decimal half-adder. Such a device is shown in Fig. 9-7. The input numbers must be encoded in some BCD representation, making the total number of Boolean inputs at least 8. The design of such a device is at least an eight-input, five-output combinational design problem. The outputs include the BCD code for the sum digit and a single Boolean variable that defines the

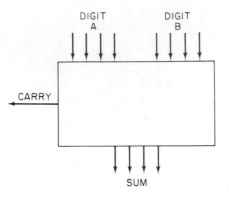

Figure 9-7. The Decimal Half-adder

carry to the next higher decade. The design of such a device is quite difficult, but computer assistance makes it possible.

EXAMPLE 9-8: Design a decimal half-adder for use with the 8421 BCD code.

Solution: The 8421 BCD half-adder can be symbolized as shown in Fig. E9-8. There are 256 input combinations, of which only 100 are fully

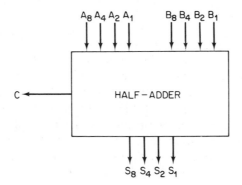

Figure E9-8

specified. The other 156 can be used as redundancies. Using the MINTE program of Sec. 4.6.5 to reduce the output functions gives

$$S_1 = A_1 \oplus B_1$$
$$S_2 = A_2 B_2 \bar{B}_1 + \bar{A}_2 \bar{A}_1 B_2 + \bar{A}_2 B_1 \bar{B}_2 B_1 + A_2 \bar{B}_2 \bar{B}_1 + A_2 \bar{A}_1 \bar{B}_2$$
$$S_4 = \bar{A}_4 B_4 \bar{B}_2 \bar{B}_1 + \bar{A}_4 \bar{A}_1 B_4 \bar{B}_2 + \bar{A}_4 \bar{A}_2 B_4 \bar{B}_1 + \bar{A}_4 \bar{A}_2 B_4 \bar{B}_2 + \bar{A}_4 \bar{A}_2 \bar{A}_1 B_4$$
$$\quad + \bar{A}_4 A_1 \bar{B}_4 B_2 B_1 + \bar{A}_4 A_2 \bar{B}_4 B_2 + \bar{A}_4 A_2 A_1 \bar{B}_4 B_1 + A_4 \bar{B}_4 \bar{B}_2 \bar{B}_1$$
$$\quad + A_4 \bar{A}_1 \bar{B}_4 \bar{B}_2 + A_4 \bar{A}_2 \bar{B}_4 \bar{B}_1 + A_4 \bar{A}_2 \bar{B}_4 \bar{B}_2 + A_4 \bar{A}_2 \bar{A}_1 \bar{B}_4$$

$$S_8 = \bar{A}_8 B_8 \bar{B}_1 + \bar{A}_8 \bar{A}_1 B_8 + \bar{A}_8 \bar{A}_2 B_8 + \bar{A}_8 \bar{A}_4 B_8 + \bar{A}_8 A_1 B_4 B_2 B_1 + A_2 B_4 B_2$$
$$+ A_2 A_1 B_4 B_1 + A_4 B_4 + A_4 A_1 B_2 B_4 + A_4 A_2 B_2 + A_4 A_2 A_1 \bar{B}_8 B_1$$
$$+ A_8 \bar{B}_8 B_1 + A_8 \bar{B}_8 \bar{B}_2 + A_8 \bar{B}_8 \bar{B}_4 + A_8 \bar{A}_1 \bar{B}_8$$
$$C = A_4 A_2 A_1 B_8 B_1 + A_8 B_8 + A_8 A_1 B_4 B_2 B_1$$

The decimal half-adder is fine for the least-significant decade, but a decimal *full* adder is required for all other decades. Its design involves at least nine input variables. An example, using the 8421 BCD code once again, follows.

EXAMPLE 9-9:   Design an 8421 BCD full adder.

Solution: The full adder is defined as shown in Fig. E9-9.

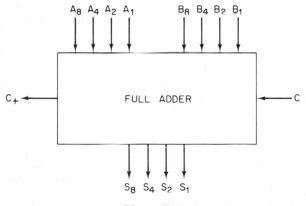

Figure E9-9

There are 512 input combinations, of which only 200 are fully specified. Computer-aided minimization gives

$$S_1 = \bar{A}_1 \bar{B}_1 C + \bar{A}_1 B_1 \bar{C} + A_1 \bar{B}_1 \bar{C} + A_1 B_1 C$$
$$S_2 = \bar{A}_2 \bar{B}_2 B_1 C + \bar{A}_2 B_2 B_1 \bar{C} + \bar{A}_2 \bar{A}_1 B_2 \bar{C} + \bar{A}_2 \bar{A}_1 B_2 \bar{B}_1 + \bar{A}_2 A_1 \bar{B}_2 B_1$$
$$+ A_2 \bar{B}_2 \bar{B}_1 \bar{C} + A_2 \bar{A}_1 \bar{B}_2 \bar{C} + A_2 \bar{A}_1 \bar{B}_2 \bar{B}_1$$
$$S_4 = \bar{A}_4 \bar{B}_4 B_2 B_1 C + \bar{A}_4 B_4 \bar{B}_2 \bar{B}_1 \bar{C} + \bar{A}_4 \bar{A}_1 B_4 \bar{B}_2 \bar{C} + \bar{A}_4 \bar{A}_1 B_4 \bar{B}_2 \bar{B}_1$$
$$+ \bar{A}_4 \bar{A}_2 B_4 \bar{B}_1 \bar{C} + \bar{A}_4 \bar{A}_1 B_4 \bar{B}_2 + \bar{A}_4 \bar{A}_2 \bar{A}_1 B_4 \bar{C} + \bar{A}_4 \bar{A}_2 \bar{A}_1 B_4 \bar{B}_1$$
$$+ \bar{A}_4 A_1 \bar{B}_4 B_2 C + \bar{A}_4 A_1 \bar{B}_4 B_2 B_1 + \bar{A}_4 A_2 \bar{B}_4 B_1 C + \bar{A}_4 A_2 \bar{B}_4 B_2$$
$$+ \bar{A}_4 A_2 A_1 \bar{B}_4 C + \bar{A}_4 A_2 A_1 \bar{B}_4 B_1 + A_4 \bar{B}_4 \bar{B}_2 \bar{B}_1 \bar{C} + A_4 \bar{A}_1 \bar{B}_4 \bar{B}_2 \bar{C}$$
$$+ A_4 \bar{A}_1 \bar{B}_4 \bar{B}_2 \bar{B}_1 + A_4 \bar{A}_2 \bar{B}_4 \bar{B}_1 \bar{C} + A_4 \bar{A}_2 \bar{B}_4 \bar{B}_2 + A_4 \bar{A}_2 \bar{A}_1 \bar{B}_4 \bar{C}$$
$$+ A_4 \bar{A}_2 \bar{A}_1 \bar{B}_4 \bar{C} + A_4 \bar{A}_2 \bar{A}_1 \bar{B}_1 \bar{C}$$

$$S_8 = \bar{A}_8 B_4 B_2 B_1 C + \bar{A}_8 B_8 \bar{B}_1 \bar{C} + \bar{A}_8 \bar{A}_1 B_8 C + \bar{A}_8 \bar{A}_1 B_8 \bar{B}_1 + \bar{A}_8 \bar{A}_2 A_8$$
$$+ \bar{A}_8 \bar{A}_4 B_8 + \bar{A}_8 A_1 B_4 B_2 C + \bar{A}_8 A_1 B_4 B_2 B_1 + A_2 B_4 B_1 C + A_2 B_4 B_2$$
$$+ A_2 A_1 B_4 C + A_4 A_1 B_4 B_1 + A_4 B_2 B_1 C + A_4 B_4 + A_4 A_1 B_2 C$$
$$+ A_4 A_1 B_2 B_1 + A_4 A_2 \bar{B}_8 B_1 C + A_4 A_2 B_2 + A_4 A_2 A_1 \bar{B}_8 C$$
$$+ A_4 A_2 A_1 \bar{B}_8 B_1 + A_8 \bar{B}_8 \bar{B}_1 \bar{C} + A_8 \bar{B}_8 \bar{B}_2 + A_8 \bar{B}_8 B_4 + A_8 \bar{A}_1 \bar{B}_8 \bar{C}$$
$$+ A_8 \bar{A}_1 \bar{B}_8 \bar{B}_1$$
$$C_+ = A_4 A_2 B_8 B_1 C + A_4 A_2 A_1 B_8 C + A_4 A_2 A_1 B_8 B_1 + A_8 B_4 B_2 B_1 C + A_8 B_8$$
$$+ A_8 A_1 B_4 B_2 C + A_8 A_1 B_4 B_2 B_1$$

Combinational half-adders and full adders can be designed for use with any BCD code, although the design work is quite tedious if computer assistance is not available. Because of this difficulty, most decimal arithmetic has been performed by using one of two BCD codes that are closely related to natural binary numbers, the 8421 BCD and XS3 codes. The decimal numbers are added just as if they are normal binary numbers and corrections are made whenever the "binary" outputs disagree with the proper decimal results.

9.3.1. 8421 BCD Addition.

Two 8421 BCD digits can also be added by using four binary full adders, as shown in Fig. 9-8. If the sum of the two

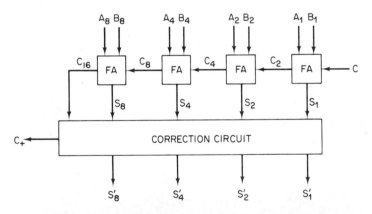

Figure 9-8. The "Binary" Addition of Two 8421 BCD Digits

digits and the incoming carry is 9 or less, then the output from the full adders is correct. Sums of 10 or more will be incorrect, however, since the four full adders act in modulo-16, rather than modulo-10, form. Thus, whenever a binary value of 10 or more appears at the sum outputs, a correction must be made.

The most common correction technique acts to add a binary 6 to the sum whenever the sum is 10 or more. This converts the decimal addition process into modulo-16 form, making the output from the full adders have the correct decimal value. The correction circuit receives the four sum bits and the carry out of the high-order full adder. It generates the correct value (by adding 6 to the five-bit sum) and generates the carry signal that goes to the next higher stage.

EXAMPLE 9-10:   Design the 8421 BCD correction circuit shown in Fig. 9-8.

Solution: The operation of the correction circuit is defined by the following table:

	Sum Input					Corrected Output				Carry
C_{16}	S_8	S_4	S_2	S_1		S_8'	S_4'	S_2'	S_1'	C_+
0	0	0	0	0	(0)	0	0	0	0	0
0	0	0	0	1	(1)	0	0	0	1	0
0	0	0	1	0	(2)	0	0	1	0	0
0	0	0	1	1	(3)	0	0	1	1	0
0	0	1	0	0	(4)	0	1	0	0	0
0	0	1	0	1	(5)	0	1	0	1	0
0	0	1	1	0	(6)	0	1	1	0	0
0	0	1	1	1	(7)	0	1	1	1	0
0	1	0	0	0	(8)	1	0	0	0	0
0	1	0	0	1	(9)	1	0	0	1	0
0	1	0	1	0	(10)	0	0	0	0	1
0	1	0	1	1	(11)	0	0	0	1	1
0	1	1	0	0	(12)	0	0	1	0	1
0	1	1	0	1	(13)	0	0	1	1	1
0	1	1	1	0	(14)	0	1	0	0	1
0	1	1	1	1	(15)	0	1	0	1	1
1	0	0	0	0	(16)	0	1	1	0	1
1	0	0	0	1	(17)	0	1	1	1	1
1	0	0	1	0	(18)	1	0	0	0	1
1	0	0	1	1	(19)	1	0	0	1	1

The sum can never exceed 19, since it results from the addition of only two decimal digits and a one-unit carry.

Solving for the minimized output functions gives

$$S_8' = S_8 \bar{S}_4 \bar{S}_2 + C_{16} S_2$$
$$S_4' = \bar{S}_8 S_4 + S_4 S_2 + C_{16} \bar{S}_2$$
$$S_2' = \bar{C}_{16} \bar{S}_8 S_2 + S_8 S_4 \bar{S}_2 + C_{16} \bar{S}_2$$
$$S_1' = S_1$$
$$C_+ = S_8 S_2 + S_8 S_4 + C_{16}$$

There are other approaches to the implementation of the correction circuit [8], but the combinational approach given above is a most direct method for its construction.

As an illustration of the technique, consider the addition of $A = 00110111 = 37_{10}$ and $B = 01101000 = 68_{10}$. The uncorrected summation of the numbers gives

$$
\begin{aligned}
A &= \quad 0011 \quad\ \ 0111 \\
B &= \quad 0110 \quad\ \ 1000 \\
\hline
S &= 01001 \quad 01111
\end{aligned}
$$

Correction of the units decade gives 0101 and a carry to the tens decade. Correcting the tens decade and including the carry from the ones decade gives 0000 and a carry to the hundreds decade. The net result is

$$
\begin{aligned}
A &= \quad 0011 \quad\ \ 0111 \\
B &= \quad 0110 \quad\ \ 1000 \\
\hline
S &= 01001 \quad 01111 \quad \longleftarrow \text{uncorrected} \\
S' &= 10000 \quad\ \ 0101 = 105_{10} \leftarrow \text{corrected}
\end{aligned}
$$

9.3.2. Excess-Three Addition.

The second BCD code that is commonly used in decimal arithmetic operations is the excess-three code. Excess-three digits can be added by using a combinational full adder as was described earlier in this section and illustrated, for the 8421 BCD code, in Ex. 9-9. Two excess-three digits can also be added in normal binary fashion, as shown for the 8421 code in Fig. 9-8, with the resulting sum digit having an "excess" of 6. For example, if 7 and 5 are added by using XS3 code, the binary result is 18_{10}, since 7 is encoded as 10 and 5 is encoded as 8. The excess-six nature of the sum assures that a high-order carry will be generated whenever the true values of the two digits add to 10 or more.

The modulo-16 nature of the four-bit binary adder removes the excess of 6 whenever a carry is made. As a result, the sum must be increased by 3 in order to maintain the excess-three code at the output. If no carry is made, the sum is still excess-six and it must be *reduced* by 3.

EXAMPLE 9-11:   Show the addition of 357 and 468 in XS3 code.

   Solution:

$$
\left.\begin{aligned}
357 &= 0110 \quad 1000 \quad 1010 \\
468 &= 0111 \quad 1001 \quad 1011
\end{aligned}\right\} \quad \text{in XS3}
$$

Adding the units decade gives

$$
\begin{array}{r}
1010 \\
+\ 1011 \\
\hline
1\ \ 0101
\end{array}
$$

The carry is sent to the tens decade. The uncorrected sum is 5_{10}, which results from $8 + 7 = 15$. The modulo-16 carry removes the excess of 6. To return the units digit to XS3 code, the 0101 value must be increased by 3, giving 1000.

The addition in the tens decade gives

$$
\begin{array}{r}
1000 \\
1001 \\
+ \quad\quad 1 \leftarrow \text{carry from units decade} \\
\hline
1\ 0010
\end{array}
$$

This corresponds to $5 + 6 + 1 = 12$. Increasing the sum by 3 gives 0101. The addition in the hundreds decade gives

$$
\begin{array}{r}
0110 \\
0111 \\
+ \quad\quad 1 \\
\hline
0\ 1110
\end{array}
$$

Since no carry occurs, the sum must be reduced by 3, giving 1011 as a result. The complete process is

$$
\begin{array}{ccc}
\overset{1}{0110} & \overset{1}{1000} & 1010 \\
+\ 0111 & +\ 1001 & +\ 1011 \\
\hline
1110 & 1\ 0010 & 1\ 0101 \\
-\ 0011 & +\ 0011 & +\ 0011 \leftarrow \text{corrections} \\
\hline
1011 & 0101 & 1000
\end{array}
$$

The final result is the XS3 code for 825_{10}.

The correction of the sum can be accomplished by a combinational converter that operates much as the 8421 BCD converter that was described in Fig. 9-8 and Ex. 9-10. The logic for the XS3 converter can be simplified, in that the carry out of the high-order full adder can be used directly as the carry to the next higher decade. A three-digit arrangement for XS3 addition is shown in Fig. 9-9. The design of the correction circuit is presented in Ex. 9-12.

EXAMPLE 9-12:   Design the XS3 combinational correction circuit that is shown in Fig. 9-9.

Solution: The circuit adds 3 to the sum whenever there is a high-order carry and subtracts 3 otherwise. The input operands to the decimal adder range from 3_{10} to 12_{10} (0–9 with the excess of 3) and may include a carry.

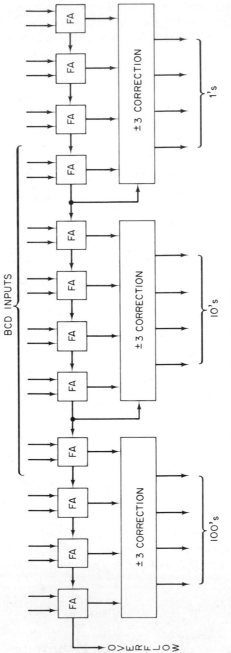

Figure 9-9. A Three-digit XS3 Adder

Thus, the possible values of the uncorrected sum range from 6_{10} to 25_{10}. The operation of the correction circuit is tabulated below. The inputs to the circuit are C, S_8, S_4, S_2, and S_1. The outputs are the bits of the corrected sum and are labeled S_8', S_4', S_2', and S_1'.

	Uncorrected Sum					Corrected Sum			
C	S_8	S_4	S_2	S_1	Correction	S_8'	S_4'	S_2'	S_1'
0	0	1	1	0	-3	0	0	1	1
0	0	1	1	1	-3	0	1	0	0
0	1	0	0	0	-3	0	1	0	1
0	1	0	0	1	-3	0	1	1	0
0	1	0	1	0	-3	0	1	1	1
0	1	0	1	1	-3	1	0	0	0
0	1	1	0	0	-3	1	0	0	1
0	1	1	0	1	3	1	0	1	0
0	1	1	1	0	-3	1	0	1	1
0	1	1	1	1	-3	1	1	0	0
1	0	0	0	0	$+3$	0	0	1	1
1	0	0	0	1	$+3$	0	1	0	0
1	0	0	1	0	$+3$	0	1	0	1
1	0	0	1	1	$+3$	0	1	1	0
1	0	1	0	0	$+3$	0	1	1	1
1	0	1	0	1	$+3$	1	0	0	0
1	0	1	1	0	$+3$	1	0	0	1
1	0	1	1	1	$+3$	1	0	1	0
1	1	0	0	0	$+3$	1	0	1	1
1	1	0	0	1	$+3$	1	1	0	0

Single-function, SOP minimization of the output functions gives

$$S_8' = CS_8 \mid CS_4S_1 \mid CS_4S_2 + S_8S_4 + S_8S_2S_1$$
$$S_4' = \bar{C}\bar{S}_4\bar{S}_2 + \bar{S}_4S_2\bar{S}_1 + \bar{C}S_4S_2S_1 + C\bar{S}_4S_1 + CS_4\bar{S}_2\bar{S}_1$$
$$S_2' = \bar{C}S_2\bar{S}_1 + \bar{C}\bar{S}_2S_1 + CS_2S_1 + C\bar{S}_2\bar{S}_1 = \bar{C}(S_2 \oplus S_1) + C(S_2 \odot S_1)$$
$$= C \oplus S_2 \oplus S_1$$
$$S_1' = \bar{S}_1$$

The addition of decimal numbers that are represented in other codes can be accomplished by designing a direct combinational adder or by converting the code into 8421 BCD or XS3, using the techniques given above, and converting the result back into the original code.

9.3.3. Decimal Subtraction. Decimal subtraction can be implemented by designing and using decimal full- and half-subtraction units, just as decimal full adders and half-adders were designed for decimal addition. This design process closely follows that of Exs. 9-8 and 9-9, with borrows (rather than carries) being generated. Several of the problems at the end of this chapter consider the design of such devices.

The subtraction operation brings with it the possibility of a negative result, and techniques such as signed magnitude and sign-and-complement representations must be introduced, just as they were for negative binary operands in Sec. 8.3. Decimal sign and magnitude uses one extra high-order digit to represent the sign. In fact, since the sign can only have two values, a single extra bit can represent it.

Signed magnitude operations involving decimal numbers have the same problems that signed binary has: the need for *both* addition and subtraction and the need for magnitude comparison prior to addition or subtraction, etc. As a result, signed magnitude decimal is normally used only as the encoding method for representing input and output data. Arithmetic operations are most often performed by using complementary notation.

The decimal equivalent to 2's complement encoding is the 10's complement (the radix complement). The 10's complement of an *m*-digit number is obtained by subtracting it from 10^m. Complementation is equivalent to multiplication by -1. Each number is given an extra digit, usually a 0 or 1, that represents its sign. Addition in the sign column is carried out in binary (modulo-2) form rather than in decimal.

EXAMPLE 9-13:   What is the 10's complement of $+157$?

Solution: Subtract 157 from $10^3 = 1000$. Give the result a minus sign. This gives the following result:

$$
\begin{array}{r}
1000 \\
-\ 157 \\
\hline
843 \Rightarrow 1{,}843
\end{array}
$$

EXAMPLE 9-14:   What are the 10's complements of 0,563, 1,271, 0,9576, and 1,12?

Answers: 1,437; 0,729; 1,0424; and 0,98

EXAMPLE 9-15:   Add $+436$ and -157 by using 10's complement notation.

Solution:

$$
\begin{array}{rl}
+436 = & 0{,}436 \\
-157 = & 1{,}843 \\
\hline
\text{Sum} = & 10{,}279
\end{array}
$$

Note that the sign digits are added in binary fashion, with the high-order carry being neglected. The answer is $+279$.

EXAMPLE 9-16:   Add $+157$ and -436 by using 10's complement notation:

Solution:

$$+157 = 0,157$$
$$-436 = 1,564$$
$$\overline{\text{Sum} = 1,721}$$

Converting the sum back to sign and magnitude form gives -279.

Overflow detection in the decimal case is the same as in the binary case. Since the sign digits are binary in nature, exactly the same logical techniques can be used.

Ten's complementation can be accomplished by using an algorithmic process that is similar to the 2's complementing algorithm given in Sec. 8.4.1. The 10's complement algorithm is

> Examine all digits, beginning at the least-significant
> end of the number. Keep all low-order 0's unchanged.
> Subtract the first nonzero digit from 10. Subtract all
> other digits from 9. Change the sign bit from 1 to 0 or from
> 0 to 1.

The result is the 10's complement of the original number.

EXAMPLE 9-17:   Complement $+12765380$.

Solution: The low-order 0 is kept unchanged. The 8 is subtracted from 10, giving 2. The remaining digits are subtracted from 9. The sign bit is set to a 1, indicating a negative number. The net result is 1,87234620. To verify the result, it may be added to its positive counterpart:

$$0,12765380$$
$$+1,87234620$$
$$\overline{10,00000000}$$

Since the high-order carry is neglected, the answer is, in fact, 0.

To implement the complementation process, a conditional "subtract from 10 or subtract from 9" logical circuit can be designed. This circuit can be combined with nonzero digit detection in order to implement the 10's complementing algorithm, closely following the 2's complementing circuit shown in Fig. 8-7.

A second approach obtains the 10's complement by adding one unit to its diminished-radix cousin, the 9's complement. This parallels the 1's complement + 1 method for finding the 2's complement of a binary number.

The 9's complement of a decimal number is found by subtracting each digit of the number from 9. Since each digit can never exceed 9, the subtraction will never produce a borrow, allowing each digit to be complemented without regard to its neighbor. This allows the use of nonconditional complementation circuitry that receives an encoded digit and generates the 9's complement of the digit. The addition of one unit can be accomplished by replacing the decimal half-adder in the units column of the decimal adder by a decimal full adder and forcing the low-order carry input to be TRUE whenever a 10's complementation is to be made.

EXAMPLE 9-18:   Design a combinational circuit that receives an 8421 BCD digit and a control signal, C. Whenever C is FALSE, the digit is to be passed unchanged. When C is TRUE, the digit is to be converted into its 9's complement.

Solution: This conditional 9's complement circuit may be represented by the diagram in Fig. E9-18. The operation of the device is summarized in the following table:

Control Signal C	Decimal Input				Decimal Output			
	A_8	A_4	A_2	A_1	a_8	a_4	a_2	a_1
0	0	0	0	0	0	0	0	0
0	0	0	0	1	0	0	0	1
0	0	0	1	0	0	0	1	0
0	0	0	1	1	0	0	1	1
0	0	1	0	0	0	1	0	0
0	0	1	0	1	0	1	0	1
0	0	1	1	0	0	1	1	0
0	0	1	1	1	0	1	1	1
0	1	0	0	0	1	0	0	0
0	1	0	0	1	1	0	0	1
1	0	0	0	0	1	0	0	1
1	0	0	0	1	1	0	0	0
1	0	0	1	0	0	1	1	1
1	0	0	1	1	0	1	1	0
1	0	1	0	0	0	1	0	1
1	0	1	0	1	0	1	0	0
1	0	1	1	0	0	0	1	1
1	0	1	1	1	0	0	1	0
1	1	0	0	0	0	0	0	1
1	1	0	0	1	0	0	0	0

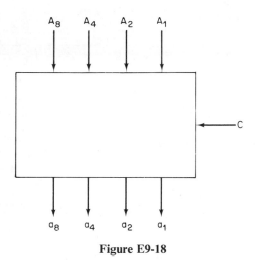

Figure E9-18

Solving for the output functions gives the following minimized SOP forms:

$$a_8 = \bar{C}A_8 + C\bar{A}_8\bar{A}_4\bar{A}_2$$
$$a_4 = \bar{C}A_4 \mid A_4\bar{A}_2 \mid C\bar{A}_4A_2$$
$$a_2 = A_2$$
$$a_1 = C \oplus A_1$$

Conditional complement circuits of the type designed in Ex. 9-18 can be used to implement a decimal adder/subtractor that subtracts by addition of complements. Figure 9-10 gives the general arrangement of such a system. The *SUBTRACT* signal causes the *B* number to be 9's complemented and forces a low-order carry. The net result is a 10's complementation. The sign of *B* is also changed, but since it is a single Boolean variable, it can be complemented via an EXCLUSIVE-OR gate.

The *A* and *B* numbers are added by decimal full adders. The binary add and correct technique can also be used. The sign bits of *A* and *B* are added by a binary full adder. The result appears in signed 10's complement form. Overflow can be detected by comparing the carries into and out of the sign bit.

Decimal addition and subtraction can be performed in the signed 9's complement form as well, with an end-around carry being required whenever there is a carry out of the sign bit. The need for the end-around carry has made the use of 9's complement less prevalent than the use of 10's complement. Nine's complement is normally used only as an intermediate step in the calculation of the 10's complement.

The 9's complementation of an XS3 digit is quite simple, as the following example illustrates.

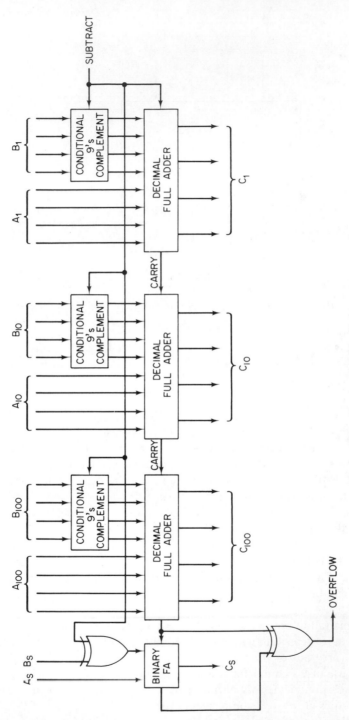

Figure 9-10. Decimal Adder/Subtractor

EXAMPLE 9-19:   Repeat Ex. 9-18 for the XS3 code.

Solution: The operation of the XS3 conditional complement circuit is tabulated below:

Control Input, C	Decimal Input				Decimal Output			
	B_8	B_4	B_2	B_1	b_8	b_4	b_2	b_1
0	0	0	1	1	0	0	1	1
0	0	1	0	0	0	1	0	0
0	0	1	0	1	0	1	0	1
0	0	1	1	0	0	1	1	0
0	0	1	1	1	0	1	1	1
0	1	0	0	0	1	0	0	0
0	1	0	0	1	1	0	0	1
0	1	0	1	0	1	0	1	0
0	1	0	1	1	1	0	1	1
0	1	1	0	0	1	1	0	0
1	0	0	1	1	1	1	0	0
1	0	1	0	0	1	0	1	1
1	0	1	0	1	1	0	1	0
1	0	1	1	0	1	0	0	1
1	0	1	1	1	1	0	0	0
1	1	0	0	0	0	1	1	1
1	1	0	0	1	0	1	1	0
1	1	0	1	0	0	1	0	1
1	1	0	1	1	0	1	0	0
1	1	1	0	0	0	0	1	1

The output equations can be found to be

$$b_0 = B_0 \oplus C$$
$$b_4 = B_4 \oplus C$$
$$b_2 = B_2 \oplus C$$
$$b_1 = B_1 \oplus C$$

The equations given in Ex. 9-19 show that the 9's complement of a digit that is encoded in XS3 BCD can be found, when desired, by changing each bit of the XS3 code to its logical opposite. This property makes 9's complementation (and its close relative, 10's complementation) quite easy to implement. The XS3 code is one of several *self-complementing* BCD codes. Other codes of this type and other properties of codes that relate to 9's and 10's complementation are discussed in Chapter 10.

The complementation process can be performed by sequential circuits as well as by purely combination circuits. For example, if a four-bit register contains a digit that is represented in XS3 code, toggling all the flip-flops that make up the register will change the encoded value into its 9's complement. A second toggling operation will return the register to its original state.

9.4 Summary

Chapter 9 covers three major topics: decimal-to-binary conversion, binary-to-decimal conversion, and decimal arithmetic. The two conversion processes can be implemented in either combinational or sequential form. The dibble-dabble and dabble-dibble algorithms can be used for these conversions; generalized techniques for using them with any decimal code are given.

BCD codes use four or more Boolean variables to represent a decimal digit. Decimal quantities that are represented in BCD code can be added by several methods. Combinational full adders and half-adders for BCD digits can be used. An alternative is to add the numbers as if they were represented in binary and correct the results where necessary. The 8421 and XS3 BCD codes are especially useful for the latter purpose.

Decimal subtraction can be performed with BCD operands. Negative results can be represented in signed decimal, 10's complement, or 9's complement notation. Decimal subtraction can be performed by adding a complemented operand. Ten's complementation for this purpose can be implemented algorithmically or as the 9's complement $+$ 1.

9.5 Did You Learn?

1. How the ten values that a decimal digit can take on can be represented by four or more Boolean variables? What BCD means?

2. How to represent a multidigit decimal number in BCD?

3. The 8421 and XS3 BCD codes?

4. How to design the combinational decimal-to-binary conversion circuits shown in Figs. 9-1 and 9-2 for *any* BCD code?

5. How to design the sequential equivalents to those same devices as per Table 9-3?

6. How to design their combinational (Table 9-4) and sequential (Table 9-6) binary-to-decimal conversion counterparts?

7. How to design a decimal half-adder and full adder for use with *any* BCD code?

8. How to add decimal numbers that are encoded in 8421 or XS3 BCD code? How to correct for errors if each decade is added in binary fashion? How to implement these operations with combinational logic?

9. How to find the 10's and 9's complements of a decimal number? Why the 9's complement is easier to find?

10. How to use the 9's complement to obtain the 10's complement?

11. How to design a conditional 9's complement circuit?

12. What *self-complementing* means? Why the XS3 code has this property? How it simplifies the complementation operation?

13. How to subtract by adding decimal complements?

9.6 References

[1] Chu, Yoahan, *Digital Computer Design Fundamentals*. New York: McGraw-Hill Book Company, Inc., 1962.

[2] Couleur, J. F., "BIDEC—A Binary-to-decimal or Decimal-to-binary Converter," *IEEE Transactions on Electronic Computers*, EC-7, No. 6 (December 1958), 313–316.

[3] Dean, K. J., "Conversion Between Binary Code and Some Binary-decimal Codes," *The Radio and Electronic Engineer* (January 1968), 49–53.

[4] Linford, John, "Binary-to-BCD Conversion with Complex IC Functions," *Computer Design*, 9, No. 9 (September 1970), 53–66.

[5] Motorola Semiconductor Products Inc., "BCD-to-binary/Binary-to-BCD Number Converter," *Technical Datasheet No. MC4001P*, Phoenix, Arizona, August 1969.

[6] Plantz, A. R., and M. Berman, "Adoption of the Octal System," *IEEE Transactions on Computers*, C-20, No. 5 (May 1971), 593–597.

[7] Rhyne, V. T., "Serial Binary-to-decimal and Decimal-to-binary Conversion," *IEEE Transactions on Computers*, C-19, No. 9 (September 1970), 808–812.

[8] Richards, R. K., *Arithmetic Operations in Digital Computers*. New York: Van Nostrand Reinhold Company, 1955.

9.7 Problems

General Concepts

9-1. Express each of the following decimal numbers in the 8421 BCD and the XS3 BCD codes.

a. 96

b. 42157

c. 35336

d. 100239

e. 5263

f. 4273618590

9-2. The following table defines one version of the 6, 3, 1, −1 BCD code. Express each of the decimal numbers given in Prob. 9-1 in this code.

Decimal Value	Code 6 3 1 −1
0	0 0 0 0
1	0 0 1 0
2	0 1 0 1
3	0 1 0 0
4	0 1 1 0
5	1 0 0 1
6	1 0 1 1
7	1 0 1 0
8	1 1 0 1
9	1 1 1 1

Note that other codes with the same 6, 3, 1, −1 "weights" can be found.

9-3. Convert each of the following 8421 BCD numbers into its decimal equivalent.

    a. 1001 0010 1000 0111

    b. 001001101001100000000100

    c. 0101 0000 0000 0011 1000

9-4. Convert each of the following XS3 numbers into its decimal equivalent.

    a. 1100 0011 0100 0101

    b. 0011 0011 1100 1010 1011 0100

    c. 011000111010110000110101

9-5. Use the 8421 BCD-to-binary conversion table given in Ex. 9-4 to convert each of the BCD numbers of Prob. 9-3 into its binary equivalent, in accordance with the dabble-dibble algorithm.

9-6. Insert the XS3 code words into Table 9-2. Using the result, convert each of the XS3 numbers given in Prob. 9-4 into its binary equivalent.

9-7. Verify the results that are given in Table 9-5. Extend the table to cover 40- and 50-bit numbers.

9-8. Example 9-6 contains a tabulation of the binary-to-8421 BCD conversion process. Use that table to convert each of the following binary numbers into its 8421 BCD equivalent.

    a. 1101101   c. 01011010

    b. 10000000   d. 11111111

9-9. Insert the XS3 code words into Table 9-4 and use the result to convert the binary numbers given in Prob. 9-8 into their XS3 equivalents.

9-10. Add the following pairs of 8421 BCD numbers by using the corrected binary summation technique that is diagrammed in Fig. 9-8. Give both the uncorrected result and the final answer. Convert the input operands into their decimal equivalents and verify the correctness of the addition operations.

a. $A = 100101010011$ $B = 001001110100$
b. $A = 0010100000110110$ $B = 0111000101100100$
c. $A = 011001100111$ $B = 100110000111$
d. $A = 010101000011001000001$ $B = 10000111011001010100$

9-11. Add each of the following pairs of XS3 numbers by using the corrected binary summation technique. Give both the uncorrected result and the final answer. Convert the input operands into their decimal equivalents and verify the correctness of the addition operations.

a. $A = 110011001100$ $B = 010001000100$
b. $A = 0100010101100111$ $B = 0011001111001100$
c. $A = 0110101010111000$ $B = 1001010101001001$

9-12. Express each of the following signed decimal numbers in 10's complement, 9's complement, 10's complement 8421 BCD, and 10's complement XS3. For example,

Signed Decimal	10's Comp.	9's Comp.	10's Comp. 8421	10's Comp. XS3
−92	1,08	1,07	1,00001000	1,00111011

Consider both the algorithmic and the 9's complement + 1 techniques for performing 10's complementation.

a. $+2753$ d. -1001
b. -2753 e. $+53$
c. $+19999$ f. -67

9-13. Convert each of the following 10's complement numbers into signed decimal.

a. 1,0000   c. 0,12345
b. 1,95623   d. 1,010209

9-14. Consider the numbers given in Prob. 9-13 to be in 9's complement, rather than 10's complement, form. Convert them into signed decimal.

9-15. The sign bit of a 2's complement or 1's complement binary number can be assigned a negative weight that is dependent on the number of magnitude bits in the binary word. Can this be done for 10's complement numbers?

For 9's complement numbers? If so, give a general formula for computing the weight of the sign bit for 9's and 10's complement operands that have N digits in their magnitude field. Then, by totaling the weights of the sign bit and the magnitude field, verify the results of the 10's and 9's complementation of the negative numbers given in Probs. 9-12, 9-13, and 9-14.

9-16. Add each of the following pairs of 10's complement numbers. Indicate overflow, should it occur.

    a. $A = 0{,}957$ $B = 1{,}253$
    b. $A = 1{,}056$ $B = 1{,}247$
    c. $A = 1{,}127$ $B = 0{,}627$
    d. $A = 1{,}999$ $B = 0{,}995$

9-17. For each of the pairs of 10's complement numbers given in Prob. 9-16, compute $A - B$ by adding the 10's complement of the B number. Repeat by adding the 9's complement of $B + 1$. Verify the results by converting A and B into signed decimal and using direct subtraction.

9-18. Figure 9-0 shows the data conversion sequence for a typical digital system that has decimal input and output but uses binary arithmetic. Show the sequence of data conversions and arithmetic operations that are involved in the addition of $A = +257$ and $B = -126$.

Functional Design Problems

*9-19. Design a combinational device that will convert from 8421 BCD into XS3 BCD. Repeat for the XS3-to-8421 conversion. Give logic equations and show an RTL and a TTL logic diagram. Assume double-rail inputs and provide positive-TRUE outputs.

*9-20. Design a sequential 8421-to-XS3 converter. Use four flip-flops. Assume that a valid 8421 code has been loaded into the flip-flops and that, upon receiving a clock pulse, the flip-flops are to "count" into the XS3 code having the same decimal value. Thus, the 8421 code words form the possible present states and the XS3 code words form the next states. Use J-K flip-flops. After completion of the design, investigate the behavior of the counter/converter from its six redundant starting states. Draw a complete state diagram.

*9-21. Repeat Prob. 9-19 for an 8421 to 6, 3, 1, -1 converter.

*9-22. Design a combinational converter that receives an 8421 BCD encoded digit and decodes its value into one-out-of-ten form.

*9-23. Design a combinational converter that changes one-out-of-ten code (ten Boolean signals, say A to J) into XS3 BCD code. Since this is a ten-variable reduction problem with 1014 redundancies, an intuitive design approach is recommended.

*9-24. Design the combinational converter that implements the halving operation required by the dabble-dibble algorithm for each of the following codes. Follow Table 9-2 and Ex. 9-4.

   a. XS3 code

   b. 6, 3, 1, -1 code, as given in Prob. 9-2

   c. Two-out-of-five code, as tabulated below:

Decimal Value	Code				
	A	B	C	D	E
0	0	0	0	1	1
1	0	0	1	0	1
2	0	1	0	0	1
3	1	0	0	0	1
4	0	0	1	1	0
5	0	1	0	1	0
6	1	0	0	1	0
7	0	1	1	0	0
8	1	0	1	0	0
9	1	1	0	0	0

Note that part c involves a total of six input variables. Tabular or computer-aided reduction is advised.

*9-25. Design the counter/converter used for sequential conversion of each of the following BCD codes into binary. Follow Table 9-3, Fig. 9-3, and Ex. 9-5. Use J-K flip-flops.

   a. XS3

   b. 6, 3, 1, -1, as given in Prob. 9-2

   c. Two-out-of-five code

*9-26. Design a combinational circuit that will convert a four-bit binary number into its BCD equivalent. Consider each of the following BCD codes.

   a. 8421 BCD

   b. XS3

   c. 6, 3, 1, -1 BCD

   d. Two-out-of-five code

*9-27. Repeat Prob. 9-26a and b for five-bit binary input and six-bit binary input.

*9-28. Design the combinational double-add circuit that is used to convert from binary to each of the following BCD codes in accordance with the dabble-dibble algorithm. Follow Table 9-4 and Ex. 9-6.

   a. XS3

   b. 6, 3, 1, -1 BCD

   c. Two-out-of-five code

*9-29. Design the combinational double-add circuit for use in converting from binary to the biquinary code given below:

Decimal Value	Code Word						
	A	B	C	D	E	F	G
0	1	0	1	0	0	0	0
1	1	0	0	1	0	0	0
2	1	0	0	0	1	0	0
3	1	0	0	0	0	1	0
4	1	0	0	0	0	0	1
5	0	1	1	0	0	0	0
6	0	1	0	1	0	0	0
7	0	1	0	0	1	0	0
8	0	1	0	0	0	1	0
9	0	1	0	0	0	0	1

*9-30. Design a sequential binary-to-BCD counter/converter for use with each of the following codes. Follow Table 9-6, Fig. 9-6, and Ex. 9-7. Use J-K flip-flops.

    a. XS3

    b. 6, 3, 1, -1 BCD

    c. Two-out-of-five code

*9-31. Design an XS3 decimal half-adder and an XS3 decimal full adder. Give an RTL logic diagram for each device.

*9-32. Repeat Prob. 9-31 for the 6, 3, 1, -1 BCD code.

*9-33. Design a decimal half-subtractor and full subtractor for use with 8421 BCD. Give a TTL logic diagram for each device.

Systems Design Problems

*9-34. Design a signed decimal to 10's complement converter for use with 8421 BCD. Each digit is passed through a conditional converter that performs the following operations:

    a. If a control input, C, is FALSE, the input is sent directly to the output.

    b. If C is TRUE, then the input is complemented in accordance with a second control input, D, as follows:

    c. If D is FALSE, then the input is 9's complemented.

    d. If D is TRUE, then the input is 10's complemented.

The overall logic of the device is as follows. The sign bit of the signed decimal input is never changed. If the sign bit is FALSE, the number is positive and the magnitude digits are not changed. This is accomplished by making the C input to all the conversion circuits FALSE when the sign bit is FALSE. If the sign bit is TRUE, then the magnitude digits are 10's complemented in accordance with the algorithm given in Sec. 9.3.3. The control logic of the complement circuits should provide the following conditions:

   a. All low-order decades that have zero inputs should not be complemented; i.e., their C input should be FALSE.
   b. The first decade that has a nonzero input should be 10's complemented; i.e., its CD inputs should be 11.
   c. All other higher-ordered decades should be 9's complemented; i.e., $CD = 10$.

Design the complete system, including the decade converters and the control logic that interrelates them. Give a complete RTL logic diagram for a converter that receives signed, three-digit input numbers. The complexity of this system points to the use of the 9's complement $+$ 1 technique for implementing 10's complementation.

  9-35. Can the device that is designed in Prob. 9-34 be used for 10's complement to signed decimal conversion? Can this process cause overflow? If so, how can it be detected and how can the converter be modified so as to give an approximately correct answer?

*9-36. Design a digital accumulator system that receives a signed, two-digit decimal number and three control signals:

CLEAR	causes the accumulator to be cleared
ADD	causes the input number to be added to the current value in the accumulator
SUBT	causes the input number to be subtracted from the current value in the accumulator

The system uses the data conversion sequence that is shown in Fig. 9-0. The system provides a signed decimal representation of the accumulated total and an *OVERFLOW* signal. Overflow causes the system to halt until a *CLEAR* signal is given. Give a complete TTL system diagram.

*9-37. Design a sequential binary-to-BCD conversion system. The system is to accept a ten-bit, signed binary number and convert it into its signed, four-digit 8421 BCD equivalent. The controller is to load the binary number into a shift register, initiate the conversion sequence, and halt after the proper number of clock periods. The conversion is to be initiated by a *START* signal.

*9-38. Design an XS3 accumulator. The system is to receive a signed, two-digit XS3 encoded number. Accumulation is implemented by using a gated-carry binary accumulator. Then, each decade of the accumulator is corrected by ± 3 in accordance with the XS3 addition technique. Design each decade accumulator so that the accumulation is completed in three clock periods: two clock periods for the gated-carry accumulation and a third clock period during which each decade is sequentially corrected by ± 3. Design the complete system, including an appropriate sequential controller. Detect and indicate overflow, halting the controller when it occurs. A *CLEAR* signal should be provided in order to restart the accumulation process. Give a complete RTL logic diagram.

10 Codes and Their Properties

10.0 Introduction

The use of a group of Boolean variables to represent, store, transmit, or otherwise relate to a multivalued item of information or quantity is commonly called *coding* or *encoding*. Each of the different informative or quantitative values is assigned a particular combination of the Boolean variables, usually in one-to-one correspondence. The combinations of variables are called *code words*, and the encoding process relates each code word to a particular informative or quantitive value.

Different coding techniques offer different advantages and disadvantages. For example, XS3 code offers easy complementation (since it is self complementing), but it is more difficult to use in binary/decimal conversions than is 8421 BCD code. Binary encoding of numerical quantities is useful and efficient for arithmetic operations and data handling, but it is not too desirable for data input and output when human interpretation is involved.

This chapter considers four useful classes of codes: BCD codes, weighted codes, unit-distance codes, and codes for error detection and correction.

10.1 BCD Codes

As is mentioned in Sec. 9.1, the term *BCD* is generally interpreted as binary-coded decimal, although Boolean-coded decimal is more applicable. The 8421, XS3, two-out-of-five, and 6, 3, 1, −1 BCD codes have already been introduced. All BCD codes must have at least four bits, and at least six redundant code words must be present.

BCD codes are used to represent the digits of a decimal number. Typical operations involving BCD code words include binary/decimal conversion and decimal addition or subtraction. Since decimal subtraction is often performed by addition of complements, the ease with which a given BCD code word can be converted to the code word for its 9's complement is of concern.

For *self-complementing* BCD codes this conversion is quite easy: the bits of the code word are logically inverted in order to obtain the 9's complement. The XS3 code is a self-complementing BCD code that also has good arithmetic properties, as is shown in Sec. 9.3.2. The 6, 3, 1, -1 code given in Prob. 9-2 is another self-complementing code. Table 10-0 lists three more self-complementing BCD codes, one of which is a second variation of the 6, 3, 1, -1 code. The bits of the third code, labeled code III, cannot be assigned numerical values, making it an *unweighted code*, as is discussed in Sec. 10.2.

Table 10-0: Some Self-complementing BCD Codes

Decimal Value	6, 3, 1, -1 *Code*				2421 *Code*				Code III			
	6	3	1	-1	2	4	2	1				
0	0	0	1	1	0	0	0	0	1	0	0	1
1	0	0	1	0	0	0	0	1	1	0	0	0
2	0	1	0	1	0	0	1	0	0	1	0	1
3	0	1	1	1	0	0	1	1	0	0	0	0
4	0	1	1	0	0	1	0	0	1	1	0	0
5	1	0	0	1	1	0	1	1	0	0	1	1
6	1	0	0	0	1	1	0	0	1	1	1	1
7	1	0	1	0	1	1	0	1	1	0	1	0
8	1	1	0	1	1	1	1	0	0	1	1	1
9	1	1	0	0	1	1	1	1	0	1	1	0

The self-complementing property can be identified by comparing each of the codes for the decimal values from 0 to 4 with its 9's complemented partner. If the code words in each pair are bit-for-bit opposites, then the code has the self-complementing property. Any self-complementing BCD code can be conditionally complemented by using only EXCLUSIVE-OR gates, as shown in Fig. 10-0.

A second property of a BCD code that makes 9's complementation easy to implement is *reflection*. A reflected BCD code word can be 9's complemented by changing only *one* of its bits. Table 10-1 lists two reflected BCD codes. The *reflection bit* is bit A for code I and bit X for code II. Reflection can be identified by comparing the upper five code words with the lower five. If they are mirrored except for one bit, then the code is reflected. A reflected BCD code can be 9's complemented by using only *one* EXCLU-

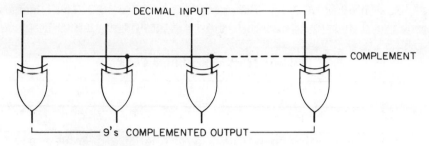

Figure 10-0. 9's Complement Circuit for a Self-complementing BCD Code

SIVE-OR gate. A conditional 9's complement circuit for use with code I, Table 10-1, is shown in Fig. 10-1.

Table 10-1: Reflected BCD Codes

Decimal Value	Code I				Code II			
	A	B	C	D	W	X	Y	Z
0	0	0	0	0	1	0	0	0
1	0	0	0	1	0	0	1	1
2	0	0	1	0	0	0	1	0
3	0	0	1	1	1	0	1	1
4	0	1	0	0	0	0	0	0
5	1	1	0	0	0	1	0	0
6	1	0	1	1	1	1	1	1
7	1	0	1	0	0	1	1	0
8	1	0	0	1	0	1	1	1
9	1	0	0	0	1	1	0	0

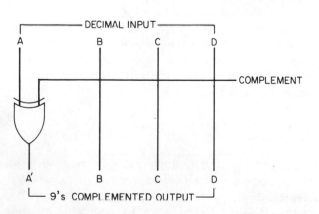

Figure 10-1. 9's Complement Circuit for a Reflected BCD Code

Several other BCD codes are presented in Table 10-2. The biquinary code uses three extra bits, but it has useful error-detection properties.

Table 10-2:   Additional BCD Codes

Decimal Value	5421 Code				5311 Code				7421 Code				Biquinary						
	5	4	2	1	5	3	1	1	7	4	2	1	A	B	C	D	E	F	G
0	0	0	0	0	0	0	0	0	0	0	0	0	0	1	0	0	0	0	1
1	0	0	0	1	0	0	0	1	0	0	0	1	0	1	0	0	0	1	0
2	0	0	1	0	0	0	1	1	0	0	1	0	0	1	0	0	1	0	0
3	0	0	1	1	0	1	0	0	0	0	1	1	0	1	0	1	0	0	0
4	0	1	0	0	0	1	0	1	0	1	0	0	0	1	1	0	0	0	0
5	0	1	0	1	1	0	0	0	0	1	0	1	1	0	0	0	0	0	1
6	0	1	1	0	1	0	0	1	0	1	1	0	1	0	0	0	0	1	0
7	0	1	1	1	1	0	1	1	1	0	0	0	1	0	0	0	1	0	0
8	1	0	1	1	1	1	0	0	1	0	0	1	1	0	0	1	0	0	0
9	1	1	0	0	1	1	0	1	1	0	1	0	1	0	1	0	0	0	0

Many, many other BCD codes are in use. An infinite number are possible, assuming that more than four variables can be used. The self-complementing and reflected properties are useful for complementation, but these codes, with the exception of the XS3 code, are difficult to use in decimal addition. A decimal full adder can be defined for each code, however, by using the techniques given in Sec. 9.3. An alternative approach is to conditionally complement the self-complementing or reflected code word and then to convert it into 8421 or XS3 code prior to the decimal addition process.

Binary-to-decimal and decimal-to-binary conversion can be implemented for any of these (or other) BCD codes by following the techniques given in Chapter 9. The combinational and/or sequential design procedures given in Chapters 3–6 permit the use of any of these BCD codes in other ways. Code-to-code conversion to drive a seven-segment decimal display, magnitude comparison, and up- or down-counter design are typical of these applications. The problems given at the end of Chapter 9 and at the end of this chapter cover some of these design situations.

10.2 Weighted Codes

The use of Boolean variables to represent numerical quantities generally involves a one-to-one correspondence between a given quantity and the code word that represents it. In many cases, the quantity can be determined directly from the TRUE/FALSE values of the code word by using Eq. (10-0):

$$N = \sum_{i=1}^{n} w_i a_i + C \tag{10-0}$$

where N is the quantity, the a_i coefficients are determined from the n bits of the code word ($a_i = 1$ if the ith bit is TRUE; $a_i = 0$ otherwise), the w_i coefficients are the *weights* that are assigned to the bits of the code, and C is the *bias constant* of the code. If Eq. (10-0) holds for all code words that have specified quantitative values, then the code is said to be a *weighted code*. Equation (10$=$0) does not apply to code words that are redundancies, i.e., that are not assigned a numerical equivalent.

For an n-bit code Eq. (10-0) allows $n + 1$ variables: the n weights and the bias constant. Thus, given $n + 1$ code word/quantity combinations, a set of linear, simultaneous equations can be formed and solved, giving a set of weights and a bias constant that will satisfy those $n + 1$ combinations. For the code to be weighted, however, the same set of weights and bias constant must satisfy *all* the code word/quantity combinations that are specified.

EXAMPLE 10-0:   Determine whether or not the following code is weighted. If it is, determine the weight that corresponds to each bit, and the bias constant.

	Code Word		
Quantity	A	B	C
0	0	0	1
1	0	1	1
2	0	0	0
3	0	1	0
4	1	0	1
5	1	1	1
6	1	0	0
7	1	1	0

Solution: The first four code words may be used to solve for a set of weights and a bias constant. Inserting these code words into Eq. (10-0) gives

$$0 \cdot w_A + 0 \cdot w_B + 1 \cdot w_C + C = 0$$
$$0 \cdot w_A + 1 \cdot w_B + 1 \cdot w_C + C = 1$$
$$0 \cdot w_A + 0 \cdot w_B + 0 \cdot w_C + C = 2$$
$$0 \cdot w_A + 1 \cdot w_B + 0 \cdot w_C + C = 3$$

The third equation gives $C = 2$. Solving for the weights gives $w_B = 1$ and $w_C = -2$. These values satisfy all four equations. The value of w_A cannot be determined from these four equations since w_A never appears in them. The fifth code word does contain a TRUE value for A, giving

$$1 \cdot w_A + 0 \cdot w_B + 1 \cdot w_C + C = 4$$

Inserting the values for w_C and C gives $w_A = 4$. The test for the weighted property continues by verifying that these weights satisfy Eq. (10-0) for the remaining code words, corresponding to 5, 6, and 7:

$$w_A + w_B + w_C + C = 4 + 1 - 2 + 2 = 5$$
$$w_A + \qquad\qquad C = 4 + 2 = 6$$
$$w_A + w_B + \qquad C = 4 + 1 + 2 = 7$$

Since the same weights and bias constant satisfy all the specified code word/numerical value combinations, the code is a weighted code. It may be referred to as 4, 1, −2 code with a bias constant of +2.

EXAMPLE 10-1: Determine whether or not the following code is weighted:

	Code Word		
Quantity	A	B	C
0	0	0	0
1	0	0	1
2	0	1	0
3	1	0	0
4	1	0	1
5	1	1	1

Solution: Inserting each of the first four code words into Eq. (10-0) gives the following expressions:

$$C = 0$$
$$w_C = 1$$
$$w_B = 2$$
$$w_A = 3$$

Using the weights and the fifth code word gives

$$w_A + w_C + C = 3 + 1 + 0 = 4$$

which is correct. The last code word gives an invalid result, however:

$$w_A + w_B + w_C + C = 3 + 2 + 1 + 0 = 6 \neq 5$$

so that the code is not weighted.

Some authors do not include a bias constant in their definitions of a weighted code (Refs. [1] and [3]), making a code such as the XS3 BCD code be "unweighted." The more general definition given in Eq. (10-0) is preferable, since it extends the definition of weighted code to include many codes that otherwise cannot be considered as weighted.

The weighted property of a code is of great use in many applications. A major example is the need for a weighted code when performing digital-to-analog conversion. This is discussed in more detail in Chapter 11. A second area wherein the weighted property is useful is in numerical display. A quantity that is expressed in a weighted code can be displayed by using a set of ON/OFF indicators, one for each bit. Each of these indicators can be labeled with the weight of the bit that it represents. The observer can read the weights of the ON indicators, sum them, add (or subtract) the bias constant, and obtain the quantity that the indicated code word represents.

The weighted property may be "overlapped" with the self-complementing property that was described in Sec. 10.1. The XS3 code is an example of a code that is both weighted and self-complementing.

EXAMPLE 10-2: What are the weights and the bias constant of the XS3 code?

Answer: 8, 4, 2, 1, and $C = -3$.

A useful combination of these two properties occurs whenever the weights for a given code sum to a value of 9 and the bias constant is 0. Any weighted code that has this property can be made to be self-complementing. The self-complementing property is not guaranteed, however. The code words must be chosen to assure its presence.

EXAMPLE 10-3: A 6, 3, 1, -1 code can be made to be self-complementing since its weights sum to 9. Give an example of a 6, 3, 1, -1 code that *is* self-complementing and a 6, 3, 1, -1 code that is not.

Answer:

Numerical Value	Self-complementing				Not Self-complementing			
	6	3	1	−1	6	3	1	−1
0	0	0	0	0	0	0	0	0
1	0	0	1	0	0	0	1	0
2	0	1	0	1	0	1	0	1
3	0	1	0	0	0	1	0	0
4	0	1	1	0	0	1	1	0
5	1	0	0	1	1	0	0	1
6	1	0	1	1	1	0	0	0
7	1	0	1	0	1	0	1	0
8	1	1	0	1	1	1	0	1
9	1	1	1	1	1	1	0	0

The second code fails to be self-complementing because the code word pairs for (0,9) and (3,6) are not bit-for-bit opposites. The weighted property is maintained throughout both codes, however.

The 2421 BCD code given in Table 10-0 is a second example of a weighted, self-complementing code. A version of the 2421 code that is *not* self-complementing is given in Table 10-3.

Table 10-3:   A 2421 BCD Code that Is Not Self-complementing

Numerical Value	Code Word 2 4 2 1
0	0 0 0 0
1	0 0 0 1
2	0 0 1 0
3	0 0 1 1
4	0 1 0 0
5	0 1 0 1
6	0 1 1 0
7	0 1 1 1
8	1 1 1 0
9	1 1 1 1

If a BCD code is used to represent a multidigit decimal number, the weights of the bits in each decade are ten times the weights of the corresponding bits of the next lower decade. For example, the weights of the bits of a three-digit number that is encoded in 8421 BCD are 800, 400, 200, 100, 80, 40, 20, 10, 8, 4, 2, and 1.

10.3 Alphameric Codes

The use of combinations of Boolean variables to represent items of information is not limited to numerical values. A common requirement for data processing and storage is the representation of the alphabetic, numerical, and punctuation symbols that make up English text. Such codes are called *alphameric* codes, since they represent both *alpha*betic and nu*meric* information. Since there are 26 letters and 10 numerals, each code word will require at least six bits. This gives 64 possible code words, enough to cover the letters (in either upper- or lowercase, but not both), the numbers, and 28 punctuation symbols and control signals.

An exception to the six-bit minimum is the teletype code, often called the *Baudot code*, after its developer, that uses only five bits per character. As a result, some code words are used to represent more than one character. This code is presented in Table 10-4.

The five-bit TTY code is extended to represent more than 32 different characters by using the special "mode-change" character. The transmitter and receiver that handle the code must begin in the same mode, usually the alphabetic mode. Changes in mode are inserted into the sequence of code

Table 10-4:　The Five-Bit Teletypewriter Code

Code Word*	Character Represented	
	Alphabetic Mode	Numeric Mode
00	Blank	Blank
01	E	3
02		
03	A	-
04		
05	S	
06	I	8
07	U	7
10		
11	D	$
12	R	4
13	J	,
14	N	
15	F	
16	C	
17	K	(
20	T	5
21	Z	
22	L	)
23	W	2
24	H	
25	Y	6
26	P	0
27	Q	1
30	O	9
31	B	
32	G	
33		
34	M	.
35	X	/
36	V	
37	Mode change	Mode change

*Octal.

**Table 10-5:   Representative Portions of the
ASCII and EBCDIC Codes**

Character	ASCII*	EBCDIC*
A	41	C1
B	42	C2
C	43	C3
D	44	C4
E	45	C5
F	46	C6
G	47	C7
H	48	C8
I	49	C9
J	4A	D1
K	4B	D2
L	4C	D3
M	4D	D4
N	4E	D5
O	4F	D6
P	50	D7
Q	51	D8
R	52	D9
S	53	E2
T	54	E3
U	55	E4
V	56	E5
W	57	E6
X	58	E7
Y	59	E8
Z	5A	E9
0	30	F0
1	31	F1
2	32	F2
3	33	F3
4	34	F4
5	35	F5
6	36	F6
7	37	F7
8	38	F8
9	39	F9
Blank	00	40

*Hexadecimal.

words whenever they are required. The operation is similar to the up-shift/
down-shift mechanism used by typewriters. The effective number of bits per
character is increased above 5, since the mode-change characters add bits to
the data that are transmitted or stored and yet do not carry any information.

Some variations of the five-bit code use two mode-change characters, one to force a transfer into alpha mode and one to force a transfer to numeric mode. This technique assures that the transmitter and receiver are in the same mode following a mode change.

The use of a six-bit code to represent alphameric information may require mode-change characters just as the five-bit code required. The transfer from uppercase to lowercase letters may be needed, for example, with special *UPSHIFT* and *DOWNSHIFT* codes being used for this purpose.

A second approach is to expand the code again, using seven or more bits, so that the upper- and lowercase forms for a given letter can be given unique code words. In fact, the need for special characters (Greek symbols and the like) and special control signals has increased the code requirements until an eight-bit code has been developed and offered as a national standard. This code, the American Standard Code for Information Interchange, or *ASCII*, is presented, in part, in Table 10-5. A portion of a second eight-bit code (it seems that no so-called standard code is ever acceptable to everyone), the Extended Binary-Coded-Decimal Interchange Code, or *EBCDIC*, is also presented in Table 10-5.

There are many other alphameric codes that are in use. Code-to-code conversion is often required when various types of digital equipment are tied together to make a data-processing system. The conversion techniques that are given in Chapter 4 are sufficient for designing the necessary converters, although computer assistance is advisable when working with codes having six or more bits. MSI devices for conversion between common codes are also available.

EXAMPLE 10-4:   Design a combinational system that receives alphameric characters that are encoded in EBCDIC and indicates whenever a vowel is present at its input. Assume that code words other than the 37 listed in Table 10-5 will never appear at the input to the device.

Solution: The eight input bits may be called $A_1, A_2, ..., A_8$. The output function is labeled V. It has 5 TRUE conditions (A, E, I, O, and U), 32 FALSE conditions, and 219 redundancies. Use of the MINTE program gives

$$V = \bar{A}_3 \bar{A}_4 \bar{A}_7 A_8 + \bar{A}_3 A_4 A_6 A_7 \bar{A}_8 + A_3 \bar{A}_4 \bar{A}_5 \bar{A}_7 \bar{A}_8$$

10.4   Unit-Distance Codes

In some applications it is desirable that a code which represents numerical information have the *unit-distance* property, meaning that the code words that represent numerical values that differ by one numerical unit will be different in only one of their Boolean variables. Codes having this property are also called *cyclic codes*, *reflected codes*, or, in the case of binary numbering, the *Gray code* [6]. The unit-distance property means that, in counting up or down in numerical sequence, a succession of code words is obtained without ever changing more than one bit at a time.

This property is extremely useful in designing electromechanical systems such as rotational shaft-position encoders. A simple encoder that converts shaft position into a two-bit binary number that indicates which quadrant the shaft is positioned in, is introduced in Sec. 4.5.1. Consider the changes that occur when the shaft rotates from the second quadrant, 01, to the third quadrant, 10. The first bit changes from FALSE to TRUE, and the second bit changes from TRUE to FALSE. The encoder cannot be made so precisely that these two changes will occur exactly at the same time. Thus, the encoder will indicate a momentary value of either 00 or 11 as it passes from 01 to 10.

The elimination of this type of "transitional" error can be accomplished in two ways. One approach uses two (or more) sets of contacts within the encoder. A single contact is chosen as the "master," with one set of contacts preceding it and one set following it. Whenever the master contact changes its logical value, the set of contacts that is selected is also changed, with the new set being the contacts that are positioned away from the area of changing value. Two common multiple-contact techniques are the U-scan and V-scan contact arrangements [3]. This technique increases the cost of the encoder because of the need for extra contacts and for the brush-selection logic.

The second approach to the elimination of the errors caused by multiple changes in logical value is to recode the sequence of values so that, when going from one quadrant to the next, only one bit will change. This type of encoding scheme is called a *unit-distance code*.

EXAMPLE 10-5:   Give a unit-distance code for an encoder having four output values. Repeat for an encoder having eight outputs (half quadrant).

Answers:

Quadrant (deg)	Code		Half-quadrant (deg)	Code
0–89	00		0–44	000
90–179	01		45–89	001
180–269	11		90–134	011
270–359	10		135–179	010
			180–224	110
			225–269	111
			270–314	101
			315–359	100

Note that the sequences of code words, including the change from the last code word to the first, never involve more than one change in logical value at a time.

EXAMPLE 10-6: Can a counter that counts in a unit-distance code be made to operate asynchronously?

Answer: No. Since only one flip-flop changes in counting from one state to the next, there can be no use of asynchronous counting logic. The transitions of one flip-flop will never overlap the transitions of another.

The unit-distance property is useful in other applications as well. Designing a control counter so that its state-to-state transitions are unit-distant will eliminate the possibility of a transient control state appearing during state changes.

10.4.1. The Gray Code. One of the most common unit-distance codes is the Gray code. This code is closely related to natural binary and is also called *reflected binary code*. The four-bit Gray code sequence is shown in Table 10-6.

Table 10-6: The Four-Bit Gray Code

Quantity	Gray Code	Binary Code
0	0 0 0 0	0 0 0 0
1	0 0 0 1	0 0 0 1
2	0 0 1 1	0 0 1 0
3	0 0 1 0	0 0 1 1
4	0 1 1 0	0 1 0 0
5	0 1 1 1	0 1 0 1
6	0 1 0 1	0 1 1 0
7	0 1 0 0	0 1 1 1
8	1 1 0 0	1 0 0 0
9	1 1 0 1	1 0 0 1
10	1 1 1 1	1 0 1 0
11	1 1 1 0	1 0 1 1
12	1 0 1 0	1 1 0 0
13	1 0 1 1	1 1 0 1
14	1 0 0 1	1 1 1 0
15	1 0 0 0	1 1 1 1

The Gray code is particularly useful because of the ease with which it can be converted to and from its natural binary equivalent. The Gray-to-binary conversion algorithm is given in Eq. (10-1):

$$b_i = q_i \oplus b_{i+1} \tag{10-1}$$

where b_i is the ith bit of the binary code word, q_i is the ith bit of the Gray code word, and b_{i+1} is the bit of the binary word that is to the left (more significant) of the ith bit. The conversion proceeds from left ($i = n - 1$) to right ($i = 0$).

EXAMPLE 10-7:   Convert 1011 from Gray to binary.

Solution: The conversion begins at the left (most significant) end of the word with $i = 3$. The bit at $i + 1$ is considered to be 0, so

$$b_3 = q_3 \oplus 0 = 1 \oplus 0 = 1$$

Similarly,

$$b_2 = q_2 \oplus b_3 = 0 \oplus 1 = 1$$
$$b_1 = q_1 \oplus b_2 = 1 \oplus 1 = 0$$
$$b_0 = q_0 \oplus b_1 = 1 \oplus 0 = 1$$

Thus, 1011 (Gray) converts into 1101 (binary). This agrees with Table 10-6.

EXAMPLE 10-8:   Convert 1011010, 110111, and 1111 from Gray to binary.

Answer: 1101100, 100101, and 1010.

A combinational Gray-to-binary converter is shown in Fig. 10-2. The converter uses $n - 1$ EXCLUSIVE-OR gates to convert an n-bit code word.

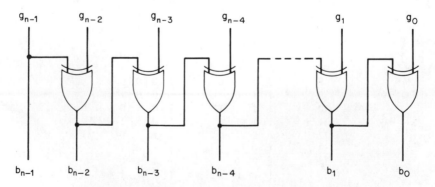

Figure 10-2. Combinational Gray-to-binary Conversion

Because of ripple, the binary output will require $n - 1$ times the delay through a single EXCLUSIVE-OR gate in order to settle, in the worst case.

A sequential Gray-to-binary converter is shown in Fig. 10-3. The Gray code word is loaded into the G shift register. The B register is initially cleared to all 0s. The *SHIFT* pulses transfer the sequence of Gray code bits into

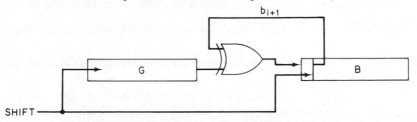

Figure 10-3. Sequential Gray-to-binary Conversion

the B register, most-significant bit first. The last bit of the binary word, b_{i+1}, is fed back and combined with the current g_i bit to produce the next value of b_i, in accordance with Eq. (10-1). The G register can be used in a circular fashion, with the g_i bits being shifted out of one end and the b_i bits being shifted into the other. This eliminates the need for a separate B register.

The reverse process, binary-to-Gray conversion, can be obtained from Eq. (10-1) by XORing b_{i+1} to both sides:

$$b_i \oplus b_{i+1} = g_i \oplus b_{i+1} \oplus b_{i+1} \qquad (10\text{-}2)$$
$$b_i \oplus b_{i+1} = g_i \qquad (10\text{-}3)$$

Equation (10-3) follows from (10-2) since $b_{i+1} \oplus b_{i+1} = 0$. Combinational and sequential forms for the binary-to-Gray conversion process can be developed by slightly modifying the logical circuits that are given in Figs. 10-2 and 10-3. This development is left to the problems at the end of this chapter.

EXAMPLE 10-9:   Convert 10110111 from binary to Gray.

Solution: It is convenient, but not necessary, to begin at the left. Since a high-order 0 is assumed, Eq. (10-3) gives

$$g_7 = b_7 \oplus 0 = 1 \oplus 0 = 1$$

Similarly,

$$g_6 = b_6 \oplus b_7 = 0 \oplus 1 = 1$$
$$g_5 = b_5 \oplus b_6 = 1 \oplus 0 = 1$$
$$g_4 = b_4 \oplus b_5 = 1 \oplus 1 = 0$$
$$g_3 = b_3 \oplus b_4 = 0 \oplus 1 = 1$$
$$g_2 = b_2 \oplus b_3 = 1 \oplus 0 = 1$$
$$g_1 = b_1 \oplus b_2 = 1 \oplus 1 = 0$$
$$g_0 = b_0 \oplus b_1 = 1 \oplus 1 = 0$$

Thus, the Gray-coded result is 11101100. This can be checked by Gray-to-binary conversion. The result is 10110111, which agrees with the original binary value.

Other properties of the Gray code and a discussion of its use in arithmetic operations are given in Ref. [6].

10.4.2. Unit-Distance BCD Codes.

The unit-distance property can be extended to codes that represent decimal numbers. This implies that the codes for digits that differ by only one unit will differ in only one bit. The codes for 0 and 9 must also be unit-distant. The unit-distance property is easily recognized by plotting the BCD code on a K-map, since code words that are unit-distant will be adjacent to each other in or at opposite ends of the same row or column. Figure 10-4 shows a unit-distance BCD code. The decimal values are plotted in the squares of the minterms that represent them.

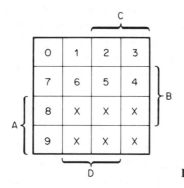

Figure 10-4. A Unit-distance BCD Code

The coding assignment shown in Fig. 10-4 maintains the next-neighbor relationship throughout the 0,1,...,8,9 sequence. Zero and 9 are also next neighbors, since they are at opposite ends of the same column. A tabulation of the code represented by the map of Fig. 10-4 is presented in Table 10-7.

Many other unit-distance BCD codes are possible. A weighted, unit-distance BCD code (or any weighted, unit-distance code, for that matter) cannot be defined, however. This is easy to prove. Each increasing numerical value represents an increase of only one unit. To be unit-distance encoded, it must also represent a one-bit change in the code word. Thus, a bit that goes from FALSE to TRUE must have a weight of $+1$ and a bit that goes from TRUE to FALSE must have a weight of -1. Since the same bit will go from TRUE to FALSE *and* from FALSE to TRUE at least once each during the counting sequence (since the sequence forms a closed path), then any bit would have to have a variable weight: $+1$ at one point and -1 at another. This rules out the possibility of a unit-distance code being weighted.

Table 10-7: A Unit-Distance BCD Code

Decimal Value	Code Word
0	0 0 0 0
1	0 0 0 1
2	0 0 1 1
3	0 0 1 0
4	0 1 1 0
5	0 1 1 1
6	0 1 0 1
7	0 1 0 0
8	1 1 0 0
9	1 0 0 0

A unit-distance BCD code can never be self-complementing, as well. This is clear from consideration of the codes for 4 and 5. To be unit-distant they must differ in only one bit, since 4 and 5 are next to each other in the decimal sequence. But, to be self-complementing, the code words for 4 and 5 must differ in *every* bit, since 4 and 5 are a 9's complemented pair. These two conditions cannot be met simultaneously.

A unit-distance BCD code can be reflected, however, since the 9's complementation of a reflected BCD code requires that only one bit be changed. This combination of properties is most easily seen by plotting the code on a K-map. Unit distance requires that the codes for the sequence of decimal values (including the 9 $\longrightarrow$ 0 change) be next to each other in or at opposite ends of the same row or column. Reflection implies that the codes for a decimal value and its 9's complement must be at mirrored locations across the TRUE/FALSE axis of the reflection bit.

A reflected, unit-distance BCD code is mapped in Fig. 10-5. The reflection variable is A. Many other codes of this type are possible.

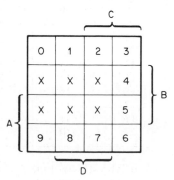

Figure 10-5. A Reflected, Unit-distance BCD Code

One problem that arises when a unit-distance BCD code is used is the maintenance of the unit-distance property over a multidigit decimal number. Consider a two-digit number that is represented by the unit-distance code given in Fig. 10-4. As the decimal value moves from 00 to 09, the unit-distance property is satisfied. The 09 ⟶ 10 transition moves from the code word 00001000 to 00010000, however, requiring that two bits change in logical value. This double change occurs whenever more than one digit changes.

A solution to this problem is to modify the usual decimal counting sequence so that only one digit changes value at a time. This type of numbering is called *cyclic decimal*. It is the decimal equivalent of the Gray code/ binary relationship. Consider a two-digit cyclic decimal counting sequence. It begins at 00 and proceeds to 09 just as normal decimal notation does. From 09 it goes to 19, however, so that only the tens digit changes. This maintains the unit-distance property.

From 19 the cyclic decimal code goes to 18,17,16,...,11,10, and from 10 it goes to 20. The two-digit cyclic decimal and normal decimal sequences for counting through the first six decades are presented in Table 10-8.

Table 10-8: Cyclic Decimal/Decimal Relationships

Decimal Value	Cyclic Equivalent	Decimal Value	Cyclic Equivalent	Decimal Value	Cyclic Equivalent
00	00	20	20	40	40
01	01	21	21	41	41
02	02	22	22	42	42
03	03	23	23	43	43
04	04	24	24	44	44
05	05	25	25	45	45
06	06	26	26	46	46
07	07	27	27	47	47
08	08	28	28	48	48
09	09	29	29	49	49
10	19	30	39	50	59
11	18	31	38	51	58
12	17	32	37	52	57
13	16	33	36	53	56
14	15	34	35	54	55
15	14	35	34	55	54
16	13	36	33	56	53
17	12	37	32	57	52
18	11	38	31	58	51
19	10	39	30	59	50

The largest two-digit cyclic decimal number is 90, corresponding to a normal value of 99; the next cyclic number is 190.

A cyclic decimal number can be converted to its decimal equivalent by using the following algorithm:

1. The left (most-significant) digit of the cyclic decimal number is used as the most-significant digit of the decimal number.

2. Each time a digit of the decimal number has an odd value, the next digit of the cyclic decimal number is 9's complemented in order to obtain its true decimal equivalent.

3. Each time a digit of the decimal number has an even value, the next digit of the cyclic decimal number is used unchanged in forming the true decimal number.

EXAMPLE 10-10:   Convert 53 from cyclic decimal to normal decimal.

Solution: The first digit is never changed, making the decimal number begin with a 5. Since 5 is odd, the next digit of the cyclic number is 9's complemented, giving a 6. Thus, 53 (cyclic) converts to 56 (decimal).

===

EXAMPLE 10-11:   Convert 109365, 243751, and 9999 into their decimal equivalents.

Answers: 190334, 243258, and 9090.

===

A combinational system for implementing the cyclic-to-decimal conversion for a four-digit number is shown in Fig. 10-6. The high-order cyclic digit passes through the converter unchanged. A combinational circuit examines the BCD code for the first digit and, if it is odd, signals a second combinational circuit to convert the BCD code for the next cyclic digit into its 9's complement. Otherwise, the next cyclic digit is unchanged. A second combinational circuit detects whether or not the D_{100} digit is odd and conditions the complement circuit for the tens decade. The process can be continued for a cyclic decimal number of any length.

If a BCD code that is *both* unit-distant and reflected is used to encode the cyclic decimal number, then each of the 9's complementation circuits shown in Fig. 10-6 can be implemented with a single EXCLUSIVE-OR gate. This gate complements the reflection bit if the previous decimal decade has an odd value.

The only additional logic that is required beyond the XOR gates is the odd/even detection circuit. The reflected, unit-distance BCD code can be chosen so as to simplify the circuitry required for this purpose, as the following example shows.

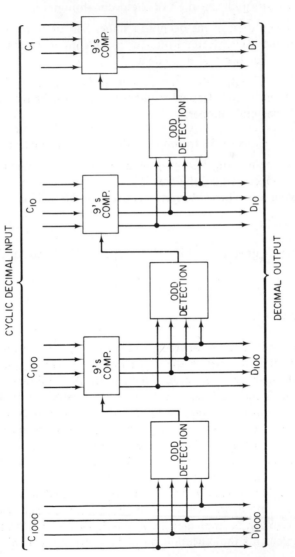

Figure 10-6. Combinational Cyclic Decimal-to-decimal Conversion

EXAMPLE 10-12: The map in Fig. E10-12a defines a reflected, unit-distance BCD code. The reflection variable is A. Design the odd/even detection logic for this code and compare its cost with that required by the similar code given in Fig. 10-5.

Solution: The detection logic considers the odd values as being TRUE and the even values as FALSE. The six unused code words can be used as redundancies. Therefore, mapping the ODD function gives the map in Fig. E10-12b, so that

$$ODD = A\bar{B} + \bar{A}B\bar{C} + \bar{A}BD + AC\bar{D}$$

for a cost of 15 c.u. A similar grouping for the code given in Fig. 10-5 yields the map in Fig. E10-12c, where

$$ODD = AB + A\bar{C}\bar{D} + ACD + \bar{A}\bar{C}D + \bar{A}\bar{B}C\bar{D}$$

for a cost of 20 c.u.

The detection/complementation circuitry for two successive decades of the cheaper code is shown in Fig. E10-12d.

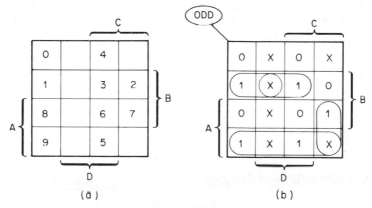

Figure E10-12a, b

Figure E10-12c

CYCLIC DECIMAL INPUT

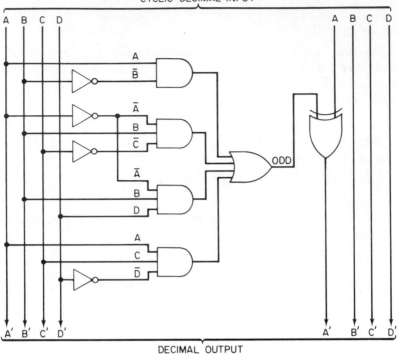

DECIMAL OUTPUT

Figure E10-12d

10.5 *Error Detection and Correction*

The final code properties that are considered in this chapter are the capabilities for error detection and error correction. When code words are transmitted from one device to another or when they are stored and later retrieved for use, errors may occur in one or more of the bits that make up the code words. Error detection is the capability of recognizing that such an error did occur and that the received or retrieved code word is not the same as the word that was transmitted or stored. The capability for error correction goes beyond that required for error detection in that it allows the bit (or bits) that are in error to be identified and corrected, thereby reconstructing the original code word, even in the face of errors.

The error-detection and error-correction capabilities of a code are closely related. This section discusses a few simplified concepts which only serve as an introduction to the general subject of error-protected encoding. Interested readers are referred to Refs. [2], [4], and [5] for further information.

10.5.1. Error Detection. Chapter 1 shows that an n-bit code can be used to represent 2^n different items of information. If each of the 2^n code words is assigned some valid meaning, error detection is impossible. Any bit or bits in error will change a given code word into some other code word, but since the error-caused code word will also be valid, there is no way to recognize the error.

Error-detection capability is obtained, at some sacrifice of the information-carrying capability of a code, by identifying some of the code as errors. A BCD code is an example, since only 10 of the 16 code words are used. If one of the other 6 code words appears, it can be recognized as an invalid, erroneous code word.

The simplest code with error-detection capability is the *parity-checked code*. If n-bit code words are to be used, then an extra bit is added to each word, forming words that are $n + 1$ bits long. The extra bit, or *parity bit*, is chosen so that the total number of TRUE variables in each code word is always even (even parity) or odd (odd parity). Since the TRUE or FALSE value of the parity bit is totally dependent on the states of the other n bits of the code word, the parity bit does not add any initial information. Thus, even though each code word has $n + 1$ bits, only 2^n different error-free code words are available. The parity-checked code gains its error-detection capability at the expense of one-half of its information-carrying capability.

Since a single bit in error will always change the number of TRUE bits in the word from even to odd, or the reverse, the "parity" of a code word containing a single bit in error will be incorrect. Thus, a parity-checked code permits all single errors to be detected.

The identification of which bit is incorrect is impossible, however, so the only allowable action is a rejection of the erroneous word; it cannot be corrected. Words containing double errors (two bits in error) cannot be detected. Thus, the smallest number of bits in error that go undetected when a parity-checked code is used is 2. This value is a useful figure of merit in comparing error-detection schemes.

The value of the parity bit can be obtained by EXCLUSIVE-ORing the other bits of the code word. The XOR of these bits will be TRUE if an odd number of them are TRUE. Including the XOR output as the parity bit will assure even parity, therefore.* Odd parity is generated by using the complement of the XOR output as the parity bit.

A simple sequential parity generation scheme uses a shift register and a single T flip-flop, as shown in Fig. 10-7. The original code word is loaded into the shift register. The *ODD/EVEN* flip-flop is cleared to the *ODD* state at that time. The *SHIFT* pulses cause the bits of the code word to shift out of the register. Each TRUE bit causes the *ODD/EVEN* flip-flop to change

---

*See Prob. 2-2f.

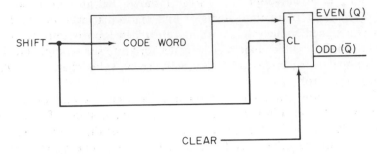

Figure 10-7. A Sequential Parity-generation System

states. Thus, after an even number of TRUE values, the $ODD/EVEN$ flip-flop will be in the ODD state, meaning that the use of the $\bar{Q} = ODD$ output as the parity bit will give an odd number of TRUE values in the total code word. An odd number of TRUE bits will leave the flip-flop in the $EVEN$ state, so that using the $Q = EVEN$ output as the parity bit will assure even parity. A similar system can be used for error detection to verify that code words that are received have the proper parity.

10.5.2. Error Correction.

The single parity bit system can be expanded to include an error-correction capability if a sequence of code words is treated as a block of bits. A common technique is the *horizontal/vertical* (H/V) *parity system* or *double parity*. Each code word in a given block of words is assigned its own parity bit, the *horizontal parity bit*. When a complete block of data has been transmitted, an additional character, the *vertical parity character*, is sent. This character is generated so that each "column" of the block has a specified odd or even parity. This system requires some special control characters such as an end-of-block character, or *EOB*, to signal the receiver that the vertical parity character is coming next.

If a single error occurs during the transmission of the entire block of data, the *EOB* character, and the vertical parity character, then horizontal parity will be incorrect in one character (row), and vertical parity will be incorrect in one column. The erroneous bit will be located at the intersection of the row and column that had the wrong parity. That bit can be complemented in order to remove the error from the block. Thus, single errors can be detected *and* corrected.

Double errors can also be detected, but they cannot be corrected. If both errors occur in the same word, then two columns will be in error, but the erroneous character will be unlocatable, since the double error restores its row to the correct parity. Similarly, double errors in the same column will mark two words as being incorrect but will not indicate which bits of the

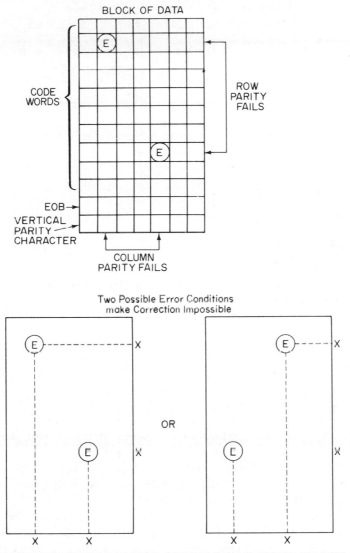

Figure 10-8. Double Errors in the Horizontal/Vertical Parity Checked Code

words are incorrect. Double errors in different words and columns will also be uncorrectable, as Fig. 10-8 shows.

Another troublesome problem arises whenever a triple error occurs. Figure 10-9 shows that three bits in error can appear to be the same as a single error that occurs in another place. Thus, if the receiver is designed to correct

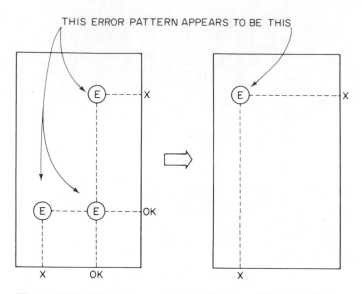

Figure 10-9. Triple Errors in the Horizontal/Vertical Parity Checked Code

the bit in all situations that appear as single bits in error, a triple error may go undetected. In fact, the triple error may cause a fourth bit to be "corrected," giving a supposedly correct block of code words that actually contains four incorrect bits.

The last situation points up the relationship between error detection and error correction. The double parity system will detect all single, double, and triple errors, plus many combinations of more than three errors. The fewest bits in error that can go completely undetected is a rectangular arrangement of four. The system will also permit the correction of single errors, but not simultaneously with the detection of triple errors. A given system can have one capability or the other, but not both.

The state-of-the-art of error correction encoding goes far beyond the H/V parity scheme. Hamming codes and the wide variety of cyclic error-correcting codes are examples. Peterson [4] is a suggested reference for those seeking more information on this topic.

10.6 Summary

The representation of information by groups of Boolean variables is called *coding*. Different codes offer different useful properties. Self-complementing or reflected BCD codes make 9's complementation very easy to implement. Some codes that represent quantitative information are *weighted*, meaning that fixed values, or weights, can be associated with each bit of the code and that the sum of the weights of the TRUE bits in a given code word will always equal the numerical quantity that is represented by the code word.

Unit-distance coding is useful in some applications, since it assures that a pair of code words which represents numerical quantities that differ by one unit will also differ in only one bit. Thus, only one bit changes at a time when counting up or down in a unit-distance code. The Gray code is a unit-distance code that is closely related to natural binary. Conversion between Gray and binary is easily accomplished by using the EXCLUSIVE-OR function. Unit-distance BCD codes are also available, with cyclic decimal numbering being used to maintain the unit-distance property over a multi-digit decimal number.

Codes that offer the capability of error detection and/or correction are also useful, especially when data are transmitted or stored. These capabilities are obtained at some sacrifice in the information-carrying capacity of the code. The parity-checked code is the simplest error-detecting code. The double parity coding system allows some error correction as well.

10.7 Did You Learn?

1. Why a BCD code must have at least four Boolean variables and six redundancies?

2. What a self-complementing BCD code is? How to identify the self-complementing property? How such a code can be 9's complemented both combinationally and serially?

3. What a reflected BCD code is? How it can be identified? How it can be 9's complemented?

4. What a weighted code is? What effect the bias constant has? How to find the weights and the bias constant?

5. How a code can be both weighted and self-complementing? The special relationship between these two properties for a code whose weights sum to 9?

6. How codes can represent both alphabetic and numerical information?

7. How a code's information-carrying capability can be extended by using mode-change characters?

8. What the unit-distance property is? How it can eliminate momentary states during a sequence of changing code words?

9. How to convert from Gray to binary and vice versa?

10. How to identify a unit-distance code by plotting it on a K-map?

11. How cyclic decimal numbering maintains the unit-distance property over a multidigit number?

12. How a unit-distance BCD code can also have the reflection property? How this is useful in converting from cyclic decimal to normal decimal numbering?

13. What error detection and error correction are?

14. Why information-carrying capability must be sacrificed in order to obtain either error detection or correction?

15. How error correction may cause certain classes of errors to go undetected?

16. How the single and double parity schemes provide error detection and error correction?

10.8   References

[1] Booth, Taylor L., *Digital Networks and Computer Systems*. New York: John Wiley & Sons, Inc., 1971.

[2] Gallager, Robert G., *Information Theory and Reliable Communication*. New York: John Wiley & Sons, Inc., 1968.

[3] Kintner, P. M., *Electronic Digital Techniques*. New York: McGraw-Hill Book Company, 1968.

[4] Peterson, W. W., *Error Correcting Codes*. New York: John Wiley & Sons, Inc., 1961.

[5] Peterson, W. W., "Error Correcting Codes," *Scientific American*, 215, No. 3 (February 1962), 96–110.

[6] Walker, Monty, "Decipher the Gray Code," *Electronic Design* (February 15, 1970), 70–74.

10.9   Problems

General Concepts

10-1.   Complete each of the following BCD codes so as to satisfy the self-complementing property.

a.	Decimal Value	Code Word	b.	Decimal Value	Code Word	c.	Decimal Value	Code Word
	0	1101		0			0	0000
	1	1111		1			1	
	2	0110		2			2	0010
	3	0011		3			3	
	4	0101		4			4	0100
	5			5	0101		5	
	6			6	0110		6	0110
	7			7	0111		7	
	8			8	1100		8	1000
	9			9	1111		9	

10-2. Can the following partial BCD code be completed so as to be self-complementing? Explain.

Decimal Value	Code Word
0	0000
1	0001
2	0010
3	0011
4	1111
5	
6	
7	
8	
9	

10-3. Complete each of the following BCD codes so as to satisfy the reflection property. Identify the reflection bit. Plot the code on a K-map and verify the reflection property by inspection of the map.

a.	Decimal Value	Code Word	b.	Decimal Value	Code Word	c.	Decimal Value	Code Word
	0	0010		0	1000		0	0000
	1	0011		1	1010		1	
	2	0111		2	0000		2	0001
	3	1111		3	0010		3	
	4	1110		4	0100		4	0010
	5			5			5	
	6			6			6	0111
	7			7			7	
	8			8			8	1111
	9			9			9	

10-4. Determine whether or not each of the following codes is weighted. If so, give the weights and the bias constant. Indicate whether or not the code is also self-complementing.

Decimal Value	Code I				Code II				Code III				Code IV					Code V		
	A	B	C	D	E	F	G	H	I	J	K	L	M	N	P	Q	R	S	T	U
0	0	0	0	0	0	0	1	0	0	0	0	1	0	0	0	0	0	0	0	0
1	0	0	0	1	0	0	1	1	0	0	0	0	0	0	0	0	1	0	0	1
2	0	0	1	1	0	1	0	0	0	0	1	1	0	0	1	1	1	0	1	0
3	0	1	0	0	0	1	0	1	0	0	1	0	1	1	0	0	0	0	1	1
4	1	0	0	0	1	0	0	0	1	1	0	1	1	1	0	0	1	1	0	0
5	0	1	1	1	1	0	0	1	1	1	0	0	1	0	0	1	0	1	0	1
6	1	0	1	1	1	0	1	1	1	1	1	1	1	0	0	1	1	1	1	1
7	1	1	0	0	1	1	0	0	1	1	1	0	1	0	1	1	0	1	1	0
8	1	1	1	0	1	1	1	0	1	0	0	1	1	0	1	0	0			
9	1	1	1	1	1	1	1	1	1	0	0	0	1	0	1	0	1			

10-5. Prove that a reflected BCD code cannot be weighted.

10-6. List the complete character string that is required to transmit the following message by using the five-bit teletypewriter code given in Table 10-4.

THE SUM IS $12.568.

10-7. Convert each of the following binary numbers into Gray code. Then convert them from Gray code back into binary.

a. 1010101010
b. 1111111111
c. 0111011101

d. 11001100111
e. 10101
f. 100001111

10-8. The K-map in Fig. P10-8 defines a unit-distance code for the decimal values from 0 to 15. It is similar to the Gray code. List the sequence of code words and verify the unit-distance property. Then, develop a second unit-distance code via the K-map process.

10-9. Convert each of the following cyclic decimal numbers into their normal decimal equivalents.

a. 123456789
b. 102030405
c. 99999999

d. 22222222
e. 122333445
f. 987652311

10-10. Write the sequence of cyclic decimal numbers that is generated in counting from 563 to 625 (natural decimal).

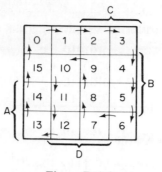

Figure P10-8

10-11. Develop an algorithm for converting from decimal into cyclic decimal. Use it to check the results of Prob. 10-9.

10-12. Determine the value of the parity bit that will give even parity to each of the following code words.

a. 1010101
b. 1111111
c. 1100000

d. 100100100100
e. 110011001111
f. 1000000

10-13. The following code words are received from a digital transmitter. They should have odd parity. Mark any erroneous words.

a. 1100111
b. 1010011
c. 1011111

d. 0000000
e. 1111111
f. 1010111

10-14. The following code words are to be transmitted by a vertical parity system. Horizontal parity is to be odd; vertical parity is to be even. An all-1s character marks the end of the block (EOB). Calculate the necessary parity bits and list the complete bit sequence that is involved in the transmission.

$$
\begin{array}{ccccccc}
1 & 0 & 1 & 0 & 1 & 1 & 0 \\
0 & 1 & 0 & 1 & 1 & 1 & 1 \\
1 & 1 & 0 & 0 & 0 & 0 & 0 \\
0 & 0 & 1 & 1 & 0 & 1 & 0 \\
1 & 1 & 0 & 1 & 0 & 1 & 0 \\
1 & 1 & 1 & 1 & 1 & 1 & 1 \\
\end{array}
$$

10-15. Repeat Prob. 10-14, using the ASCII code words for the following characters as the block of data. A *period* acts as the *EOB* character. The codes are given in Table 10-5.

ABXC,1259

10-16. The following block of data is received from the transmission system described in Prob. 10-14. An all-1s *EOB* is used.

```
0101010 0 ←— horizontal parity bits
1010101 1
0100110 1
0011111 0
1010001 0
1100110 1
1111111 0
```
vertical parity character → 0011110 1

a. Find any parity failures.
b. Correct the error, if possible.
c. Assuming that there are three bits in error, show several possible locations.

Functional Design Problems

*10-17. Design a combinational converter that will convert code III, Table 10-0, into 8421 BCD. Show an RTL logic diagram.

*10-18. Design a 6, 3, 1, −1 up-counter. Use the 6, 3, 1, −1 code that is given in Table 10-0. Show a TTL logic diagram. Give a complete state diagram. Verify that if the outputs from the counter are all complemented, the counter will count *down* the 6, 3, 1, −1 sequence.

*10-19. Design a 2421 BCD up/down-counter (see Table 10-0). Let the control signal be *U*, with up-counting being enabled when *U* is TRUE. Use J-K flip-flops.

*10-20. Design an up-counter for code II, Table 10-1. Show that the counter can be made into a down-counter by complementing its *X* output.

*10-21. Design a combinational converter that will convert the alphabetic characters of the five-bit teletypewriter code of Table 10-4 into their ASCII equivalents, as given in Table 10-6.

*10-22. Design a combinational converter that receives a three-digit cyclic decimal number that is encoded in 2421 BCD code (Table 10-0) and converts it to natural decimal. Show an RTL circuit diagram.

*10-23. Repeat Prob. 10-22 for code II, Table 10-1.

Systems Design Problems

*10-24. Design a digital receiver that will receive the five-bit teletypewriter code, Table 10-4, and detect the mode-change characters. The outputs from the receiver are the five-bit character and a mode control signal, *ALPHA*.

A *CLEAR* signal places the system in the *ALPHA* mode. Thereafter, mode changes force $ALPHA = 0$ and $ALPHA = 1$ in toggling fashion. The system should delete the mode-change character from the character string that appears at its output.

*10-25. A digital transmission system receives seven bits of information that are to be serially transmitted over a communications link. Each transmission begins with a TRUE bit, the *START* bit. The seven data bits then follow. A parity bit that forces even parity is generated and transmitted immediately after the last data bit. This transmission is followed by at least one period during which the output is kept FALSE. The system receives the seven input signals from toggle switches. The transmission cycle is initiated by a *SEND* control signal. The controller halts after the proper number of clock periods. The clock should be adjusted to give a transmission rate of 500 bits per second. Design the complete system.

*10-26. Design a digital receiver that will operate with the transmitter of Prob. 10-25. The *START* bit will signal the beginning of each code word. The receiver should accept the word and present it in parallel at its output. Erroneous parity should be indicated.

11 Analog/Digital Conversion

11.0 Introduction

The concept of a generalized digital system is introduced in Chapter 1 (see Fig. 1-0). The input and output signals that are received and/or generated by such a system are shown to be of two types: digital signals and analog signals. Some digital systems deal only with digital signals. Data-processing computers are a common example. Their input data come from punched cards, paper tape, digital magnetic tape, or some other digital media. Computed output data are printed or punched or recorded in some other digital fashion.

In many digital systems, however, nondigital (analog) information must be received, processed, and returned to the user. Typical analog data include temperature, pressure, force, speed, or other continuous descriptors of real-world conditions. These types of signals are first converted into an appropriate form by transducers. Electrical transducers that convert their input variable into a voltage or current are used in conjunction with electronic digital systems. Other types of transducers are in use, however, such as those having a pressure output that are used with fluidic digital systems.

The output signals from the transducers are converted into digital form by *analog-to-digital converters* (ADCs). The digital code words that are produced by this conversion describe the input conditions to the digital system. These code words are processed by the system in order to produce the desired output data.

In many situations output data must be converted back into a continuous form. Typical examples of this requirement are systems whose digital outputs must actuate a servo system, drive a meter or an X-Y recorder, or operate a cathode-ray-tube display unit. The devices that perform this conversion are called *digital-to-analog converters* (DACs).

This chapter discusses the basic concepts of digital-to-analog and analog-to-digital conversion. Voltage-oriented converters are considered. The major emphasis is upon the logic aspects of operating these devices and of using them in conjunction with digital systems that have been designed à la Chapters 1–7. The design of DACs is really more electronic than digital in nature. As a result, the section on digital-to-analog conversion is brief, with electronic considerations kept to a minimum. References to more detailed information sources are provided.

It is also significant that a wide variety of DACs and ADCs are now available from commercial sources, with all the necessary electronics and logical circuitry already designed and conveniently packaged. These devices may therefore be considered as predesigned components rather than as subsystems that must be designed and constructed. This situation is similar to the effect that the widespread availability of flip-flops, gates, and other logical devices has had upon digital design, in that the component-level design of these devices is no longer required, or expected, of a digital systems designer.

11.1 Digital-to-Analog Conversion

Figure 11-0 is a general diagram of a voltage-output DAC. The digital code word that is to be converted into analog form is applied to the input of the

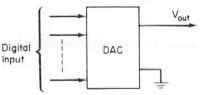

Figure 11-0. A Voltage-output Digital-to-Analog Converter

DAC. The internal circuitry of the DAC interprets the TRUE/FALSE values of the bits of the digital code word and produces the appropriate output voltage, V_{out}.

The output voltage can be related to the digital input in three ways: random, linear, or nonlinear. A random DAC must be defined by a tabulation of the coded inputs and the voltages that each should produce. An example of this type of tabulation is given in Table 11-0. A random DAC can be constructed as shown in Fig. 11-1. Each input code word is used to close a relay that connects the appropriate voltage to the DAC output. The code words are decoded by using normal combinational logic. Each of the logical signals is buffered by an appropriate relay driver circuit. Solid-state switching can be used if the DAC must operate at higher speeds.

Table 11-0: A Random DAC

Digital Input	Voltage Output (V)
000	$+10.0$
001	-10.0
010	$+5.0$
011	-3.0
100	-1.5
101	$+3.6$
110	$+7.2$
111	0.0

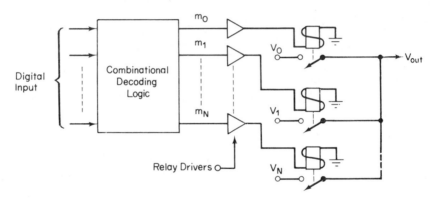

Figure 11-1. A Random DAC

A random DAC having n input bits may require as many as 2^n voltage sources and 2^n switches or relays. As the number of input bits increases, this requirement increases exponentially.

The random DAC may exhibit erratic behavior during the period just following a change in its digital input. An asynchronous change may cause the brief appearance of transition states at the input, and several relays or switches may be activated in rapid sequence. This will cause a rapid variation in V_{out}. Also, if two or more switches are closed at the same time, the short circuit that results may damage the switches or the voltage sources. Thus, the use of fully synchronous input changes or a unit-distance coding technique is generally advisable with a random DAC.

A linear DAC differs from the random DAC in that its output voltage is directly related to the magnitude represented by its digital input. Linear DACs can also be made to operate with signed numbers, with the polarity of the output voltage changing in accordance with the sign of the input code word.

A linear DAC can be characterized by two parameters. One is the *unit voltage increment*, which is the change in voltage that corresponds to increasing the digital input by one unit. This is the smallest change that the output voltage can make; it is also called the *step voltage*, the *unit voltage*, or the *quantization voltage*. The second parameter is the number of bits that the DAC receives at its input. An *n*-bit DAC can receive 2^n different code words. If the unit voltage is represented by v, then the range of the output voltage of the DAC must be $2^n v$ or less. It will be less than this value if not all the 2^n code words are allowed, as when BCD code is used as the input.

The voltage output from a linear DAC can begin at 0 and increase to $V_{max} \leq 2^n v$, or it can be biased to work between a negative lower limit and a positive upper limit. A tabulation describing a typical linear DAC is given in Table 11-1. The unit voltage for this DAC is 0.1 V.

Table 11-1: A Three-Bit Binary DAC

Binary Input	Voltage Output (V)
000	0.0
001	0.1
010	0.2
011	0.3
100	0.4
101	0.5
110	0.6
111	0.7

A linear DAC can be constructed according to the schematic given for a random DAC in Fig. 11-1. A better approach, however, is to use a weighted code to represent the input quantity and to implement the DAC operation by using a voltage summing amplifier. This type of arrangement is shown, in generalized form, in Fig. 11-2. Only *n* switches, $n + 1$ fixed-gain voltage

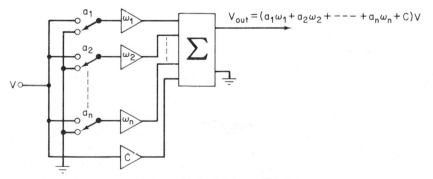

$$V_{out} = (a_1 \omega_1 + a_2 \omega_2 + --- + a_n \omega_n + C)V$$

Figure 11-2. A Linear DAC

amplifiers (one extra to allow for the bias constant, if the weighted code has one), and one voltage summing amplifier are required. Each TRUE bit of the input code word closes a switch, thereby applying a voltage that is proportional to the weight of that bit to the summing amplifier. Input bits that are FALSE leave their switches tied to 0. The summing amplifier adds the various voltages that it receives, including any fixed-bias voltage, and produces the output voltage.

EXAMPLE 11-0: Show a DAC that will operate with one digit of 8421 BCD code.

Answer: See Fig. E11-0.

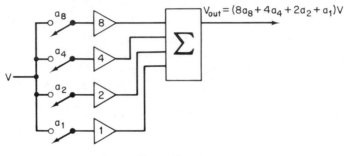

Figure E11-0

EXAMPLE 11-1: Show a DAC for use with one digit of the 6, 3, 1, -1 code.

Answer: See Fig. E11-1.

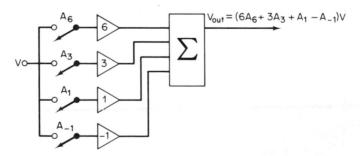

Figure E11-1

EXAMPLE 11-2: Show a DAC for use with one digit of the XS3 code.
Answer: See Fig. E11-2.

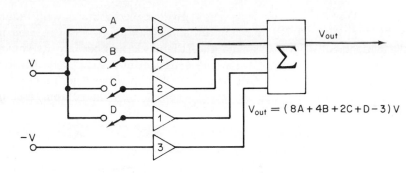

$$V_{out} = (8A + 4B + 2C + D - 3)V$$

Figure E11-2

EXAMPLE 11-3:   Show two ways to arrange a two-digit DAC for use with 8421 BCD code.

Answers: See Fig. E11-3a and b.

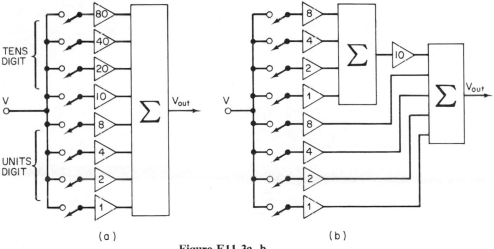

(a)   (b)

Figure E11-3a, b

A DAC of the type illustrated in Exs. 11-2 and 11-3 can be designed for any weighted code. The summing amplifier can be constructed from an operational amplifier. The various weighting voltages are then obtained by using a single voltage source and varying the gain that each input receives during the voltage summing process. This can be accomplished by using differing input impedances between the switched voltage sources and the summing amplifier.

Also, special resistance networks, notably the *weighted resistor network* and *ladder decoder networks*, can be used to perform both the weighting and voltage summing operations. Details on all of these facets of the design of DACs are given in Refs. [2], [3], and [4].

Linear DACs can be made to operate with signed codes as well. Figure 11-3 shows how an analog inversion amplifier (gain of -1) can be used to implement the conversion of a sign-and-magnitude number into its analog equivalent. The conversion of a sign-and-complement number is even easier, since the sign bit has a numerical weight just like the other bits. Figure 11-4 shows a five-bit DAC that operates with 2's complement binary input. A similar arrangement can be used with signed decimal numbers.

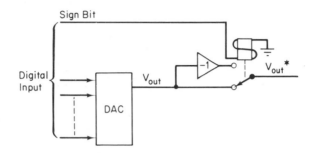

Figure 11-3. A DAC for use with Sign-and-Magnitude Code

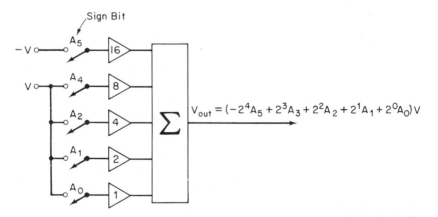

$$V_{out} = (-2^4 A_5 + 2^3 A_3 + 2^2 A_2 + 2^1 A_1 + 2^0 A_0) V$$

Figure 11-4. A 2's Complement DAC

The nonlinear DAC generates an output voltage that is nonlinearly related to the magnitude of its digital input. Typical input/output relationships include logarithmic and square-root converters. These types of DACs are used in special-purpose situations. Their design is not discussed herein.

Of major concern to the digital designer is the widening availability of DACs that are preengineered and packaged as complete conversion systems (Refs. [1] and [5]). These devices offer low cost and standardization. Most commercially available DACs are designed for use with binary or 8421 BCD input. This may require some code conversion prior to the input to the DAC. The component parts of the DAC, e.g., the voltage switches, the weighting resistance networks, or the summing amplifier, are also available as predesigned components. Thus, a designer can assemble off-the-shelf components into a "customized" DAC.

11.2 Analog-to-Digital Conversion

The analog-to-digital conversion process is closely related to its inverse, digital-to-analog conversion. In fact, almost all ADCs use a DAC as a component part of their operational hardware. A general diagram for an ADC is shown in Fig. 11-5. The voltage that is to be converted, or *digitized*, to use

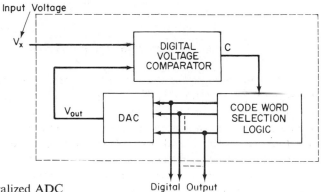

Figure 11-5. A Generalized ADC

the common jargon, is applied to the ADC. This voltage, V_x, is compared with the output from a DAC, V_{out}. The comparison device is a *digital voltage comparator*, which is usually a high-gain differential amplifier whose logical output, C, assumes one of the two logical voltage levels when $V_x < V_{out}$ and assumes the other voltage level when $V_x > V_{out}$.

The logical control system of the ADC uses the output signal from the comparator to determine how to change the digital code word that is driving the DAC. This code is increased or decreased until the DAC output agrees with the voltage input within some tolerance. This tolerance can be no better than the quantization unit of the DAC. Thus, a more accurate ADC will require a more complex DAC and will probably take more time to make its conversion.

The key to the operation of the ADC is the way in which its control logic seeks to find the digital code that most closely represents the input voltage. There are several common techniques, each offering a trade-off between operational usefulness and hardware complexity. The basic organization of each of these types is that shown in Fig. 11-5.

11.2.1. The Ramp ADC.
The block diagram for a *ramp* ADC is shown in Fig. 11-6. The DAC is driven by an up-counter that counts in the code that is compatible with the DAC. The conversion process begins with the counter being reset to 0. The system clock is then used to increment the counter, with the counting continuing until the output of the DAC becomes

greater than the input voltage. This will be indicated by a change in the logical value of the comparator output.

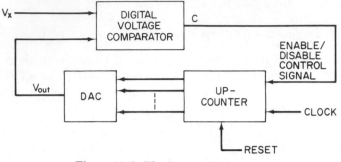

Figure 11-6. The Ramp ADC

The control logic for this type of converter is quite simple. One disadvantage of this technique is the variable period that the conversion process takes. The conversion time is linearly related to the magnitude of the input voltage. The counting speed is determined by the rate at which the DAC input can be changed, and the time required for its output voltage and the comparator output to settle to their correct values.

A second problem inherent in the ramp conversion technique arises if the input voltage changes during the conversion process. If V_x decreases, the counting process may stop, but it cannot reverse its counting sequence in order to reach the lower value. This problem can be solved by using a sample-and-hold circuit [2] that stores the value that the input voltage has at the beginning of the conversion and keeps that value constant during the conversion.

The ramp ADC must be reset in order to start a second conversion cycle. A second type of ADC eliminates the need for this reset, in that it attempts to follow any changes in the input voltage.

11.2.2. The Tracking ADC. The tracking ADC replaces the up-counter shown in Fig. 11-6 by an up/down-counter. Up-counting is enabled whenever the comparator indicates that $V_x > V_{out}$. Down-counting is enabled when $V_x < V_{out}$. The counter will count up or down until the value of V_{out} that is closest to the current value of V_x is reached. It will then cycle about that value, counting in an up-down-up-down-$\cdots$ sequence that causes V_{out} to range above and below V_x.

If V_x changes in value, then the counter will increment or decrement in order to follow the change. This is the so-called "tracking" mode. The converter will be limited to a maximum rate of change, the *slew rate*, that is defined by its counting rate and the unit voltage of the DAC. This slew rate is

$$\frac{\Delta V}{\Delta t} = v \times f \quad \text{V/sec} \tag{11-0}$$

where v is the step voltage and f is the counting frequency in pulses per second. If the input voltage changes faster than the rate defined by Eq. (11-0), then the tracking ADC will not be able to follow the change. It will continue to count up or down toward the new value of V_x, however, and will eventually reach the stable condition, provided that V_x does not continue to change.

EXAMPLE 11-4: A tracking ADC has a clock frequency of 10,000 pps and a DAC whose step voltage is 1 mV. Assuming that the ADC has stabilized at $V_x = 1$ V, how long will it take to stabilize if the input voltage changes instantaneously to 5 V?

Solution: The change in voltage is 4 V. The slew rate of the ADC is $10^4 \times 0.001 = 10$ V/sec. Thus, the ADC will take 0.4 sec to stabilize.

11.2.3. The Successive-Approximation ADC. The successive-approximation ADC is the most common of all ADC types. The conversion hardware is similar to that used by the ramp and tracking ADCs (Fig. 11-6) in that a DAC is used to generate a test voltage that is compared to the input voltage. The successive-approximation ADC uses a different control sequence, however. Its control logic uses the output signal from the digital comparator to guide the ADC through an organized search for the digital equivalent of the input voltage. Unlike the ramp and tracking ADCs, the conversion period of the successive-approximation ADC is the same for all values of the input voltage.

The successive-approximation conversion technique can be used only with a restricted set of weighted codes. The requirements that the code must meet are

1. Its weights must all be positive.*

2. Its weights must monotonically decrease from a maximum value down to a minimum value.

These restrictions are necessitated by the search procedure that is used during the successive-approximation process. They are met by natural binary and 8421 BCD code, as well as by XS3, 4221, 5321, and many other BCD codes. They are not met by the 6, 3, 1, -1 code or by the Gray code, for example.

The successive-approximation ADC uses a data register, rather than a counter, to generate and store the results of the conversion process. This register is initially set to a value of $100\ldots00$, corresponding to the most-significant bit of the code being TRUE. After a period sufficient for the DAC output and the comparator to settle, the logical output from the comparator will indicate whether or not V_{out} is greater than the input voltage, V_x. If it is,

---

*A negatively weighted sign bit is allowable, as is explained later in this section.

then the weight of the most-significant bit is too large to be included in the digital code word and that bit must be reset. Otherwise, the MSB is left TRUE.

The requirements that were given earlier assure that the decision as to whether or not to keep the highest-weighted bit TRUE, once it is made, will never be reversed. This applies for all lower-weighted bits, as well.

Once the TRUE/FALSE value of the most-significant bit is known, the successive-approximation process moves to the second bit. It is enabled, the DAC output is compared to V_x, and the decision as to whether or not to keep the second bit TRUE is made. The process continues to the least-significant bit. Thus, the conversion will take one clock period for each bit of the digital code word that is generated. The length of each period is determined by the settling speed of the DAC and the comparator.

EXAMPLE 11-5:   A successive-approximation ADC uses an eight-bit binary DAC whose unit voltage is 0.1 V. Describe the sequence through which the converter digitizes an input voltage of 20.5 V.

Solution: The ADC has an input voltage range from 0.0 to $(2^8 - 1)$ $\times$ 0.1 = 25.5 V. The conversion process begins with the data register of the ADC set to 10000000, thereby giving a DAC output of 12.8 V. Since V_x is greater than 12.8 V, the most-significant bit is kept TRUE.

Enabling the next bit gives 11000000, corresponding to $V_{out} = 19.2$ V. Since this, too, is less than V_x, the second bit is kept TRUE. Enabling the third bit gives 11100000 and $V_{out} = 22.4$ V. V_x is now exceeded, and the third bit is reset. The complete conversion process is summarized in the following table:

Clock Period	Digital Code	V_{out}	Comparator Output*	Decision Made
1	10000000	12.8	1	Keep TRUE
2	11000000	19.2	1	〃
3	11100000	22.4	0	Reset
4	11010000	20.8	0	〃
5	11001000	20.0	1	Keep TRUE
6	11001100	20.4	1	〃
7	11001110	20.6	0	Reset
8	11001101	20.5	1	Keep TRUE

*Comparator output assumed to be TRUE when $V_x > V_{out}$.

The final result is $V_x = 11001101_2 \times 0.1$ V. The decimal equivalent of the binary code word is 205_{10}, which corresponds to the correct result.

The key to the successive-approximation process is the assurance that a given bit, once its value has been decided, will never need to be changed.

The following example shows why this process cannot be used with some weighted codes.

EXAMPLE 11-6:   Consider a one-digit ADC that is to use the 6, 3, 1, -1 BCD code. Assume the unit voltage is 1.0 V and that the input voltage is 8.0 V. Show why the successive-approximation process fails.

Solution: The conversion begins with the data register set to 1000, giving $V_{out} = 6.0$ V. Since this value is less than V_x, the first bit is kept TRUE. The second bit is then enabled, giving 1100 and $V_{out} = 9$ V. This exceeds V_x, so that the new bit is reset. Continuing to the next bit gives $1010 = 7.0$ V, which is less than V_x. When the last bit is tried, the result is $1011 = 6.0$ V, and it, too, will be kept, even though it makes the answer even more in error. The decision that was made concerning the second bit was incorrect, since the correct result should be $1101 = 8.0$ V. The presence of the negative weight causes the successive-approximation technique to fail.

Conversion of a voltage into its 6, 3, 1, -1 encoded equivalent can be accomplished by code-to-code conversion of the output of an 8421 BCD ADC.

It should also be noted that while the successive-approximation process will work with BCD codes having more than one way to represent a given quantity, such as the 4221 code, the conversion process will give a specific set of answers that may differ from the desired code words. Properties such as self-complementation may not be maintained.

EXAMPLE 11-7:   An ADC produces one digit of 4221 BCD. If the unit voltage is 1.0 V, what are the ten possible code words that the successive approximation process will give?

Answer:

Input Voltage (V)	Coded Result
0.0	0000
1.0	0001
2.0	0100
3.0	0101
4.0	1000
5.0	1001
6.0	1100
7.0	1101
8.0	1110
9.0 or more	1111

The coded results do not exhibit the self-complementing property.

The control logic for a successive-approximation ADC is quite easy to implement. A state counter that provides n states (one for each bit of the encoded output) must be provided. This can be implemented as an n-bit ring counter or as an m-bit counter, where $m \geq \log_2 n$.

The conversion can be initiated by a *START* signal that forces the controller into its beginning state, say P_1, sets the first bit of the data register, say D_1, and resets all the other flip-flops that make up that register. The controller combines the state signals and the logical output from the comparator in order to execute the successive-approximation sequence. The resetting of the bit that is under consideration can be combined with the setting of the next bit. Thus, the control logic for resetting the most-significant bit and for setting the next bit can be

$$K_1 = \bar{C} P_1 \cdot CL \qquad J_2 = P_1 \cdot CL \qquad (11\text{-}1)$$

where C is TRUE when $V_x > V_{\text{out}}$. The control logic given in Eq. (11-1) will reset the most-significant bit, D_1, if the comparator indicates that V_{out} is greater than V_x. The next bit, D_2, will be set at the same time.

The conversion process is continued by repeating the control expression given in Eq. (11-1):

$$K_2 = \bar{C} P_2 \cdot CL \qquad J_3 = P_2 \cdot CL \qquad (11\text{-}2)$$
$$K_3 = \bar{C} P_3 \cdot CL \qquad J_4 = P_3 \cdot CL \qquad (11\text{-}3)$$

or, in general,

$$K_i = \bar{C} P_i \cdot CL \qquad J_{i+1} = P_i \cdot CL \qquad (11\text{-}4)$$

The last flip-flop is set at the end of the $(n-1)^{\text{th}}$ clock period. It is conditionally reset at the end of the nth period by

$$K_n = \bar{C} P_n \cdot CL \qquad (11\text{-}5)$$

The control logic given in Eqs. (11-1)–(11-5) can be used for conversion of the unknown voltage into any weighted code that meets the requirements that were given earlier. Only the DAC must be changed; it must provide the proper values of V_{out} for the code that is being used.

Figure 11-7 shows the logic diagram for a five-bit binary ADC that uses the successive-approximation method. The state sequence is provided by a ring counter. The *START* signal is synchronized to the system clock by a pulse-catcher. It then sets the ring counter into P_1, sets D_1, and resets all the other flip-flops. The successive-approximation process continues until P_5 is reached. After the value of the least-significant bit has been determined, the ring counter leaves P_5 and goes into the all-zero state. It remains in this condition until the next conversion cycle is initiated by the *START* signal.

The conversion of bipolar voltages can also be accomplished by using the successive-approximation technique, providing that a sign-and-complement notation is used. This type of coding allows the sign bit to be assigned a negative weight as shown in Chapters 8 and 9. The DAC that is used by

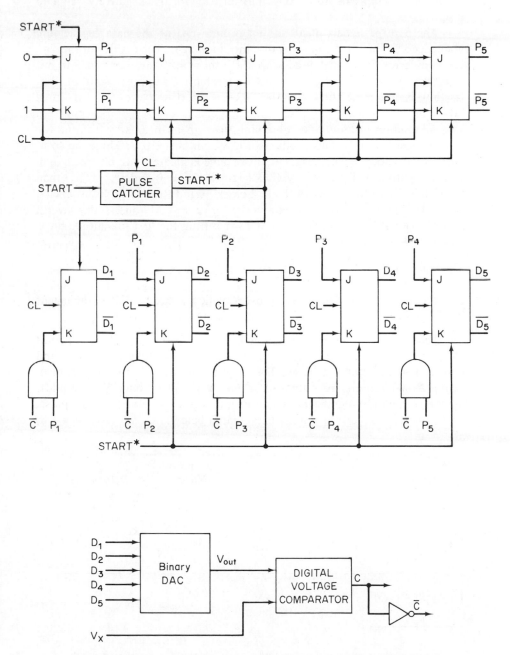

Figure 11-7. A Five-bit Binary ADC (Successive Approximation)

the successive-approximation ADC must provide for this negatively weighted input.

The control signals of all the magnitude bits of the data register are unchanged by the extension to bipolar input voltages. The control of the sign bit is different from that used with the other bits, however, since it has a negative weight. The sign bit must be treated as the most-significant bit. Its value is determined first, with the TRUE/FALSE values of the magnitude bits being determined thereafter.

A common technique for obtaining the sign bit begins by clearing the data register to all 0s. This produces a DAC output of 0 V. Thus, the comparator will indicate whether the input voltage is greater than or less than 0. If it is negative, as indicated by $C = 1$ (since $V_{out} = 0$ and $V_x < V_{out}$) then the sign bit is set at the end of the first clock period. If the sign bit is labeled as D_0 and the state counter is modified to give an additional state, say P_0, then the control logic for the sign and for setting the first magnitude bit is

$$J_0 = CP_0 \cdot CL \qquad J_1 = P_0 \cdot CL \qquad (11\text{-}6)$$

Thereafter, Eqs. (11-1)–(11-5) can be used.

EXAMPLE 11-8: Show the operations by which a five-bit, 2's complement ADC digitizes an input of -9 V. Assume a unit voltage of 1.0 V.

Solution: The sign bit has a weight of -16 V. The initial clock period compares $V_x = -9$ V to $V_{out} = 0$ V. Since V_x is negative, $V_x < V_{out}$ and $C = 1$. This causes D_0 to be set. The largest magnitude bit is then enabled, giving 1,1000 and a DAC output of $-16 + 8 = -8$ V. Since V_x is still less than V_{out}, the most-significant magnitude bit is reset by the next clock pulse. The complete conversion process is tabulated below:

Clock Period	Code Word	V_{out} (V)	Comparator Output	Decision
0	0,0000	0	1	Set Sign Bit
1	1,1000	-8	0	Reset
2	1,0100	-12	1	Keep TRUE
3	1,0110	-10	1	"
4	1,0111	-9	1	"

The digital result is $1,0111 = -9_{10}$ V.

Successive-approximation ADCs are available as prepackaged functional devices, but the control logic for an ADC of this type is so simple that it is feasible for the logical portion of the ADC to be constructed from digital components. The use of a commercially available DAC and comparator is desirable, however.

11.3 Summary

Chapter 11 introduces the digital-to-analog converter, or DAC, and the analog-to-digital converter, or ADC. These devices perform the conversions between continuous signals and their numerical counterparts. The emphasis of the chapter is upon the logical aspects of these devices rather than their electronic characteristics.

The electronic DAC is shown to be a device for converting a digital code word into a voltage that closely approximates the numerical value of the code word. The *unit voltage* is the increment of voltage that the DAC relates to a one-unit change in its numerical input.

An ADC converts an unknown voltage into the digital code word that most closely represents the value of the voltage. An ADC usually consists of a DAC, a voltage comparator, and a logical control section that seeks out the proper code word. Several types of ADC control techniques are explained.

11.4 Did You Learn?

1. What an analog variable is?

2. Why the output from a DAC is only approximately continuous?

3. What the unit voltage of a DAC is? How to determine the voltage range of a DAC?

4. How the use of a weighted code simplifies the design of a DAC? How to use variable-gain voltage amplifiers and a voltage summing amplifier to implement the DAC operation for a weighted code?

5. How to construct a multidigit decimal DAC? A signed DAC?

6. What a voltage comparator does?

7. The general arrangement for an ADC? What function a DAC has in this arrangement? What the comparator does for the ADC?

8. How the ramp and the tracking ADCs operate? What the limitations on their conversion speed are?

9. How a successive-approximation ADC operates?

10. Why a successive-approximation ADC cannot operate with some weighted codes?

11. How to design the control logic for an *n*-bit successive-approximation ADC, with and without a sign bit?

11.5 References

[1] Baasch, Tom, "Monolothic IC D/A Converters Give Wide Choice of Design Concepts," *Electronic Products*, 12, No. 2 (July 1970), 56–59.

[2] Cadzow, James A., and Hinrich R. Martens, *Discrete-time and Computer Control Systems*. Englewood Cliffs, N.J.: Prentice-Hall, Inc., 1970.

[3] Johnson, Clarence L., *Analog Computer Techniques*. New York: McGraw-Hill Book Company, 1956.

[4] Malmstadt, H. V., and C. G. Enke, *Digital Electronics for Scientists*. Reading, Mass.: W. A. Benjamin, Inc., 1969.

[5] Marcott, C. M., "Single Chip D/A Converter," *Electronic Products*, 12, No. 1 (June 1970), 46–50.

11.6 Problems

11-1. Design a DAC for use with each of the following codes. Follow Exs. 11-0 through 11-3.

    a. 2421 BCD (Table 10-0)
    b. 5421 BCD (Table 10-1)
    c. 5311 BCD (Table 10-1)
    d. 7421 BCD (Table 10-1)
    e. Three-digit, 2421 BCD

*11-2. Give a complete system design for a five-bit binary ADC that uses the ramp control technique. Show a complete RTL logic diagram. Assume that a five-bit DAC with a unit voltage of 0.1 V is available, along with a digital voltage comparator that has two input terminals, *A* and *B*. When $V_A < V_B$, the comparator output is $+3.6$ V. When $V_B < V_A$, the output is 0 V.

*11-3. Modify the design of Prob. 11-2 to produce a tracking ADC.

*11-4. Using the comparator described in Prob. 11-2 and an appropriate DAC, design a complete two-digit, 8421 BCD ADC.

11-5. Show why the successive-approximation technique cannot be used with 2431 BCD.

*11-6. Modify the ramp ADC of Prob. 11-2 and the tracking ADC of Prob. 11-3 so that they will work with 2's complement binary code.

11-7. Design a complete ADC system that produces a two-digit, 10's complement output in 5421 BCD code. Use the comparator described in Prob. 11-2.

Appendix

```
28          SUBROUTINE   MINTE (N,C)
       C
       C    ******************************************************************
       C    ******************************************************************
       C    *                                                                *
       C    *                ---- MINTE -- MINTE ----                        *
       C    *        BOOLEAN MINIMIZATION PROGRAM. ALL PRIME IMPLI-          *
       C    *        CANTS AND ESSENTIAL IMPLICANTS FOUND FOLLOWING          *
       C    *        THE PROGRAMMED PROCEDURE GIVEN BY                       *
       C    *                                                                *
       C    *        CARROLL, C. C., "A FAST ALGORITHM FOR BOOLEAN           *
       C    *            FUNCTION MINIMIZATION," TECHNICAL REPORT            *
       C    *            NO. AU-T-3, PROJECT THEMIS, AUBURN UNIV.,           *
       C    *            AUBURN, ALABAMA, DECEMBER, 1968.                    *
       C    *                                                                *
       C    *        CYCLIC PRIME-IMPLICANT TABLES, SHOULD THEY              *
       C    *        ARISE, ARE REDUCED BY FOLLOWING THE APPROXI-            *
       C    *        MATE METHOD GIVEN BY                                    *
       C    *                                                                *
       C    *        BOWMAN, R. M., AND E. S. MC.VEY, "A METHOD FOR          *
       C    *            THE FAST APPROXIMATE SOLUTION OF LARGE PRIME        *
       C    *            IMPLICANT CHARTS," IEEE TRANS. ON COMPUTERS,        *
       C    *            VOL. C-19, NO. 2, FEB., 1970, PP. 169-173.          *
       C    *                                                                *
       C    *        THIS PROGRAM IS WRITTEN IN STANDARD FORTRAN             *
       C    *        WITH THE EXCEPTION OF THE "IAND" FUNCTION. IT           *
       C    *        IS DESCRIBED AT THE END OF THE LISTING. THE             *
       C    *        SIMPLE OPERATION OF THE IAND FUNCTION MUST BE           *
       C    *        PROVIDED, USUALLY BY AN ASSEMBLY LANGUAGE SUB-          *
       C    *        ROUTINE THAT IS CALLED FROM THE MAIN SUBROU-            *
       C    *        TINE.                                                   *
       C    *                                                                *
       C    *        THIS VERSION OF THE SUBROUTINE IS DIMENSIONED           *
       C    *        TO HANDLE 10 VARIABLES. THE CALL IS                     *
       C    *                                                                *
       C    *            CALL MINTE(N,C)                                     *
       C    *                                                                *
       C    *        WHERE N IS THE NUMBER OF VARIABLES AND                  *
       C    *        C IS AN ARRAY CONTAINING INTEGER VALUES                 *
       C    *        OF -1, 0, 1, FOR REDUNDANCY, FALSE, AND                 *
       C    *        TRUE MINTERMS. C MUST BE DIMENSIONED AT                 *
       C    *        LEAST 2**N.                                             *
       C    *                                                                *
       C    ******************************************************************
       C    ******************************************************************
29           INTEGER X(10),X1/' X  '/,X2/' 1  '/,X3/' 0  '/,DIF,
            * XE(3)/'ESSE','NTIA','L   '/,XC(3)/'CHOS','EN  ',
            * '    '/,Y(10)/' A  ',' B  ',' C  ',' D  ',' E  ',
            * ' F  ',' G  ',' H  ',' I  ',' J  '/,C(1024),MINT(1024)
            * ,NUM(1024),PRIMI(512),PRIMJ(512),COST(512),T(512),
            * TABI(1025),TABLE(12500)
30           REAL TX(1024),TY(512)
31           A2=ALOG(2.0)
32           NT=2**N
33           MK=0
34           M=0
35           DO 125 I=1,NT
36           TABI(I)=1
37           IF(C(I).EQ.0) GO TO 125
38           M=M+1
```

```
39              MINT(M)=I-1
40      125     CONTINUE
41              WRITE(6,995) N
42              MP=0
43              DO 3 I=1,NT
44      3       NUM(I)=I-1
45              WRITE(6,700)(NUM(I),C(I),I=1,NT)
        C
        C       SOLVE FOR PRIME IMPLICANTS
        C
46              DO4 I=1,M
47              II=MINT(I)
48              M1=M-I+1
49              DO 5 J=1,M1
50              NN=0
51              JJ=MINT(M-J+1)
52              IF (IAND(II,JJ).NE.II) GO TO 5
53              CIST=1.0
54              DIF=JJ-II
55              IF(C(II+1).GT.0 ) NN=NN+1
56              IF (C(JJ+1) .GT.0)   NN=NN+1
57              IF(DIF.EQ.0) GO TO 9
58              ICNT=1
59              NUM(1)=II
60              DO 6 K=1,N
61              IZ=IAND(2**(K-1),DIF)
62              IF(IZ.EQ.0) GO TO 6
63              IC2=ICNT
64              DO 8 L=1,IC2
65              ICNT=ICNT+1
66              NUM(ICNT)=NUM(L)+IZ
67              LL=NUM(ICNT)+1
68              IF(C(LL))8,5,30
69      30      NN=NN+1
70       8      CONTINUE
71       6      CONTINUE
72              CIST=ICNT+1
73      9       IF(NN.EQ.0) GO TO 5
74              IF(MP.EQ.0) GO TO 10
75              DO11 L=1,MP
76              IF(PRIMJ(L).LT.JJ) GO TO 11
77              IF(IAND(PRIMJ(L),JJ).NE.JJ) GO TO 11
78              IF(IAND(PRIMI(L),II).EQ.PRIMI(L)) GO TO 5
79       11     CONTINUE
80       10     MP=MP+1
81              PRIMI(MP)=II
82              PRIMJ(MP)=JJ
83              T(MP)=0
84              ICOST =ALOG(CIST)/A2
85              COST(MP)=N-ICOST
86       5      CONTINUE
87              IF(C(II+1).LT.0) GO TO 4
88              TABI(I)=MK+1
89              DO 13 L=1,MP
90              IF(PRIMJ(L).LT.II) GO TO 13
91              IF(IAND(PRIMI(L),II).NE.PRIMI(L)) GO TO 13
92              IF(IAND(PRIMJ(L),II).NE.II) GO TO 13
93              MK=MK+1
94              TABLE(MK)=L
95       13     CONTINUE
```

```
 96              TABI(I+1)=MK+1
 97              IC=MK-TABI(I)
 98              IF(IC.EQ.0) GO TO 15
 99              A=IC
100              TX(I)=1.0/A
101              GO TO 4
      C
      C     SET ESSENTIAL TERM
      C
102   15       J=TABLE(TABI(I))
103              T(J)=-1
104              NUM(1) = PRIMI(J)
105              LL=NUM(1)+1
106              C(LL)=-1
107              ICNT=1
108              DIF = PRIMJ(J)-PRIMI(J)
109              DO 16 K=1,N
110              IZ=IAND(2**(K-1),DIF)
111              IF(IZ.EQ.0) GO TO 16
112              ICT2=ICNT
113              DO 17 L=1,ICT2
114              ICNT=ICNT+1
115              NUM(ICNT)=NUM(L)+IZ
116              LL=NUM(ICNT)+1
117   17       C(LL)=-1
118   16        CONTINUE
119    4        CONTINUE
      C
      C  ALL PRIME IMPLICANTS ARE FOUND--CK FOR CYCLIC CHART
      C
120              LL=0
121              DO 27 L=1,M
122              IF(C(MINT(L)+1)) 27,27,25
123   25       LL=LL+1
124              NUM(LL)=L
125   27        CONTINUE
126              IF(LL.EQ.0) GO TO 31
      C
      C  SOLVE CYCLIC CHART
      C
127              BIX=0.0
128              DO 28 I=1,LL
129              IF(BIX.GE.TX(NUM(I))) GO TO 28
130              BIX=TX(NUM(I))
131              II=NUM(I)
132   28        CONTINUE
133   48       DO 49 I=1,MP
134   49       TY(I)=0.0
135              DO 50 I=1,LL
136              IT=TABI(NUM(I))
137              JT=TABI(NUM(I)+1)-1
138              DO 50 J=IT,JT
139              TY(TABLE(J))=TY(TABLE(J))+TX(NUM(I))
140   50        CONTINUE
141   55       BIY=0.0
142              IT=TABI(II)
143              JT=TABI(II+1)-1
144              DO 60 J=IT,JT
145              IF(TY(TABLE(J))-BIY) 60,56,57
146   56       IF(BIY.EQ.0.0) GO TO 60
```

```
147            IF(COST(JJ).LE.COST(TABLE(J))) GO TO 60
148     57     BIY=TY(TABLE(J))
149            JJ=TABLE(J)
150     60     CONTINUE
        C
        C      CHOOSE PRIME IMPLICANT
        C
151            T(JJ)=1
152            BIX=0.0
153            DO 65 I=1,LL
154            IJ=MINT(NUM(I))
155            IF (PRIMJ(JJ).LT.IJ) GO TO 61
156            IF(IAND(PRIMJ(JJ),IJ).NE.IJ) GO TO 61
157            IF(IAND(PRIMI(JJ),IJ).NE.PRIMI(JJ)) GO TO 61
158            TX(NUM(I))=0.0
159     61     IF(BIX.GE.TX(NUM(I))) GO TO 65
160            BIX=TX(NUM(I))
161            II=NUM(I)
162      65    CONTINUE
163            IF(BIX.NE.0.0) GO TO 48
164     31     CONTINUE
        C
        C      WRITE OUTPUT
        C
165            WRITE(6,1000) (Y(K),K=1,N)
166            DO 40 J=1,MP
167            I3=PRIMI(J)
168            K3=PRIMJ(J)-I3
169            DO 35 K=1,N
170            K1=K-1
171            X(N-K1)=X3
172            IF(IAND(I3,2**K1).NE.0)X(N-K1)=X2
173            IF(IAND(K3,2**K1).NE.0)X(N-K1)=X1
174      35    CONTINUE
175            IF (T(J)) 38,37,39
176         37 WRITE(6,1003) J,COST(J),(X(I),I=1,N)
177            GO TO 40
178         38 WRITE(6,1003) J,COST(J),(X(I),I=1,N),XE
179            GO TO 40
180         39 WRITE(6,1003) J,COST(J),(X(I),I=1,N),XC
181      40    CONTINUE
182            WRITE(6,2001)
183            IF(LL.EQ.0)  RETURN
184            WRITE(6,1004)
185            DO 70 I=1,M
186            IT=TABI(I)
187            JT=TABI(I+1)-1
188            IF(C(MINT(I)+1).GT.0)  WRITE(6,1002) MINT(I),(TABLE(J)
              *  ,J=IT,JT)
189      70    CONTINUE
190            WRITE(6,753)
191        995 FORMAT('1',T21,'*** BOOLEAN MINIMIZATION PROGRAM ***',
              * /// T16,'THIS FUNCTION CONTAINS',I5,' VARIABLES',//
              * T16, 'A LISTING OF THE INPUT DATA FOLLOWS',/ T16,'TR'
              *,'UE MINTERMS = 1',/ T16,'FALSE MINTERMS = 0',/ T16,
              * 'REDUNDANT MINTERMS (DON''T CARES) = -1',/)
192        700 FORMAT('0',( T6,5(I5,' = ',I2,', ')))
193       1000 FORMAT('1',T6,'THE FOLLOWING IS A LIST OF THE PRIME '
              *,'IMPLICANTS OF THE',/ T6,'MINIMIZED FUNCTION.',//
              * T10,'ESSENTIAL PRIME IMPLICANTS ARE SO LABELED, AND'
```

```
                 * ,/ T10,'PRIME IMPLICANTS SELECTED FROM A CYCLIC CHA'
                 *,'RT ARE',/ T10,'LABELED AS CHOSEN.',// 5X'    NO.',
                 * 3X,'COST',7X,'PRIME IMPLICANTS',/ 22X,10A4,/)
194      1003 FORMAT(' ',5X,I4,2X,I4,6X,13A4)
195      2001 FORMAT(' ',/ 5X,'X INDICATES A MISSING VARIABLE,0 IN'
                 *,'DICATES A COMPLEMENTED',/ 5X,'VARIABLE AND 1 INDIC'
                 *,'ATES A TRUE VARIABLE.',/ 5X,'THE FUNCTION IS REPRE'
                 *,'SENTED BY THE SUM OF BOTH THE',/ 5X,'ESSENTIAL AND'
                 *,' THE CHOSEN PRIME IMPLICANTS.')
196      1004 FORMAT('1',//////,10X,' CONSTRAINT   TABLE',// T8,'COV'
                 *,'ERED',4X,'COVERING PRIME',/ T8,'MINTERM',6X,'IMPLI'
                 *,'CANTS.')
197      1002 FORMAT(' ',T9,I5,T16,15I6,9(/,7X,15I6))
198       753 FORMAT('1')
199            RETURN
200            END
         C

201            FUNCTION IAND (I1,I2)
         C
         C          THIS FUNCTION COMBINES THE BITS IN I1 AND I2
         C          ACCORDING TO THE BOOLEAN "AND" FUNCTION. FOR
         C          EXAMPLE
         C
         C              I1= 0101100101110101
         C              I2= 1001011101010010
         C
         C          IAND= 0001000101010000
         C
         C
         C          THE FOLLOWING STATEMENTS ARE FOR USE WITH THE
         C          W A T F I V FORTRAN COMPILER ONLY.
         C
202            EQUIVALENCE (I,A)
203            A=AND( I1,I2)
204            IAND=I
205            RETURN
206            END

         //$DATA
```

*** BOOLEAN MINIMIZATION PROGRAM ***

THIS FUNCTION CONTAINS   5 VARIABLES

A LISTING OF THE INPUT DATA FOLLOWS
TRUE MINTERMS = 1
FALSE MINTERMS = 0
REDUNDANT MINTERMS (DON'T CARES) = -1

```
 0 =  1,      1 =  0,      2 = -1,      3 =  1,      4 =  1,
 5 =  0,      6 =  1,      7 =  1,      8 = -1,      9 =  0,
10 =  0,     11 =  0,     12 = -1,     13 =  0,     14 = -1,
15 =  1,     16 =  0,     17 = -1,     18 =  0,     19 =  0,
20 =  0,     21 =  1,     22 =  0,     23 =  1,     24 = -1,
25 =  0,     26 =  1,     27 =  0,     28 =  1,     29 =  0,
30 =  0,     31 = -1,
```

THE FOLLOWING IS A LIST OF THE PRIME IMPLICANTS OF THE
MINIMIZED FUNCTION.

ESSENTIAL PRIME IMPLICANTS ARE SO LABELED, AND
PRIME IMPLICANTS SELECTED FROM A CYCLIC CHART ARE
LABELED AS CHOSEN.

NO.	COST	PRIME IMPLICANTS					
		A	B	C	D	E	
1	3	0	X	X	0	0	CHOSEN
2	3	0	0	X	X	0	
3	3	0	0	X	1	X	ESSENTIAL
4	3	0	X	1	X	0	
5	3	0	X	1	1	X	
6	3	X	X	1	1	1	CHOSEN
7	3	X	1	X	0	0	ESSENTIAL
8	4	1	0	X	0	1	CHOSEN
9	4	1	0	1	X	1	
10	4	1	1	0	X	0	ESSENTIAL

X INDICATES A MISSING VARIABLE,0 INDICATES A COMPLEMENTED
VARIABLE AND 1 INDICATES A TRUE VARIABLE.
THE FUNCTION IS REPRESENTED BY THE SUM OF BOTH THE
ESSENTIAL AND THE CHOSEN PRIME IMPLICANTS.

```
              CONSTRAINT   TABLE

      COVERED        COVERING PRIME
      MINTERM          IMPLICANTS.
          0           1      2
          4           1      2       4
         15           5      6
         21           8      9
         23           6      9
```

Index

561